Data Management Solutions Using SAS® Hash Table Operations

A Business Intelligence Case Study

Paul Dorfman
Don Henderson

§.sas®

sas.com/books

Contents

About This Book

What Does This Book Cover?

This book is about the **How**, the **What,** and the **Why** of using the SAS DATA step hash object. These three topics are interconnected and quite often SAS users focus on just a small part of what SAS software can do. This is especially true for the SAS hash object and hash tables. Far too many users immediately think of the use of hash tables as just a very powerful table lookup facility (a **What**), and that notion then influences their understanding of the **How** and the **Why**.

The authors have found that the SAS hash object and hash tables provide for a very robust data management and analysis facility, and we collaborated on this book to provide the insights we have discovered:

- More **What**s: e.g., data management; data aggregation, . . .
- More **Why**s: e.g., efficiency; flexibility, parameterization, . . .
- More **How**s: e.g., memory management, key management, . . .

The focus of this book is to provide the readers with a more complete understanding and appreciation of the SAS hash object. As such, we have included a number of SAS programs that illustrate this broad range of functionality. Many of the programs use features of the SAS DATA step language that many readers may not be familiar with. This book does not attempt to describe those techniques in detail; instead, the authors will expand upon traditional SAS DATA step programming techniques that are particularly relevant to the SAS object in a series of blog entries. You can access the blog entries from the author page at support.sas.com/authors. Select either "Paul Dorfman" or "Don Henderson." Then look for the cover thumbnail of this book, and select "Blog Entries."

The book is organized around a Proof of Concept (PoC) project whose goal is to convince a group of business and IT users that the SAS hash object can be used to address many of their requirements for data management and reporting.

Is This Book for You?

This book is intended for any SAS programmer who has an interest in learning more about what can be done with the SAS hash object and specifically about how to use the hash object to assist in the analysis of data to address business intelligence problems. The hash object is more than just a technique for table lookup; the point of this book is to broaden that perspective.

How to Read This Book

The book is organized into four parts. There is no requirement to read this book in order.

Part 1 focuses on the **How** of the hash object and provides a deep dive into the details of how it works. It provides a high-level overview of the hash object followed by a discussion of both table-level and item-level operations. It concludes with a more advanced discussion of item-level enumeration operations. Part 1 is probably best used by first reading Chapter 1 to get a better understanding of the kinds of tasks the hash object can be used for. The remaining Part 1 chapters can be reviewed later.

The focus of Part 2 is **What** the hash object should be used for, along with a discussion of **Why** that hash object is a good deal for many business intelligence questions. It starts with a discussion of the sample data used in the book and how the business users are interested in providing answers to business intelligence and analytical questions. It then provides an overview of common business intelligence and analytical data tasks. Part 2 also discusses the use of the SAS hash object to support the creation and updating of the data warehouse or data mart table. Following that, the discussion moves to using the hash object to support a range of data aggregation capabilities via a number of sample programs that you the reader can adapt to your business problems. Readers with some experience with DATA step programming might want focus on Part 2 after reviewing the overview chapter in Part 1.

Part 3 introduces how some more advanced features of the hash object can facilitate data-driven techniques in order to offer more flexibility and dynamic programming. It also addresses techniques for memory management and data partitioning, focusing on all three of the topics of **How**, **What,** and **Why**. Part 3 should be reviewed in detail once the reader feels comfortable with the examples presented in Part 2.

Two short case studies are included in Part 4. The first illustrates using the hash object to research alternatives metrics. The second one provides an example of using the hash object to support answering ad-hoc questions. The sample programs in Part 4 leverage the example programs from Part 2. Reviewing the examples in Part 4 can probably be done in any order by referring to the techniques used.

More details about each part, including suggestions for what to focus on, can be found in the short introductions to each of the 4 parts.

You can access a glossary of terms from the author page at support.sas.com/authors. Select either "Paul Dorfman" or "Don Henderson." Then look for the cover thumbnail of this book, and select "Glossary of Terms."

What Are the Prerequisites for This Book?

The only prerequisite for this book is familiarity with DATA step programming. Some knowledge of the macro language is desirable, but is not required.

What Should You Know about the Examples?

This book includes examples for you to follow to gain hands-on experience with the SAS hash object.

Software Used to Develop the Book's Content

All of the examples in this book apply to SAS 9.3 and SAS 9.4. Where differences exist, we have done our best to reference them. Many of the examples also work in early releases of SAS, but the examples have not been tested using those earlier releases.

Example Code and Data

The sample data for this book is for a fictitious game called Bizarro Ball. Bizarro Ball is conceptually similar to baseball, with a few wrinkles.

We have been engaged by the business users who are responsible for marketing Bizarro Ball about their interest in reporting on their game data. They currently have no mechanism to capture their data and so we have agreed to write programs to generate data that can be used in a Proof of Concept. The programs, most of which use the hash object, generate our sample data and are discussed in a series of blog entries. You can access the blog entries from the author page at support.sas.com/authors. Select either "Paul Dorfman" or "Don Henderson." Then look for the cover thumbnail of this book, and select "Blog Entries."

Selected example programs do make use of DATA step programming features which, while known by many, are not widely used. The authors plan to write blog entries (as mentioned above) about some of those techniques, and readers are encouraged to suggest programming techniques used in the book for which they would like to see a more detailed discussion.

You can access the example code and data from the author page at support.sas.com/authors. Select either "Paul Dorfman" or "Don Henderson." Then look for the cover thumbnail of this book, and select "Example Code and Data."

An Overview of Bizarro Ball

The key features of Bizarro Ball that we agreed to implement in our programs to generate the sample data include:

- Creating data for 32 teams, split between 2 leagues with 16 teams in each league.
- Each team plays the other 15 teams in their league.
- Each team plays each other team a total of 12 times; 6 as the home team and 6 as the away team. In other words, they play a balanced schedule.
- Games are played in a series consisting of 3 games each.
- Each week has 2 series for each team. Games are played on Tuesday, Wednesday, Thursday; the second series of games are played on Friday, Saturday, and Sunday. Monday is an agreed upon off-day for each team. This off-day is used when it is necessary to schedule a date for a game that was canceled (e.g., due to the weather). It was agreed that, to simplify the programs to generate our sample data, we would assume that no such *makeup* games are needed.
- Since each team plays each other team in their league 12 times, this results in a regular season of 180 games. Since each team plays 6 games a week, the Bizarro Ball regular season is 30 weeks long.
- Another simplifying assumption that was agreed to was that we could generate a schedule without regard to constraints related to travel time or rules about consecutive home or away series.
- Each game is 9 innings long, and games can end in a tie.
- If the home team (which always bats in the bottom half of an inning) is ahead going into the bottom half of the 9th inning, they still play that half-inning. The reason for that is that the tie breakers for determining who the league champion is include criteria that could adversely impact a good team if they are often ahead at the beginning of the bottom half of the 9th inning.
- Each team has 25 players and has complete control over the distribution of the positions a player can play.

- Each team would set its lineup for each game using whatever criteria they felt appropriate. We informed the business users that using the logic to implement a rules-based approach to do this did not add value to the PoC and would take significant extra time. So it was agreed we could randomize the generation of the line-up for each game.

There are a number of key differences between Bizarro Ball and baseball. Therefore, in the interests of time and focusing on how the hash object that can be used to address business problems, we agreed to a number of simplifying assumptions with our business users. Those assumptions are discussed in the blog posts mentioned above.

SAS University Edition

This book is compatible with SAS University Edition. If you are using SAS University Edition, then begin here: https://support.sas.com/ue-data .

The only requirement is to make sure to extract the ZIP file of sample data and programs in a location accessible to the SAS University Edition. Example code and data can be found on the author pages:

support.sas.com/dorfman
support.sas.com/henderson

We Want to Hear from You

SAS Press books are written *by* SAS Users *for* SAS Users. We welcome your participation in their development and your feedback on SAS Press books that you are using. Please visit sas.com/books to do the following:

- Sign up to review a book
- Recommend a topic
- Request information on how to become a SAS Press author
- Provide feedback on a book

Do you have questions about a SAS Press book that you are reading? Contact the author through saspress@sas.com or https://support.sas.com/author_feedback.

SAS has many resources to help you find answers and expand your knowledge. If you need additional help, see our list of resources: sas.com/books.

About These Authors

Paul Dorfman is an independent consultant. He specializes in developing SAS software solutions from ad hoc programming to building complete data management systems in a range of industries, such as telecommunications, banking, pharmaceutical, and retail. A native of Ukraine, Paul started using SAS while pursuing his degree in physics in the late 1980s. In 1998, he pioneered using hash algorithms in SAS programming by designing a set of hash routines based on SAS arrays. With the advent of the SAS hash object, Paul was one of the first to use it practically and to author a SUGI paper on the subject. In the process, he introduced hash object techniques for metadata-based parameter type matching, sorting, unduplication, filtering, data aggregation, dynamic file splitting, and memory usage optimization. Paul has presented papers at global, regional, and local SAS conferences and meetings since 1998.

Don Henderson is the owner and principal of Henderson Consulting Services, a SAS Affiliate Partner. Don has used SAS software since 1975, designing and developing business applications with a focus on data warehousing, business intelligence, and analytic applications. Don was one of the primary architects in the initial development and release of SAS/IntrNet® software in 1996, and he was one of the original developers of the SAS/IntrNet® Application Dispatcher. Don is the author of *SAS® Server Pages: Generating Dynamic Content, Building Web Applications with SAS/IntrNet®: A Guide to the Application Dispatcher*, and *Data Management Solutions Using SAS® Hash Table Operations: A Business Intelligence Case Study*. Don has presented numerous papers at SAS Global Forum and at regional SAS user group meetings, and he continues to be a great supporter of SAS software solutions.

Learn more about these authors by visiting their author pages, where you can download free book excerpts, access example code and data, read the latest reviews, get updates, and more:

support.sas.com/dorfman
support.sas.com/henderson

Acknowledgments

The authors would like to thank all the technical reviewers, including Paul Grant, Elizabeth Axelrod, Michele Ensor, and Grace Whiteis, who provided invaluable feedback. They are also grateful to the SAS Press staff. Both groups helped make this book a reality.

We would also like to thank the R&D team at SAS for implementing the powerful facilities of the SAS hash object.

Special thanks to the customers whose challenging business intelligence requirements led us to research how to address those requirements. Results of that research led us to write this book in order to share that knowledge. Special thanks to Rich N and Paulius M of a large health insurance company and to Tom H of a regional supermarket co-op's marketing department.

Conversations on various online communities (e.g., http://communities.sas.com) also provided much food for thought.

Paul would like to thank Don for his idea to write this book in the first place, for spearheading the effort all the way through, and for his energy and organizational skill, without which the book would never have seen the light of day. Furthermore, Paul would like to thank Don for discovering new capabilities of the hash object, for invigorating and edifying discussions, and for patience with his coauthor's quirks, which he calls "dorfmanisms". Last but not least, he would like to thank Don for introducing him to the data-rich game of baseball and taking the education process as far as treating him to a real game at Fenway Park.

Paul owes special gratitude to his wife, Doris Dorfman, for her infinite patience and support for his effort despite its immense impact on the family schedule. Finally, Paul is deeply indebted to his mother, Eugenia Kravchenko, not only for making him (and hence this book) happen but also for her relentless enthusiastic encouragement since the first day this book was conceived.

Don would like to thank several of his fellow baseball blog colleagues at the blog TalkNats.com, a forum for baseball fans who motivated him to start analyzing baseball data and provided suggestions for the rules of game Bizarro Ball: Stephen Mears, Bob Schiff (who coined the term *Bizarro Ball* to describe this game), Stephen Mahaney, and Andrew Lang. The sample data for this book is based on the fictional game Bizarro Ball.

Don would also like to thank Paul for enlightening him about all the things that could be done with the SAS hash object and indulging him to use sample data about a game like baseball. And, lest he forget, special thanks to his wife, Celia Henderson, for her patience and understanding when he would disappear for hours and days at a time using the excuse that *I need to work on the book.*

xx

Part One—The HOW of the SAS Hash Object

The goal of Part One is to describe the hash object essentials, data operations, and tools to implement them along with the concepts essential to using the SAS hash object, in particular:

- Hash object and table essentials.
- The standard general table operations commonly known as CRUD (*Create*, *Retrieve*, *Update*, *Delete*).
- The implementation of these operations with the hash object tools.
- Data exchange between the hash table and program data vector (PDV) host variables.
- Compile time and run time aspects of the hash object behavior.

The hash object tools support two categories of standard data table operations: table-level and item-level. The item-level operations are much more numerous and can be further subdivided into the (a) direct key access operations and (b) enumeration operations.

As such, Part One contains four chapters:

1. Chapter 1 provides an overview of what the SAS object is and what it can be used for.
2. Chapter 2 provides a deep dive into table-level operations.
3. Chapter 3 provides a deep dive into direct key access item-level operations.
4. Chapter 4 describes, in detail, how to enumerate hash table items, i.e., process them sequentially.

This is not intended as a treatise on the SAS hash object, nor as a reincarnation of the related parts of SAS product documentation. However, it is intended to serve as a reference to explain the **HOW** of the examples in Parts Two, Three, and Four.

Chapter 1: Hash Object Essentials

1.1 Introduction

The goal of this chapter is to discuss the organization and data structure of the SAS hash object, in particular:

- Hash object and table structure and components.
- Hash table properties.
- Hash table lookup organization.

- Hash operations and tools classification.
- Basics of the behind-the-scenes hash table structure and search algorithm.

On the whole, the chapter should provide a conceptual background related to the hash object and hash tables and serve as a stepping stone to understanding hash table operations.

Since we have two distinct sets of users who are in this Proof of Concept, this chapter would likely be of much more interest to the IT users as they are more likely than the business users to understand the details and the nuances discussed here. We did suggest that it would be worthwhile for the business users to skim this chapter, as it should give them a good overview of the power and flexibility of the SAS hash object/table.

1.2 Hash Object in a Nutshell

The first production release of the hash object appeared in the SAS 9.1. Perhaps the original motive for its development had been to offer a DATA step programmer a table look-up facility either much faster or more convenient - or both - than the numerous other methods already available in the SAS arsenal. The goal was certainly achieved right off the bat. But what is more, the potential capabilities built into the newfangled hash object were much more scalable and functionally flexible than those of a mere lookup table. In fact, it became the first in-memory data structure accessible from the DATA step that could emerge, disappear, grow, shrink, and get updated dynamically at run time. The scalability of the hash object has made it possible to vastly expand the original hash object functionality in future versions and releases, and its functional flexibility has enabled SAS programmers to invent and implement new uses for it, perhaps even unforeseen by its developers.

So, *what is the hash object*? In a nutshell, it is a dynamic data structure controlled during execution time from the DATA step (or the DS2 procedure) environment. It consists of the following principal elements:

- A *hash table* for data storage and retrieval specifically organized to perform table operations based on searching the table quickly and efficiently via its *key*.
- An underlying, behind-the-scenes *hashing algorithm* which, in tandem with the specific table organization, facilitates the search.
- A set of tools to control the very existence of the table - that is, to create and delete it.
- A set of tools to activate the table operations and thus enable information exchange between the DATA step environment and the table.
- Optional: a *hash iterator object* instance linked to the hash table with the purpose of accessing the table entries sequentially.

The terms "hash object" and "hash table" are most likely derived from the hashing algorithm underlying their functionality. Let us now discuss the hash table and its specific features and usage prerequisites.

1.3 Hash Table

From the standpoint of a user, the hash object's table is a table with rows and columns - just like any other table, such as a SAS data file. Picture the image of a SAS data set, and you have pretty much pictured what a hash table may look like. For example, let us suppose that it contains a small subset of data from data set Bizarro.Player_candidates:

Table 1.1 Hash Object Table Layout

Hash Table Variables				
Key Portion		Data Portion		
Team_SK	Player_ID	First_name	Last_name	Position_code
115	23391	Ryan	Coleman	C
158	38259	Ronald	Wright	CF
189	24603	Alan	Torres	CIF
189	59690	Gregory	Roberts	COF
193	11628	Henry	Rodriguez	MIF
259	30598	Eugene	Thompson	SP

Reminds us of an indexed SAS data set, does it not? Indeed, it looks like a relational table with rows and columns. Furthermore, we have a *composite key* (Team_SK, Player_ID) and the rest of the variables associated with the key, also termed the *satellite* data. The analogy between an indexed SAS data set and a hash table is actually pretty deep, especially in terms of the common table operations both can perform. However, there are a number of significant distinctions dictated by the intrinsic hash table properties. Let us examine them and make notes of the specific hash table nomenclature along the way.

1.4 Hash Table Properties

To make the hash table's properties stand out more clearly, it may be useful to compare them with the properties of the indexed SAS data set from a number of perspectives.

1.4.1 Residence and Volatility

- The hash table resides *completely in memory*. This is *one* of the factors that makes its operations very fast. On the flip side, it also limits the total amount of data it can contain, which consists of the actual data and some underlying overhead needed to make the hash table operations work.
- The hash table is *temporary*. Even if the table is not deleted explicitly, it exists only for the duration of the DATA step. Therefore, the *hash table cannot persist across SAS program steps*. However, its content can be saved in a SAS data set (or its logical equivalent, such as an RDBMS table) before the DATA step completes execution and then reloaded into a hash table in DATA steps that follow.

1.4.2 Hash Variables Role Enforcement

- The hash variables are specifically defined as belonging to two distinct domains: *the key portion* and *the data portion*. Their combination in a row forms what is termed a *hash table entry*.
- Both the key and the data portions are *strictly mandatory*. That is, at least one hash variable *must* be defined for the key portion and at least one for the data portion. (Note that this is different from an indexed SAS table used for pure look-up where no data portion is necessary.)
- The two portions communicate with the DATA step program data vector (PDV) differently. Namely, only the values of the data portion variables can be used to update their PDV *host variables*.
- Likewise, only the data portion content can be "dumped" into a SAS data file.
- In the same vein, in the opposite data traffic direction, *only the data portion* hash variables can be updated from the DATA step PDV variables or other expressions.

- However, if a hash variable is defined in the key portion, a hash variable with the same name can also be defined in the data portion. Note that because the data portion variable can be updated and the key portion variable with the same name cannot, their values can be different within one and the same hash item.

1.4.3 Key Variables

- Together, the key portion variables form the *hash table key* used to access the table.
- The table key is *simple* if the key portion contains one variable, or *compound* if there is more than one. For example, in the sample table above, we have a two-term compound key consisting of variables (Team_SK, Player_ID).
- A *compound key is processed as a whole*, i.e., as if its components were concatenated.
- Hence, unlike an indexed SAS table, the *hash table can be searched based on the entire key only*, rather than also on a number of its consecutive leading components.

1.4.4 Program Data Vector (PDV) Host Variables

Defining the hash table with at least one key and one data variable is not the only requirement to make it operable. In addition, in order to communicate with the DATA step, the hash variables must have corresponding variables predefined in the PDV before the table can become usable. In other words, at the time when the hash object tools are invoked to define hash variables, variables with the same exact names must already exist in the PDV. Let us make several salient points about them:

- In this book, from now on, we call the PDV variables corresponding to the variables in the hash table the *PDV host variables*.
- This is because they are the PDV locations from which the hash data variables get their values and into which they are retrieved.
- When a hash variable is defined in a hash table, it is from the existing host variable with the same name that it inherits all attributes, i.e., the data type, length, format, informat, and label.
- Therefore, if, as mentioned above, key portion and the data portion each contain a hash variable with the same name, it will have all the same exact attributes in both portions as inherited from one, and only one, PDV host variable with the same name.
- The job of creating the PDV host variables, as any other PDV variables, belongs to the DATA step compiler. It is complete when the entire DATA step has been scanned by the compiler, i.e., before any hash object action - invoked at run time - can occur.
- Providing the compiler with the ability to create the PDV host variables is sometimes called *parameter type matching*. We will see later that it can be done in a variety of ways, different from the standpoint of automation, robustness, and error-proneness.

In order to use the hash object properly, you must understand the concept of the PDV host variables and their interaction with the hash variables. This is as important to understand as the rules of Bizarro Ball if you want to play the game.

1.5 Hash Table Lookup Organization

- The table is internally organized to facilitate the *hash search algorithm*.
- Reciprocally, the algorithm is designed to make use of this internal structure.

- This tandem of the table structure and the algorithm is *sufficient and necessary* to facilitate an extremely fast mechanism of *direct-addressing table look-up* based on the table key.
- Hence, there is no need for the overhead of a separate index structure, such as the index file in the case of an indexed SAS table. (In fact, as we will see later, the hash table itself can be used as a very efficient memory-resident search index.)

For the purposes of this book, it is rather unimportant how the underlying hash table structure and the hashing algorithm work – by the same token as a car driver can operate the vehicle perfectly well and know next to nothing about what is going on under the hood. As far as this subtopic is concerned, hash object users need only be aware that its key-based operations work fast – in fact, faster or on par with other look-up techniques available in SAS. In particular:

- The hash object performs its *key-based* operations *in constant time*. A more technical way of saying it is that the run time for the key-based operations *scales* as $O(1)$.
- The meaning of $O(1)$ notation is simple: The speed of hash search does not depend on the number of items in the table. If N is the number of unique keys in the table, the time needed to either find a key in it or discover that it is not there *does not depend on N*. For example, the same hash table is searched equally fast for, say, $N=1,000$ and $N=1,000,000$.
- It still does not hurt to know how such a feat is achieved behind the scenes. For the benefit of those who agree, a brief overview is given in the last, optional, section of this chapter, "*Peek Under the Hood*".

1.5.1 Hash Table Versus Indexed SAS Data File

To look at the hash table properties still more systematically, it may be instructive to compile a table of the differences between a hash table and an indexed SAS file:

Table 1.2 Hash Table vs Indexed SAS File

Attribute		Hash Table	Indexed SAS File
Table residence medium		RAM	Disk
Time when defined/structured		Run	Compile
Key portion	Required or not	Yes	Yes
	PDV host variables	Yes	Yes
	Partial key look-up	No	Yes
Data portion	Required or not	Yes	No
	PDV host variables	Yes	No
Key search	Structure	AVL trees	Index binary tree
	Algorithm	Hash + AVL trees	Binary search
	Index / index file	No	Yes
	Scaling time	$O(1)$=constant	$O(log2\ (N))$

1.6 Table Operations and Hash Object Tools

To recap, the hash object is a *table in memory* internally organized around the hashing algorithm and tools to store and retrieve data efficiently. In order for any data table to be useful, the programming language used to access it must have *tools* to facilitate a set of fundamental standard *operations*. In turn, the *operations* can be used to solve programming or data processing *tasks*. Let us take a brief look at the hierarchy comprising the *tasks*, *operations*, and *tools*.

1.6.1 Tasks, Operations, Environment, and Tools Hierarchy

Whenever a data processing task is to be accomplished, we do not start by thinking of *tools* needed to achieve the result. Rather, we think about accomplishing the task in terms of the *operations* we use as the building blocks of the solution. Suppose that we have a file and need to replace the value of variable VAR with 1 in every record where VAR=0. At a high level, the line of thought is likely to be:

1. *Read.*
2. *Search* for records where VAR=0.
3. *Update* the value of VAR with 1.
4. *Write.*

Thus, we first think of the *operations* (read, search, update, write) to be performed. Once the plan of operations has been settled on, we would then search for an environment and *tools* capable of performing the *operations*. For example, we can decide whether to use the DATA step or SQL environment. Each environment has its own set of tools, and so, depending on the choice of environment, we could then decide which *tools* (such as statements, clauses, etc.) could be used to perform the *operations*. The logical hierarchy of solving the problem is sequenced as *Tasks->Operations->Environment>Tools*.

In this book, our focus is on the SAS hash object environment. Therefore, it is structured as follows:

*   Classify and discuss the hash table operations.
*   Exemplify the hash object tools needed to perform each operation.
*   Demonstrate how various data processing tasks can be accomplished using the hash table operations.

1.6.2 General Data Table Operations

Let us consider a *general data table* - not necessarily a hash table, at this point. In order for the table to serve as a programmable data storage and retrieval medium, the software must facilitate a number of *standard table operations* generally known as CRUD - an abbreviation for *Create, Retrieve, Update, Delete*. (Since the last three operations cannot be performed without the *Search* operation, its availability is tacitly assumed, even though it is not listed.) For instance, an indexed SAS data set is a typical case of a data table on which all these operations are supported via the DATA step, SQL, and other procedures. A SAS array is another example of a data table (albeit in this case, the SAS tools supporting its operations are different). And of course, all these operations are available for the tables of any commercial database.

In this respect, a SAS hash table is no different: The hash object facilitates all the basic operations on it and, in addition, supports a number of useful operations dictated by their specific nature. They can be subdivided into two levels: One related to the table *as a whole*, and the other - to the *individual table items*. Below, the operations are classified based on these two levels:

Table 1.3 Hash Table Operations Classification

Level	Operation	Function
Table	Create	Create a hash object instance and define its structure.
	Delete	Delete a hash object instance.
	Clear	Remove all items from the table without deleting the table.
	Output	Copy all or some data portion variable values to a file.
	Describe	Extract a table attribute.
	Search	Determine if a given key is present in the table.
Item	Insert	Insert an item with a given key and associated data.
	Delete All	Delete *all items* (the group of items) with a given key.
	Retrieve	Extract the data portion values from the item with a given key.
	Update All	Replace the data portion values of *all items* with a given key.
	Order	Permute same-key *item groups* into an order based on their keys.
	Enumerate by Key (Keynumerate)	Retrieve the data from the item group with a given key sequentially.
	Selective Delete	Delete specific items from the item group with a given key.
	Selective Update	Update specific items in the item group with a given key.
	Enumerate All	Retrieve the data from all or some table items sequentially, without using a key. Requires the *hash iterator object*.

1.6.3 Hash Object Tools Classification

The hash table operations are implemented by the *hash object tools*. These tools, however, have their own *classification*, *syntactic form*, and *nomenclature*, different from other SAS tools with similar functions. Let us consider these three distinct aspects.

The hash object tools fall into a number of distinct categories listed below:

Table 1.4 Hash Object Tools Classification

Hash Tool	Purpose
Statement	Declare (and, optionally, create an instance of) a hash object.
Operator	Create an instance of a hash object.
Attribute	Retrieve hash table metadata.
Method	Manipulate a hash object and its data.
Argument tag	Specify or modify actions performed by operators, and methods.

Generally speaking, there exists *no one-to-one correspondence* between a *particular hash tool* and a *particular standard operation*. Some operations, such as search and retrieval, can be performed by more than one tool. Yet some others, such as enumeration, require using a combination of two or more tools.

1.6.4 Hash Object Syntax

Unlike the SAS tools predating its advent, most hash tools are invoked using the *object-dot syntax*. Even though it may appear unusual at first to those who have not used it, it is not complicated and is easy to learn from code examples (abundant in this book) and from the documentation, as well as online forums, such as *communities.sas.com*. Since the beginning of SAS 9, the DATA step compiler has been endowed with the facility to parse and recognize this syntax as valid. In fact, this is the only way the compiler looks at code that uses the hash object tools: Syntax checking is the only thing done with the hash object at compile time. Everything else is done by the object itself at execution (run) time.

1.6.5 Hash Object Nomenclature

The key words used to call the hash object tools to action have their own naming conventions, *more or less* reflecting the nature of actions they perform and/or operations they support.   However, their nomenclature is conventional in the sense that it adheres to using the standard SAS names.

This is also true of the name given to the hash object itself when it is declared. Since the DATA step compiler views such a name as a variable (albeit of a *non-scalar* data type different from that of numeric or character), it must abide by all the SAS variables' naming conventions - including the rules imposed by the value of the VALIDVARNAME system option currently in effect. Thus, for example, if VALIDVARNAME=ANY, the hash object can be named # by referencing it in code as "#"n; however, then all subsequent references to it must follow this exact form.

1.7 Peek Under the Hood

As noted earlier, it is not really necessary to be aware of the hash object's underlying table structure and its hashing look-up mechanism in order to use it productively. However, a degree of such awareness is akin to that of inquisitive drivers who know, at a high level, how their cars work. Likewise, a more inquisitive hash object user is better equipped than one who is oblivious to what is going on beneath the surface – particularly in certain situations (some of which are presented later in this book) when the object is utilized on the verge of its capacity.

We provided this peek at the request of one of the IT users who were most interested in understanding the ins/outs of the SAS hash object. So, as just stated, the details that follow are targeted to more advanced IT users and programmers.

1.7.1 Table Organization and Unindexed Key Search

When there is a need to rapidly search a large table, creating an index on the table key may be the first impulse. After all, this is how we look for a keyword in a book. The idea that the *table organization itself* coupled with a suitable look-up strategy (and not relying on a separate index) can be used for searching just as successfully may not spring to mind just as naturally.

And yet, it should be no surprise. For instance, let us think how we look for a book page given its number. Using the fact that the pages are ordered, we open the book somewhere in the middle and decide in which half our page must be located. To wit, if the page number being sought is greater than the page number the book is opened to, we continue to search only the pages above the latter; and if the opposite is true, we continue to search below it. Then we proceed in the same manner with the remaining part of the book and repeat the process until the page we need is finally located. In this process of *binary search*, the book is nothing more than a table keyed by the page number. Effectively, taking advantage of the page order as the table organization, we use a divide-and-conquer algorithm to find the page we need. In doing so, we need no index to find it, relying solely on the table structure (i.e., the fact that it is ordered) itself.

In the same vein, an *ordered* table with N unique arbitrary keys (not necessarily serial natural numbers like book pages) can be searched in $O(log2(N))$ time using the binary search. In this case, the key order is the table's organization, and the binary search is the algorithm. The $O(log2(N))$ is an example of the so-called *"big O" notation*. $O(log2(N))$ means that every time N doubles, the maximum number of the comparisons between the keys in the table and the search key needed to find or reject it (let us denote this number as Q) increases merely by 1. By contrast, with the linear ("brute-force") search, Q is *proportional* to N, which is why in "*big O*" notation its search behavior is described as $O(N)$.

Thus, the binary search (though more complex and computationally heavy) scales much better with the table size. For example, at N=4, Q=2 for the linear search and Q=3 for the binary search. But already at N=16, the respective numbers are Q=8 and Q=5, while at N=1024, they diverge as far from each other as Q=512 and Q=11, correspondingly.

However, while the binary search is obviously very good (and consequently widely used), it has a couple of shortcomings. First, it requires the table to have been sorted by the key. Second, Q still grows as N grows, albeit more slowly. The tandem of the hash table organization coupled with the hashing algorithm rectifies both flaws: (a) it does not require presorting, and (b) it facilitates searching with $O(1)$ running time. Let us look more closely how it works in the hash object table.

1.7.2 Internal Hash Table Structure

The hash table contains a number of AVL (*Adelson-Volsky and Landis*) search trees, which can also be thought of as "buckets". An AVL tree is a *balanced binary tree* designed in such a way and populated by such a balancing mechanism that its search run time is always $O(log2(N))$ – i.e., the same as the binary search - regardless of the uniformity or skewness of the input key values. (Balancing the tree while it is being populated requires a bit of overhead; however, in the underlying SAS software it is coded extremely tightly and efficiently.) Visually, the structure (where N is the number of unique keys and H is the number of trees) can be viewed schematically as follows:

Table 1.5 Hash Object Table Organization Scheme

Tree (bucket) Number	1	2	…	H
Keys, ~N/H per tree	Tree#1 keys	Tree#2 keys	Tree#… keys	Tree#H keys

The number of the AVL trees in the hash object table (denoted as H above) is controlled by the value assigned to the argument tag HASHEXP, namely, H=2**HASHEXP. So, the number of buckets can be only a power of 2: 1, 2, 4, and so on up to the maximum of 2**20=1,048,576. If the HASHEXP argument tag is not specified, HASHEXP:8, i.e., H=256 is assumed by default. Any HASHEXP value over 20 is auto-truncated to 20. We will return to the question of picking the right number of trees later after discussing the central principle of the hashing scheme.

1.7.3 Hashing Scheme

The central idea behind hashing is to insert each key loaded into the table into its own tree in such a clever way that *each tree receives about the same number of keys regardless of their values and without any prior knowledge of these values' actual distribution.*

If that were possible, we would have N/H keys in each tree. Suppose we have 2**20 (about 1 million) keys to load. If we loaded them in a single tree (HASHEXP:0, H=1), using the binary tree search to find or reject a search key would require 21 comparisons between the search key and some keys in the tree.

However, if the keys were *uniformly* loaded into H=1024 trees (HASHEXP:10), each would contain only about 1024 keys. Given a search key, we would know the number of the bucket where it is located. Searching among 1024 keys in that tree via the binary search would then require only 11 key comparisons to find or reject the search key; i.e., the look-up speed would be practically doubled.

1.7.4 Hash Function

This divide-and-conquer strategy is not difficult to envision. But how do we make sure that each bucket receives an *approximately equal number of input keys* if we know nothing about their values *a priori*? If our input keys were, say, natural numbers from KEY=1 to KEY=1024 and we had 8 trees to fill, we would just divide the keys in 8 equal ranges, with exactly 4 keys per tree. But what to do if we do not even know anything about the input key values? And what if the keys are not merely natural numbers but arbitrary character keys or, even worse, mixed-type compound keys?

Luckily, there exist certain transformations called *hash functions*. A *good hash function* has three fundamental properties:

1. It can accept N *arbitrary keys with arbitrary values* as arguments and return a natural number TN (tree number) from 1 to H: *TN = hash_function (KEY)*.
2. Each TN number from 1 to H will be assigned to approximately N/H keys, *with very little or no variation* between different TN numbers.
3. For any given key-value, it can return one and only one value of TN. In other words, there will be no situation when the same key is assigned to two different trees.
4. It is reasonably fast to compute. Indeed, if it were very slow to calculate and hence take an inordinately long time to find to which tree a key belongs, it would defeat the very purpose of rapid and efficient search in the first place.

To see how such a feat can be pulled off, let us form a composite key from the pair of variables (Team_SK, Player_ID) from data set Bizarro.Player_candidates. Now let us plug all of its 10,000 distinct values into the cocktail of the nested functions used as a hash function below to distribute the resulting TN values into numbers from 1 to 8. Key-indexed array *Freq* below is used to obtain the frequency on the number of TN values placed into each bucket:

Program 1.1 Hash Function Bucket Distribution

```
data _null_ ;
  set bizarro.Player_candidates end = LR ;
  TN = 1 + mod (rank (MD5 (cats (Team_SK, Player_ID))), 8) ;
  array Freq [8] (8*0) ;
  Freq[TN] + 1 ;
  if LR then put Freq[*] ;
run ;
```

From the values of *Freq[TN]* printed in the SAS log, we get the following picture:

Table 1.6 Hash Function Bucket Distribution

TN	1	2	3	4	5	6		7	8
Freq[TN]	1242	1234	1286	1234	1270	1276		1216	1242

The reason the keys are distributed so evenly is that the MD5 function that is supplied by SAS is a hash function itself. Above, it consumes the composite key (Team_SK, Player_ID) converted into a character string by the CATS function and generates a 16-byte character string (so-called *signature*). The RANK function returns the position of the first byte of the signature in the collating sequence. Finally, the MOD function uses the divisor of 8 to distribute the position number (ranging from 0 to 255) between the values of TN from 1 to 8.

While there is about 4 percent variability between the most and least filled buckets, for all practical intents and purposes using the binary tree search within any of these buckets would be equally fast. As opposed to binary-searching all 10,000 keys, it would save about 3 key comparisons per binary search in the fullest bucket. Since comparing keys is usually the most computationally taxing part of searching algorithms, distributing the keys among the trees may well justify the overhead of computing TN by using the hash function above.

The expression given above is merely one example of a decent hash function used just to illustrate the idea. It works well because MD5 is itself a hash function. Though the internal hash function working for the hash object behind the scenes is different, conceptually it serves the same goal of distributing the keys evenly across the allocated number of the AVL trees.

1.7.5 Hash Table Structure and Algorithm in Tandem

Now that we know that having a good hash function is possible, we can spell out how the hash table's internal structure and the hashing algorithm work in tandem. Given a key to search for:

- The key is plugged into the hash function.
- The hash function returns a tree number TN from 1 to H. The tree with that specific value of TN is the only tree where the key can be possibly found.
- If the tree is empty, the key is not in the table.
- Otherwise, the tree is binary-searched, and the key is either found or rejected.

Further actions depend on the concrete task. For instance, all we may want to know is whether the key is in the table, so no further action is necessary. Alternatively, if the key is not found, we may want to load it, in which case it will be inserted into the tree, whose number is TN. Or if the key is found, we may want to remove it from the table if necessary.

1.7.6 The HASHEXP Effect

To recap, the value of the argument tag HASHEXP determines the number of AVL trees H the hash object creates for its hash table. It follows from the nature of the algorithm that the fewer keys there are in each tree, the speedier the search. But let us look at it more closely:

- Suppose that we have $N=2^{**}24$ (about 16.7 million) keys in the table. With HASHEXP:20 and hence $H=2^{**}20$ (about 1 million), there will be $2^{**}4=16$ keys hashing, on the average, to one bucket. Searching for a key among the16 keys within any AVL tree requires about log2(16)+1=5 key comparisons.

- Now if we had HASHEXP:8 (the default), i.e., H=2**8=256, there would be 2**16=65,536 keys, on the average, hashing to one bucket tree. That would require 17 comparisons to find or reject a search key in a given tree. And indeed, a simple test can show that HASHEXP:20 results in searching about twice as fast as HASHEXP:8.

- The penalty of increasing HASHEXP and H comes in the form of the amount of a certain part of *base memory* required to allocate 2**20 trees versus 2**8. However, base memory is not memory needed to hold the actual keys and data but rather memory required to support the table infrastructure. In other words, it is memory occupied by the empty table. It is static; i.e., once allocated, it does not change regardless of the amount of actual data loaded in the table. And the penalty is severe: For example, on the Windows 64-bit platform, a table with two numeric variables (one per portion) needs about 17 MB with HASHEXP:20 and about 1 MB with HASHEXP:8. Despite the 17 times difference, in the modern world of gigantic cheap memories, the 16 MB static difference is of little consequence.

- Thus, HASHEXP:20 can be coded safely under any circumstances to trade faster execution for a few megabytes of extra memory.

- Even so, it still hardly makes sense to allocate more trees H than the number of unique keys N and waste any memory on empty trees. Yet valuing HASHEXP with a value less than 8 (default) does not make much sense either because the base hash memory difference between HASHEXP:8 (256 trees) and HASHEXP:0 (1 tree) is practically negligible.

The real limitation on the hash object memory footprint comes from the length of its entry multiplied by the number of hash items. It is crucially important in a number of data processing situations when using the hash object appears to be the best or only solution but requires large data volumes to be stored in a hash table. In this book, a separate chapter is dedicated to the ways and means of reducing hash memory usage.

1.7.7 What Is in the Name?

It is an easy guess that the SAS hash object is called the hash object because its built-in hash function and hashing algorithm underlie its functionality and performance capabilities. The name has been in use for almost two decades and irrevocably internalized by the SAS community.

To those who are interested in SAS lore it may be interesting to know that the first name for this wonderful addition to the SAS arsenal was "associative array". Perhaps the idea was to take the attention off the mechanism underlying the object and refocus it on what it does from the perspective of usage.

An associative array is an abstract data type resembling an ordinary array, with the distinction that (a) it is dynamic (i.e., grows and shrinks as items are inserted and deleted) and (b) can be subscribed, not only using an integer key, but any key, including a composite key with character components. It is easy to perceive that the hash object possesses both properties. So, calling the hash object an associative array initially made sense, for most SAS programmers are well familiar with arrays and could relate to the new-fangled capability from this angle.

Chapter 2: Table-Level Operations

2.1 Introduction

In this chapter, we will discuss the hash table operations pertaining to the table as a whole entity. It means that these operations are concerned not with particular items in the table but with its existence, properties, and all its items at once.

2.2 CREATE Operation

The *Create* operation creates an operable initialized instance of a hash object (and the associated hash table). It involves the following compile time and run-time stages:

- Compile time:
  1. Declare a hash object with a given name.
  2. Define a PDV variable of type hash with the same name.

- Run time:
  1. Create a hash object instance.
  2. Generate a unique value of type hash to identify the instance.
  3. Assign this value to the PDV variable defined above.
  4. Define the key portion hash variables.
  5. Define the data portion hash variables.
  6. Validate the syntax used at stages 1, 4, 5.
  7. Check that for each defined hash variable, a host variable with the same name exists in the PDV (i.e., check for *parameter type matching*).
  8. Initialize the hash object instance.

The following code snippet is a simple example of implementing this plan:

Program 2.1 Chapter 2 Create Operation Template.sas

```
data _null_ ;
   declare hash H ;
   H = _new_ hash() ;
   H.defineKey  ("K") ;
   H.defineData ("D") ;
```

```
      H.defineDone () ;
      stop ;
      K = . ;
      D = "" ;
run ;
```

Using this step as a base template, let us take a look at the different stages of the *Create* operation.

2.2.1 Declaring a Hash Object

The hash tool for declaring a hash object is the DECLARE statement. Despite its outward simplicity, it would benefit us to dwell on a number of points:

- The DECLARE statement can be abbreviated as DCL.
- DECLARE|DCL and HASH are *keywords*, and so the DATA step compiler checks them for syntactic validity.
- The keyword HASH must be followed by a name given to the object. In the example above, the object is named H. However, from the standpoint of syntax, it can be any *valid* SAS variable name within the constraints of the VALIDVARNAME system option currently in effect.
- In the form shown above (i.e., with no parentheses after the object name - we will dwell more on it later), the DECLARE|DCL statement is a *compile-time only* directive. It means that, at run time, the statement is ignored. (In this sense, it is similar, for example, to the ARRAY statement.)
- The name assigned to the object - in the step above, H - is the name of a PDV *variable* the compiler creates when it parses the DECLARE statement. Its future purpose is to hold a *non-scalar value of type hash* identifying a concrete hash object instance. Correspondingly, variable H is a *non-scalar* variable *of type hash*.
- Since variable H is non-scalar, i.e., not numeric or character, it cannot coexist in the same DATA step with a numeric or character variable with the same name.
- Likewise, it cannot coexist with a SAS array with the same name. Although from the standpoint of the compiler, the array name is also a reference to a non-scalar variable, its data type is different from type hash.
- Hence, if a numeric or character variable or array with this name is already present in the PDV, it cannot be used to name a hash object. For the same reason, this variable cannot be assigned to a scalar variable, nor can a scalar variable be assigned to it. Failure to observe these safety rules will create a data type conflict and generate a compilation error. In short, an attempt to use such a variable as if it were numeric or character may result in an error message that a scalar cannot be converted to an object of type hash (or vice versa).
- The best practice in this regard is to ensure that the compiler *does not see* the variable name used to name the object *anywhere in the step* other than as a *valid* reference to the hash object it denotes.
- In particular, it means that the name of any scalar (numeric or character) variable or of an array cannot be the same as a hash object name, and vice versa.
- With respect to the scalar variables, this rule encompasses not only the variables *overtly* referenced in the DATA step by name, but any variable present in the PDV. Thus, it includes the variable coming from any data set or view referenced in the SET, MERGE, UPDATE, or MODIFY statements.

It must be also mentioned that a given DATA step can contain only a single DECLARE statement with a given hash object name. Repeating the DECLARE statement with the same object name, such as:

```
data _null_ ;
  dcl hash H1 ;
  dcl hash H2 ; *No conflict here;
  dcl hash H1 ; *Compile-time error: H1 is already defined;
run ;
```

results in a compile-time error and prevents the step from being executed. An error message is written to the log indicating that variable H1 has been already defined, as a non-scalar variable cannot be declared more than once. (Not coincidentally, the compiler reacts the same way and with the same error message to an attempt of defining an array with the same name twice.)

2.2.2 Creating a Hash Object Instance

Before hash table variables can be defined, it is not enough to merely *declare* the hash object. In addition, an *instance* of the declared object must be created. In the step above, it is done by using the _NEW_ operator:

```
H = _new_ hash() ;
```

First, let us make a few observations about the *syntax* of this statement:

- _NEW_ and HASH are keywords, and so their syntax ought to be strictly observed.
- The receiving variable named on the left side of the assignment (in this case, H) must have been already compiled as type hash. In other words, this statement must be *preceded* by a valid DECLARE statement defining H as type hash.
- It *cannot* be done in reverse. If the compiler sees the statement with the _NEW_ operator first, it will, by default, set variable H as numeric (i.e., scalar), and the DECLARE statement trying to define H as type hash will create a data type conflict.
- The blank space between the parentheses following the HASH keyword is intended for argument tags. They may have a profound impact on the hash object operations and are discussed elsewhere in the book. When they are left out, as above, they are assigned default values.

Now let us see what *actions* this deceptively simple statement implies:

- Create a *new instance* of hash object H.
- Generate a distinct *non-scalar* value *of type hash* to identify the newly created hash object instance. One way to think of it is of a pointer to the location in memory where the object instance resides.
- Assign this pointer value to PDV variable H of type hash.

The PDV value of H is what the program uses to identify the *concrete* hash object instance when the object is referenced by name - in this case, H. Any hash tool, such as a method or attribute, referenced by H, works on the instance pointed at by the *current* PDV value of H. In other words, the hash object instance whose identifying value is currently stored in variable H is *active*. Yet another way to express it is to say that the current PDV value of H *surfaces* the hash object instance it identifies.

Understanding this concept is quite important because, as we will see later in the book, for a given object name (such as H) defined in the DECLARE statement, more than one hash object instance can be created and used. In this case, we need to know how to tell the program to use a concrete instance at a given execution point according to program logic; and this is done by making the instance we want the program to work on *active* (or, which is the same, by *surfacing* it).

Another takeaway from this section is that the program is intended to use only a single instance of any given hash object, and measures must be taken to prevent the statement that creates a new instance from being executed more than once. We will discuss these measures later in this chapter.

2.2.3 Combining Declaration and Instantiation

The two statements that declare a hash object and then create a new instance of it:

```
declare hash H ;
H = _new_ hash() ;
```

can be combined into a single statement:

```
declare hash H() ;
```

Note that the only syntactic addition to the DECLARE statement used before is the *pair of parentheses* after the hash object name. In a *single* line of code, this statement:

At compile time:

1. Declares a hash object named H.
2. Creates PDV variable H of type hash to hold object instance identifying values.

At run time:

1. Creates a new instance of hash object H.
2. Generates a distinct type hash value of H to identify it.
3. Assigns this value to variable H, thereby making the instance active.

Note that since the combined DECLARE statement is not an *overt* assignment statement, it may *seem* that it does not assign anything to variable H. However, this is not true: It does assign, behind-the-scenes, the newly created value identifying the instant to variable H, just like the overt assignment statement with the _NEW_ operator.

Also note that the space between the parentheses following the object name can be filled with argument tags and their arguments in exactly the same manner as when the _NEW_ operator is used in a separate statement.

The compound DECLARE statement, for most intents and purposes, is equivalent to the two separate statements it replaces. However, combining the compile time part with the run-time part in a single statement results in a certain loss of flexibility. In particular, if there is a need to create and use two instances of hash object H at different points in the DATA step program, it is perfectly okay to code:

```
dcl hash H ;
H = _new_ hash() ; *Create instance #1;
...
H = _new_ hash() ; *Create instance #2;
```

However, the same purpose cannot be achieved by coding:

```
dcl hash H() ;
...
dcl hash H() ; *Compile time error;
```

because the second statement will result in a compile time error. Seeing the *first* statement, the compiler interprets its compile time declarative part as a directive to define variable H of type hash and creates it. But when the compiler sees the same declarative part in the *second* statement, it stops compilation with an error, since it cannot define a non-scalar variable more than once, and generates an error message stating that variable H is already defined.

2.2.4 Defining Hash Table Variables

Now an instance of hash object H has been created, and the PDV value of variable H has been assigned a unique value that identifies the instance as active (i.e., surfaced). However, at this point, no hash table associated with the instance is yet defined. To do so, at the next stage of the *Create* operation we need to provide the object constructor with the names of hash variables for the key and data portions of the table.

This is done by calling the DEFINEKEY and DEFINEDATA methods, respectively. In the sample DATA step above, the key portion is defined with a single variable K, and the data portion - with a single variable D - as follows:

```
H.defineKey  ("K") ;
H.defineData ("D") ;
```

Any method call generates a numeric return code indicating if the call has succeeded (if the return code is zero) or failed (if it is not zero). In this book, the method call style shown above is termed *unassigned* since its return code is *not assigned* to a separate variable. The other style, termed *assigned*, captures the return code by assigning it to a numeric variable, so that it can be examined later. For example, in the following assigned calls (equivalent to the unassigned calls above) the return codes are captured in variable RC:

```
rc = H.defineKey  ("K") ;
rc = H.defineData ("D") ;
```

As a side note, in this case calling the methods assigned or unassigned is merely a matter of style preference because these methods always succeed (i.e., return a zero code) if their syntax is formally correct. If there is anything wrong with them otherwise (e.g., if a variable has an invalid name, a conflict with another variable, etc.), the error will be caught later on at the next stage of the *Create* operation.

Let us now look at some details associated with the DEFINEKEY and DEFINEDATA methods:

- The calls do not have to follow the order shown above, i.e., either the DEFINEKEY method or the DEFINEDATA method can be called first or vice versa.
- In these method calls, the hash variable names are defined using *character literal constants* - in other words, quoted strings with a fixed value. As a side note, single and double quotes are equally valid, and their choice is a matter of the style one prefers - unless, of course, the content between them is a macro variable reference (in which case double quotes must be used to resolve it).
- Due to the dynamic nature of the hash object, the method argument defining a hash variable does not have to be just a character literal. In fact, it can be any DATA step *character expression* (unquoted), as long as it resolves to the character value representing the name of the variable we

want to define (such as, in this case, "K" or "D". This is a valuable feature further expounded upon in this chapter and also used in other chapters to create dynamic code.

- The hash variables being defined inherit all their attributes, such as the length, data type, format, etc., from the like-named host variables present in the PDV at the time of call. In our example, they are variables K and D placed into the PDV during the DATA step compilation phase as a result of parsing the assignment statements following the STOP statement.

2.2.5 Omitting the DEFINEDATA Method

It is possible to omit the DEFINEDATA method and call the DEFINEKEY method only. At times, it can be useful to do so; however, the user ought to be mindful of the following:

- It *does not mean* that the table will have *only the key portion* and the data portion will be absent. As already noted, a SAS hash table cannot exist without both.
- So, if the table definition includes a DEFINEKEY call only, all the variables defined by it to the key portion will be *automatically included in the data portion* as well.
- It should be kept in mind, especially when the key portion is defined with numerous and/or long variables, that including them automatically in the data portion by omitting a DEFINEDATA method call automatically doubles the overall hash entry length and increases the hash table memory footprint.
- Though it is possible to call DEFINEKEY without calling DEFINEDATA, *the reverse is not true*! At least one valid DEFINEKEY call must be always included in the definition. Failure to do so will be detected by the DEFINEDONE method call and result in a run-time error with the log message that the keys are "uninitialized".

For example, suppose that in our sample step above we omitted the DEFINEDATA call and included only

```
H.defineKey ("K") ;
```

Then it would be exactly equivalent to calling both methods with variable K as an argument:

```
H.defineKey ("K") ;
H.defineData("K") ;
```

It means that now both the key portion and the data portion contain variable K. And if there were more than one variable defined by a stand-alone DEFINEKEY call, all of them, in the same order, would end up in the data portion as well.

2.2.6 Wrapping Up the Create Operation

The last stage of the *Create* operation is executed by calling the DEFINEDONE method, either unassigned or assigned:

```
H.defineDone() ; /* unassigned call */
```

or

```
rc = H.defineDone() ; /* assigned call */
```

The DEFINEDONE method is responsible for the following actions:

- Validate the internal syntax of the DEFINEDATA and DEFINEKEY method calls.

- Make sure that host variables with the same exact names exist in the PDV.
- If either condition above is not satisfied, the DEFINEDONE call will fail, return a non-zero code, generate an error message in the log, and stop the DATA step.
- Otherwise, initialize the hash object instance.

2.2.7 PDV Host Variables and Parameter Type Matching

As it has been repeatedly stated, for every variable name defined in the hash table entry, a like-named variable must exist in the PDV. A SAS programmer new to hash object programming might at first get the impression that when the DATA step compiler parses hash variable definition statements, such as:

```
H.definekey ("K") ;
H.defineData ("D") ;
```

it will infer from them the type of K and D as numeric and place both into the PDV. This is what the compiler does, for example, when it parses code unrelated to the hash object and encounters a new variable name as part of a SAS expression.

Yet, with respect to the hash object *this is not the case* at all. While parsing code related to the hash object, the compiler performs only two actions:

1. *Validates the syntax.* For example, if in the DEFINEKEY call above, the period were missing, or the method name were mistyped, or the parentheses and quotes were unbalanced, etc., a compilation error would occur.
2. *Validates the reference variable.* The compiler checks if the variable by which the object is referenced - such as variable H above - has been already properly defined as a variable of type hash. Thus, if variable H of type hash were not defined in a valid DECLARE statement before H is referenced, it would also lead to a compilation error.

As long as these two items check out, the compiler's job is done as far as hash object code is concerned. It does not see the hash variable names passed to the DEFINEDONE and DEFINEDATA methods, does not check if they are valid, and therefore does not create the corresponding variables in the PDV.

This is why a means must be provided for the compiler to create variable named K and D (in this case) in the PDV, replete with their attributes, at compile time. This is the purpose of the two assignment statements coded last in the sample DATA step after the STOP statement:

```
K = . ;
D = "" ;
```

To wit, their goal is to place *host variables* named K and D, corresponding to the hash variables also named K and D, into the PDV and make them available for the hash object at run time later. The procedure can be described as follows:

- During the *compile phase*, the DATA step compiler parses the assignment statements.
- It infers from the literals assigned to K and D that a numeric variable K and a character variable D with length 1 be placed into the PDV.
- This way, by the time the DEFINEDONE method is called during the *execution phase*, variables K and D are already in the PDV, along with their respective attributes.

- Seeing that *hash variable* K has been defined in the key portion, the DEFINEDONE method searches the PDV for a *host variable* with the name K. Since it is there, hash variable K passes the check.

- The same actions are then performed with respect to the hash variable D and host variable D.

- If both K and D pass muster, they are initialized for use in the hash table. This wraps up the *Create* operation.

Pre-defining host variables in the PDV at compile time, so that the hash object can use them later at run time, is also termed *parameter type matching*. Note that placing parameter type matching statements (in this case, the assignments) after the STOP statement is optional; i.e., they can appear anywhere in the DATA step. Above, it is done primarily to highlight the *temporal separation* between the compilation phase (during which the host variables are created in the PDV) and the execution phase (during which they are relied upon by the hash object operations).

From the standpoint of the *Create* operation, the location of parameter type matching statements in the step is irrelevant. If parameter type matching is their only purpose - that is, they are not intended to be executed at run time - the program has to be structured accordingly. Above, this is done by placing them after the STOP statement. Another way is to place them in a block of code preceded by an IF condition that is *always false*, such as:

```
IF 0 then do ;
  <parameter type matching code>
end ;
```

2.2.8 Other Ways of Hard-Coded Parameter Type Matching

Needless to say, it does not necessarily have to be done via assignment statements. Any valid block of code letting the compiler populate the PDV with variables with the same names as the defined hash variable names will also work. For example, instead of using the assignment statements above, a LENGTH statement could be used to achieve exactly the same parameter type matching effect. The only purpose of the MISSING call routine below is to avoid the pesky "uninitialized" warning in the log if one of the variables is not valued.

```
length K 8 D $ 1 ;
call missing (K, D) ;
```

Or, alternatively, the RETAIN statement could be used as well, with the same result:

```
retain K . D "" ;
```

2.2.9 Dynamic Parameter Type Matching via File Reference

The parameter type matching techniques shown above suffer from the same basic flaw: They are essentially hard-coded. It is okay if the DATA step in question is where the variables defined by these techniques are created in the first place. However, more often than not, the values with which a hash table is eventually populated come from reading a SAS data file. In this case, hard-coding presents a problem, and here is why.

Suppose that we have a data set containing variables K and D. For example:

Program 2.2 Chapter 2 HashValues Sample Data set.sas

```
data hashValues ;
   input K D:$1. ;
cards ;
1 A
2 B
3 C
run ;
```

Suppose further that we want to use K as the hash table key and D as its data portion variable - for example, suppose we want to insert the (K,D) value pairs from file hashValues into the table later on. If we decided to use hard-coding for parameter type matching, we would first need to ensure that the data types and lengths of the hard-coded variables match those in the file. In turn, it means that we would need to find out what those attributes are by doing, for example, any of the following:

- Locate the original code used to create the file (it may not be even available).
- Query dictionary.columns or sashelp.vcolumn.
- Run the CONTENTS procedure and look at the output.
- Take a look at the file properties via the SAS viewer or another interface.

Doing any of those things runs counter to the principles of robust automated programming. Worse still, after finding the attributes of K and D in the file, they would need to be hard-coded in the program *correctly*, so as to avoid conflicts with the like-named variables in the file. Such practice is quite problematic for two reasons:

1. Errare humanum est ("*to err is human*").
2. The more variables are involved in the process, the more laborious it gets and the more ominous the truth encapsulated by the adage above becomes.

Therefore, it is much less labor-intensive and much less error-prone just to let the compiler itself read the descriptor of the data set in question (in this case, hashValues) and place the needed variables along with their attributes into the PDV. Moreover, it is easy to do because the compiler performs this action anytime the name of the data set in question is referenced by a file-reading statement, such as SET, MERGE, UPDATE, or MODIFY. Given that, there are three distinct cases:

1. One of these statements referencing the file in question is already present somewhere in the DATA step, and the requisite variables are kept, i.e., not eliminated by the KEEP= or DROP= data set option.
2. Same as above, but the requisite variables are *not* kept.
3. None of these statements referencing the file in question is present anywhere in the step.

In case #1, parameter type matching occurs automatically, and so no other action to achieve it is required. In case #2, all that is required to achieve parameter type matching is to recode the DROP= or KEEP= variable list (or omit it altogether) in order to ensure that the requisite variables are kept. Once it is done, this case becomes no different from case #1. We will delve more into these two cases later after concentrating on case #3.

2.2.10 Parameter Type Matching by Forced File Reference

In case #3, the simplest way to attain the goal is to include a *non-executable* SET statement referencing the file in question (and keeping the requisite variables) anywhere in the DATA step. Making it non-executable ensures that the file is seen only at compile time, and no data is read from it at run time (thus preventing the statement from possibly compromising the rest of the program).

If the DATA step contains the unconditional STOP statement (as in the step above), any statement following it is non-executable. Hence, in this case the parameter type matching SET statement can be simply included after STOP, e.g.:

```
  stop ;
  ...
  SET hashValues (keep = K D) ;
  ...
run ;
```

This parameter type matching technique operates as follows:

- Since SET is coded after the STOP statement, it reads no actual data from data set hashValue*s* at run time.
- However, at compile time the compiler reads the descriptor of the data set and places variables K and D and their attributes into the PDV.
- It initializes PDV variables K and D to the missing values of the appropriate data types, thus avoiding "uninitialized" warnings in the SAS log.

If the step does not already contain an unconditional STOP statement, the same parameter matching effect can be achieved by coding, somewhere in the step, the SET statement preceded by an *obviously false* condition to make it non-executable. For example:

```
IF 0 then SET hashValues (keep = K D) ;
```

Because the condition above is always false, it prevents the SET statement from being executed at run time, yet still exposes it to the compiler at compile time. Its actions are exactly the same as of the SET statement placed after STOP.

In the ensuing chapters, this robust parameter type matching method will be used widely in both variations. Note that the technique allows for a number of modifications depending on the situation and need. For example, if variables K and D were not in the same file but in two different files - say, hashValuesK and hashValuesD, respectively - the issue can be addressed simply by recoding the SET statement as their concatenation, i.e.:

```
IF 0 then SET hashValuesK(keep = K) hashValuesD(keep = D) ;
```

Incidentally, the MERGE statement can be used instead of SET to the same effect, regardless of whether it references one file or more.

2.2.11 Parameter Type Matching by Default File Reference

Under a number of realistic scenarios, no special measures to ensure parameter type matching are needed at all. This happens in two cases.

1. When the same file, from which we want the compiler to obtain the host variables, is already referenced explicitly in order to read the actual data from it. For example, consider this variation of our sample DATA step:

```
data _null_ ;
   dcl hash H()
   H.defineKey  ("K") ;
   H.defineData ("D") ;
   H.defineDone () ;
   do until (lr) ;
      SET hashValues (keep = K D) end = lr ;
      <e.g.: code to insert(K,D) values into table H>
   end ;
   <more code>
   stop ;
run ;
```

Since the compiler sees the SET statement referencing hashValues, it is unnecessary to reference it again elsewhere in the step for the purpose of parameter type matching. This is because from the standpoint of the compiler reading the data set descriptor, it does not matter whether the SET statement is run-time executable or not.

2. When the host variables with the same names as the intended hash variables occur naturally as part of the DATA step program, and so the compiler places them and their attributes into the PDV during the compilation phase. Consider, for instance, the following snippet:

```
data _null_ ;
   dcl hash H() ;
   H.defineKey ("K") ;
   H.defineData ("D") ;
   H.defineDone () ;              ❷
   do K = 1 to length ("ABCDEF") ;  ❶
      D = char ("ABCDEF", K) ;     ❷
      <code to insert current (K,D) pair into table H>
   end ;
   <more code>
   stop ;
run ;
```

❶   The compiler places variable K into the PDV as numeric as an effect of parsing the DO statement where K is used as the loop index.

❷   Then it parses the next statement and creates host variable D as character 1 because this is the type and length the function CHAR returns.

❸   Thus, by the time program control hits the DEFINEDONE call at run time, the host variables for hash variables K and D are already in the PDV with the attributes required, and so no extra measures are needed to make it happen.

2.2.12 Defining Multiple Hash Variables

So far, we have dealt with the case of a single hash variable in the key and data portion. However, in most real-life situations, the key portion or data portion or both comprise more than one variable. Therefore, we need a way to tell the DEFINEKEY and DEFINEDATA methods how to include them all. For example, consider the variables in data set Bizarro.Player_candidates with the following attributes:

Figure 2.1 Player_candidates Data Set Metadata Sample

Column Name	Type	Length
Player_ID	Number	8
Team_SK	Number	8
First_Name	Text	12
Last_Name	Text	12
Position_Code	Text	3

Now suppose that we need to create a hash table H with composite key (Player_ID,Team_SK) and the data portion containing the rest of the variables - for example, intending to use table H as a lookup table downstream. The simplest (but not necessarily the best) way of doing it is to pass comma-separated lists of the respective variable names as character literals to the DEFINEKEY and DEFINEDATA methods as arguments:

Program 2.3 Chapter 2 Define Multiple Hash Variables.sas

```
data _null_ ;
   dcl hash H() ;
   H.defineKey ("Player_ID","Team_SK") ;
   H.defineData("First_name","Last_name","Position_code") ;
   H.defineDone() ;
   stop ;
   set bizarro.Player_candidates ;
run ;
```

The SET statement facilitates parameter type matching by letting the compiler examine the descriptor of Bizarro.Player_candidates and place all its variables in the PDV. To reiterate, the order of the DEFINEKEY and DEFINEDATA calls is irrelevant, and they can be swapped.

To date, this technique of defining multiple hash variables as hard-coded variable lists has been used predominantly - in particular because the SAS documentation neither offers or suggests any other way. However, it can also be done differently. Namely, each variable can be defined using its own method call. For instance, the two calls above can be replaced, without changing the final result whatsoever, with the following series of individual calls, each comprising a single variable name:

```
   H.defineKey ("Player_ID") ;
   H.defineKey ("Team_SK") ;
   H.defineData("First_name") ;
   H.defineData("Last_name") ;
   H.defineData("Position_code") ;
```

Moreover, delimited-list calls and individual calls can be combined without contradicting each other. For example, to define the data portion, we can list First_name and Last_name in one call and leave Position_code for another:

```
H.defineData("First_name", "Last_name") ;
H.defineData("Position_code") ;
```

The calls, either individual or combined, can be issued in any order: It will only alter the sequence in which the variables are placed into the corresponding portions of the hash entry.

2.2.13 Defining Hash Variables as Non-Literal Expressions

The ability to define hash variables one at a time shown above raises the question: Why would it make sense to define them one at a time in separate method calls if it can be done by listing them in a single call? The answer is that it makes no sense as long as the variable names are *hard-coded as character literal constants*, i.e., fixed quoted values, such as "Player_ID", "Team_SK", etc.

However, it starts making sense as soon as we realize that, generally speaking, any argument to the DEFINEKEY or DEFINEDATA method represents a generic SAS *character expression*. A character literal constant is merely the most basic character expression (and it is also the most static since it represents a fixed value).

Suppose that at the time of a DEFINEKEY or DEFINEDATA call, we have a PDV character variable _hVarName of length $32 valued with the name of a hash variable we need to define. For example, imagine that we want to call DEFINEKEY to define hash variable Player_ID; and somewhere in the step before the method call we have the statements:

```
retain _kVarList "Player_ID Team_SK" ;
length _hVarName $ 32 ;
_hVarName = scan(_kVarList,1) ;
```

It means that _hVarName is populated with the value "Player_ID". But _hVarName, being a character variable, is a character expression, too. Therefore, in this case, we can pass it to the method, unquoted, instead of hard-coding a literal constant. That is, instead of coding:

```
H.defineKey ("Player_ID") ;
```

we can code:

```
H.defineKey (_hVarName) ;
```

Note that though in the above snippet _hVarName is populated with "Player_ID" via the SCAN function, the concrete way by which it receives the value is irrelevant. For example, as we will see later on, a variable similar to _hVarName can come from a data set populated with the names of the hash variables to be defined.

Developing the idea of using non-literal expressions further, let us observe that the SCAN function expression is a character expression in its own right. Hence, instead of creating an intermediate variable (such as _hVarName), the entire expression can be passed to the DEFINEKEY method call directly:

```
H.defineKey (scan(_kVarList,1)) ;
```

In sum, any valid character expression can be passed to the DEFINEKEY and DEFINEDATA methods as arguments as long as it resolves to the value representing the name of the hash variable we need to define. Needless to say, the value must be a valid SAS variable name and have a like-named counterpart host variable in the PDV.

2.2.14 Defining Hash Variables Dynamically One at a Time

Now it should be easy to understand why using non-literal expressions to define hash variables one at a time can actually shorten a program and make it tidier. Suppose that we have a list of numeric *D1-D100* to be defined in the data portion of table H. Passing the variable names as character literals to the DEFINEDATA method, we would have to code:

```
h.defineData ("D1","D2",...,"D99","D100") ;
```

Coding this kind of argument list is tedious, messy, and error-prone - it's easy to accidentally mistype a name or miss a quote or a comma. A more astutely lazy programmer could write a macro or a separate preliminary DATA step to assemble the requisite list with all the requisite quotes and commas and pass it to the method as a macro variable reference. For example:

```
data _null_ ;
  length arg $ 32767 ;
  do x = 1 to 100 ;
    arg = catx (",", arg, quote (cats ("D", x))) ;
  end ;
  call symputx ("arg", arg) ;
run ;
```

And then downstream in the DATA step where DEFINEDATA is called:

```
H.defineData (&arg) ;
```

However, neither jumping through these sorts of hoops nor hard-coding is necessary if we take into account the dynamic character expression nature of the DEFINEDATA arguments. Instead, we can simply call the method repeatedly in a DO loop for each hash variable one at a time in the *same* DATA step where DEFINEDATA is called:

```
array DD D1-D100 ;
do over DD ;
  H.defineData(put(vname(DD),$32.)) ;
end ;
```

Above, in each iteration of the loop, the character expression passed to DEFINEDATA automatically resolves to the name of the individual hash variable inferred from the corresponding array element and passes it to the method call. The final result of adding one hash data variable to the data portion one at a time in this manner will be exactly identical to hard-coding (if done correctly) or resolving the macro variable reference. Of course, the same is true if we should need to add a long list of hash variables to the key portion of H by calling the DEFINEKEY method.

2.2.15 Defining Hash Variables Using Metadata

As noted above, most of the time the key and data values loaded into a hash table come from variables in a SAS data file. In such cases, programming logic almost always dictates that the hash variables be defined with the same names as the names of the data set variables the key and data values come from. The names of these variables are already stored in the dictionary table Dictionary.Columns or in the view

Sashelp.Vcolumn in the $32 character variable Name. Since variable Name is just a case of a character expression, using its values to define the names of our hash variables we need to do only the following:

1. Read the dictionary table or view and filter it to suit our needs.
2. For each row read from the filtered table or view, call the DEFINEKEY and/or DEFINEDATA method (depending on whether the respective value of Name is designated for the key or data portion) and pass Name to the method call.

To illustrate the concept, let us suppose that we intend, down the line, to load data from data set Bizarro.Player_candidates into hash table H. Correspondingly, we want to define its hash variables as named after the variables in the data set. More specifically, we want to:

- Define the hash variables in the key portion using the data set variable names ending in "_ID" and "_SK" (for example, we may know that together, the variables with such suffixes form a unique composite key).
- Define the hash variables in the data portion using the names of the rest of the variables in the data set.

A primitive, static, and error-prone way of doing this is to eyeball the metadata related to the data set (for instance, in the SAS viewer) and then hard-code the arguments to the DEFINEKEY and DEFINEDATA method calls based on the findings. A more advanced, dynamic, and robust approach is to exploit the system dictionary tables as outlined above. The dictionary view Sashelp.Vcolumn makes it possible to define the requisite hash variables dynamically right within the DATA step where the rest of the *Create* operation is performed:

Program 2.4 Chapter 2 Define Hash Variables Selectively Metadata.sas

```
data _null_ ;
 dcl hash H() ;
 do until (lr) ;                                                        ❶
   set sashelp.vcolumn (keep=memname libname Name) end=lr ;
   where libname="BIZARRO" and memname="PLAYER_CANDIDATES" ;  ❷
   isKey = scan (upcase (Name), -1, "_") in ("ID","SK") ;     ❸
   if isKey then H.defineKey(Name) ;                          ❹
   else         H.defineData(Name) ;
 end ;
 H.defineDone () ;                                                      ❺
 stop ;
 set bizarro.Player_candidates ;                                       ❻
run ;
```

The hash variable definition plan executed above (after the hash table is declared) is as follows:

❶ Use an explicit DO loop to read a subset of sashelp.vcolumn view one record at a time.

❷ Subset sashelp.vcolumn view to the rows related to data set Bizarro.Player_candidates only. Column Name read from it contains, as its values, the names to be defined to either the key portion or data portion.

❸ If Name ends in "_ID" or "_SK", set Boolean variable isKey to 1; else set it to 0.

❹ If isKey=1, use expression Name (a variable *is* an expression) as the argument to the DEFINEKEY call. Otherwise, use it as the argument to the DEFINEDATA call.

❺ At this point, all the rows from sashelp.vcolumn subset have been read, and for each value of *Name* coming from them, either DEFINEKEY or DEFINEDONE has been called. Call DEFINEDONE to wrap up the *Create* operation.

❻ Make sure, at compile time, that all the host variables corresponding to the hash variables defined to table H reside in the PDV.

Incidentally, the IF-THEN-ELSE block ❹ above can be replaced, with the same effect, by a single statement:

```
RC = ifN (isKey, H.defineKey(Name), H.defineData(Name)) ;
```

Due to the way the IFN functions works, if isKey = 1 (i.e., evaluates as true), the DEFINEKEY method is called; otherwise if isKey=0 (i.e., evaluates false), the DEFINEDATA method is called. Note that the assignment statement here is a *dummy* statement, i.e., it is used merely as a vehicle to execute the IFN function. Respectively, RC is used merely as a dummy variable to make the assignment statement valid. (As explained earlier, capturing the return code from these method calls is unnecessary. That said, by way of coding, RC variable in this case *will* indeed receive the return code from whichever method is called - which is why it is named RC in the first place.)

Alternatively, instead of using sashelp.vcolumn view directly in the DATA step, the system table Dictionary.Columns could be used in a preliminary SQL step to create a subset related to Bizarro.Player_Candidates and containing only the fields Name and isKey. Though doing so requires an extra step, it also offers certain advantages, such as a cleaner log, better performance, and the ability to use the LIKE operator to create variable isKey.

2.2.16 Multiple Instances Issue

The following statement is a *run time* directive (unlike the DECLARE statement, which is a *compile time* directive):

```
H = _new_ hash() ;
```

It means that it is executed every time program control passes through it. Hence, if it is placed inside a loop, it will create a *new* instance of hash object H at every iteration. It will occur regardless of whether the loop is an explicit DO or the implied DATA step loop. Therefore, in the following step, the statement is *executed* twice:

```
data _null_ ;
  dcl hash H ;
  do i = 1 to 2 ;
    H = _new_ hash() ;
  end ;
 *...rest of program;
run ;
```

At run time, the DCL statement is ignored, but the assignment statement is executed twice and thus creates two separate instances of hash object H. The same happens if program control passes through the assignment statement in the implied DATA step loop. For example:

```
data _null_ ;
  set Bizarro.Player_candidates (obs=2) ;
  dcl hash H ;
  H = _new_ hash() ;
 *...rest of program...;
run ;
```

It is no different if the compound DECLARE statement is used:

```
data _null_ ;
  set Bizarro.Player_candidates (obs=2) ;
  dcl hash H() ;
 *...rest of program...;
run ;
```

In this case, the *declarative* part of the statement is ignored at run time; yet, the part that creates a new hash object instance is executed twice, and so two separate instances of H are created.

This behavior can be useful if the program *intends* to create and use multiple instances of the same hash object. However, most programs use only a single instant of every named hash object. In this case, this default behavior results in a number of undesirable side effects:

- More instances of the same hash object than needed are created.
- If an instance is not used, it needlessly consumes memory and other computer resources.
- Worse still, this overhead can be compounded if the unintended instances are numerous. For example, if in the step above, the input were not limited by the OBS=2, a separate instance of H would be created for every observation read in.

Moreover, if the loop contained the entire block of code representing the *Create* operation, every one of its run-time statements and method calls would be re-executed for each input observation and would thus add to the overhead:

```
data _null_ ;
  dcl hash H() ;
  H.definekey("Player_ID") ;
  H.definedata("Position_code") ;
  H.definedone() ;
  set bizarro.Player_candidates ;
 *...rest of program;
run ;
```

In this case, not only would another instance of H be needlessly created for each input observation, but the *Create* operation methods would be needlessly called just as many times.

Therefore, if we need only a single instance per hash object, measures must be taken to ensure that no more than one instance is created and acted upon. If multiple instances are needed, our DATA step program must make sure that each instance can be referenced. Section 2.2.18 suggests several such alternatives. Chapter 9 Hash of Hashes – Looping Thru SAS Hash Objects presents a use case for creating multiple instances with the same name.

2.2.17 Ensuring Single Instance Usage

The obvious way to ensure that only a single hash object instance is created and initialized is to ensure that program control passes through the *Create* operation statements only once. Generally speaking, there are two techniques to achieve it:

1. Execute them only on the condition of _N_=1.
2. Use the DO loop to take explicit control of reading the input.

These two approaches are exemplified in the exhibit below, where:

- Input file Player_candidates is a WORK library copy of file Bizarro.Player_candidates.
- Variable LR used with the END=LR option is initially automatically set to LR=0. The SET statement sets it to LR=1 when it reads the last input record.
- Note that the name "LR" is an abbreviation denoting the "last record". It is used in this context below and throughout the book.

Table 2.1 Ensuring a Single Hash Object Instance

1. Using _N_=1 Condition	2. Using Explicit DO Loop
```data _null_ ; if _N_ = 1 then do ; dcl hash H() ; H.definekey("Player_ID") ; H.definedata("Position_code") ; H.definedone() ; end ; if LR = 1 then do ; *code after last record; end ; set Player_candidates end=LR ; *code for each input record; run ;```	```data _null_ ; dcl hash H() ; H.definekey("Player_ID") ; H.definedata("Position_code") ; H.definedone() ; do until (LR = 1) ; set Player_candidates end=LR ; *code for each input record; end ; *code after last record; stop ; run ;```

For the coding style shown on the *left*:

- The condition _N_=1 prevents the program from executing the *Create* operation code block more than once by rendering it operable only in the first iteration of the implied DATA step loop.
- The step is stopped when, in the last iteration of the implied loop, the SET statement attempts to read from the empty buffer after the last record has been read.
- If more code is needed after the last input record has been processed, the condition IF LR=1 ensures that it is executed only once. Though it *seems* logical to code this block *last* (just before RUN), it is instead placed *before* SET. Doing so ensures that it is executed even if a conditional DELETE or subsetting IF statement coded after the SET statement should evaluate true on the last record.

For the coding style on the *right*:

- The *Create* operation code block is executed *unconditionally*.
- The file is processed by reading its records explicitly in a DO UNTIL loop terminated after the SET statement reading the last record makes it LR=1.

- If more code is needed after that, it is placed, *unconditionally*, after the DO loop.
- The STOP statement terminates the step. This way, *all code* is executed only during the first iteration of the implied loop since it never iterates again.

Both techniques have their preferred uses depending on the program logic and, to some extent, preferred programming style. In this book, both styles are exemplified, the choice depending on the circumstances.

The style shown on the left is suggested in the SAS documentation. However, the style on the right is more logically straightforward, especially if file post-processing is needed, and, to a degree, more efficient. Note that this style is a version of a technique commonly known in the SAS programming community as *the DoW loop*. Above, it is used to take explicit looping control over the entire input file. Another variant of it, also exemplified in this book, is used to take control over each BY group, one at a time, read from a sorted or grouped file.

2.2.18 Handling Multiple Instances

In the previous section, we dealt with the ways to ensure that only a single instance of a given named hash object is created and used when this is what the program needs. However, under different circumstances using multiple instances of the same object is not only desirable but advantageous in terms of flexibility and dynamic code. That raises a question: If more than one instance of the same hash object is created, how do we tell the program which one to use? To answer it, suppose that we have created two instances of hash object H, as in the following schematic DATA step:

```
data _null_ ;
  dcl hash H ;
  H = _new_ hash() ; *Create instance of H #1;
 *...code block #1...;
  H = _new_ hash() ; *Create instance of H #2;
 *...code block #2...;
 *...rest of program...;
run ;
```

Each time the same statement is executed, it creates a new instance. Hence, when it is called twice, as above, the following happens:

1. When it is executed for the first time, it creates a new instance (#1) and makes it active by storing its identifying value in PDV variable H. Thus, any reference to H in the code block #1 will cause the program to work on instance #1.

2. When it is executed for the second time, it creates another new instance (#2) and makes it active by *overwriting* the PDV value of H with the pointer value identifying instance #2. Now any reference to H in the code block #2 and the rest of the program will cause it to work on instance #2.

Now let us suppose that we need the rest of the program, instead of working on instance #2, to resume working on instance #1 again. With the program as shown above, it presents a problem. Namely, the pointer value identifying instance #1 (originally stored in H) is no longer available since it is overwritten in H and not stored anywhere else. So, even though the instance exists, it can no longer be identified by the program.

The way around the problem is to create another variable of type hash and use it to save the value identifying instance #1. Then, later on, the saved value can be reassigned back to H and thus direct the program to resume working on instance #1 again:

```
data _null_ ;
  dcl hash SAVE ;
  dcl hash H ;
  H = _new_ hash() ; *Create H instance #1;
  SAVE = H ;         *Save current PDV value of H;
 *...block #1...;
  H = _new_ hash() ; *Create H instance #2;
 *...block #2...;
  H = SAVE ;
 *...rest of program...;
run ;
```

The reason we need another DECLARE|DCL statement is that the value of type hash identifying instance #1 cannot be saved in a scalar variable. Instead, we need another variable of type hash (in this case, SAVE), and the DECLARE statement is the only vehicle to create it.

After the value saved in variable SAVE is reassigned to H, instance #1 is *reactivated* since now the PDV value of H is again related to this instance. That is, any reference to object H in the rest of the program will cause it to work on instance #1.

If the program needs to use and intermittently activate more than two hash object instances, more type hash variables can be created, each in a separate DECLARE statement, to store their identifying values for later use. However, it is easy to perceive that as the number of such instances grows (and especially if it is not known beforehand), this technique can quickly become unwieldy.

Fortunately, the pointers to hash object instances can be stored in a separate hash table and retrieved from it into the PDV. This much more suitable way of activating an instance will be discussed and exemplified later in the book (especially in Chapters 6 and 9). However, regardless of the technique, the capability to surface an individual instance at will makes programs using the hash object highly flexible and dynamic.

2.2.19 Create Operation Hash Tools

- Statements: DECLARE (DCL).
- Operators: _NEW_.
- Methods: DEFINEKEY, DEFINEDATA, DEFINEDONE.

2.3 DELETE (Table) Operation

This operation serves to delete a hash object instance *altogether*, including its table and hence all of its table's content. It is useful when all the data processing the programs needs to do with the table is finished and it is no longer needed. By deleting the instance, we free up the memory occupied by both the items stored in its table and its underlying structure - in contrast to the *Clear* operation after which the underlying structure (and the memory it occupies) is preserved.

2.3.1 The DELETE Method

The only way to delete a hash object instance is to call the DELETE method. Suppose we have declared a hash object named H (and thus created a PDV variable H of type hash) and created one or more of its instances. The following call deletes the *active instance* of H - that is, the instance identified by the *current* PDV value of H:

```
rc = H.delete() ;
```

If the instance pointed at by the current PDV value of variable H exists, it will be deleted successfully, and the method will return RC=0. It will fail in two cases:

- No instances of hash object H have been created.
- The instance identified by the current PDV value of H no longer exists because it has been deleted previously.

In both cases, the DATA step will be aborted with an error message stating that object H is uninitialized.

The method can always be called *unassigned*, i.e.:

```
H.delete() ;
```

The reason for it is that capturing its return code in a separate variable offers no utility. If the DELETE method fails, the step is instantly aborted; and so no further programming action is possible, based on the return code.

2.3.2 DELETE Operation Details

A few points regarding the DELETE method deserve to be emphasized:

- It does *not* delete the PDV variable, such as H above, associated with the object. Once defined, this variable persists for the duration of the step. In this sense, it is no different from any other variable defined in the PDV.
- The method does *not* delete all instances of the hash object referenced in the call.
- It deletes only the *active* instance. Hence, if other instances need to be deleted, each of them must be made *active* first and then deleted using a separate call.

As a side note, in contrast to the CLEAR method described below, the DELETE method can successfully delete a hash object instance even if its hash table is locked by a hash iterator.

2.3.3 Delete (Table) Operation Hash Tools

- Methods: DELETE.

2.4 CLEAR Operation

This is categorized as a *table-level operation* because it deletes *all the hash table items* at once and *releases the memory* formerly occupied by them. While the operation *eliminates the items* from the table, it *preserves its entry*. In other words, it leaves the table empty, yet keeps the table itself and its defined structure.

2.4.1 The CLEAR Method

The *Clear* operation is performed by calling the CLEAR method. If the hash object whose table we need to clear is named H, the only piece of code needed to trigger the *Clear* operation is:

```
rc = H.CLEAR() ;
```

The CLEAR method can *always* be called *unassigned*, i.e., without capturing its return code in a separate variable:

```
H.CLEAR() ;
```

This is because for this method (as well as a number of others), capturing its return code is useless. There are only two reasons why this method can fail:

1. The hash object instance referenced by H does not exist. In this case, the step will be immediately aborted.
2. The table is locked by a hash iterator (discussed in detail later). In this case, the step will be immediately stopped as well.

In either case, if the method call should fail, no further statements would be executed. Thus, the return code, even if captured, could not be examined, and so there is no reason to capture it in the first place.

2.4.2 Clear Operation vs Delete (Table) Operation

The *Clear* operation is extremely valuable in the situations when a hash table is used to process one block of data after another. Most often (though not always) it happens, for example, when the table is populated during the processing of one BY group and then needs to be reinitialized in preparation for the processing of the next one. By emptying the table before every BY group, the *Clear* operation ensures that the table uses only as much memory as it needs to load the largest BY group - as opposed to the amount of memory required to load the whole file. It can also be used if a single DATA step is generated by macro language logic, which needs to clear and reload the table using, for example, a WHERE clause.

In principle, the same can be done using the *Delete (table)* operation. However, since it also deletes the table itself, it requires redoing the entire *Create* operation before each consecutive block of data being dealt with. Compared to merely purging the table of its items while keeping the table itself, recreating the table can be quite expensive. In real-life situations, where the data blocks to be processed may number in millions, the accumulated cost of multiple *Delete (table)* operations can quickly become prohibitive.

Having said that, the *Delete (table)* operation has one advantage: It can delete a table locked by a hash iterator. However, this advantage is moot because there exist simple ways (discussed later on) to unlock the table.

2.4.3 CLEAR Operation Hash Tools

- Methods: CLEAR.

2.5 OUTPUT Operation

The *Output* operation is designed to *unload* (i.e., write or copy) the data *currently* stored in a hash table to a SAS data set file indicated by the program. We classify it as *table-level* because by default (i.e., unless specifically filtered), it writes *every* hash item as an output data set observation using a single statement. Before discussing the operation in earnest, let us first note some high-level details:

- Only the data stored in the data portion hash variables is written out; the key portion variables are ignored. Hence, if the key-values are needed in the output file, they have to be defined both in the data portion and key portion.

- Every data portion hash variable becomes an output data set variable with the name and all other attributes inherited from the corresponding PDV host variable.
- The operation is executed at run time completely independently from the DATA step facilities writing data to the data sets specified in the DATA statement.

The hash object tools supporting the *Output* operation are the OUTPUT method and the DATASET argument tag used to specify the output file. Let us take a look at them first.

2.5.1 The OUTPUT Method

Suppose that we need to write the data stored in the data portion variables of hash table H to a SAS data set Work.fromHash. This is done by calling the OUTPUT method, where the name of the output data set is specified using the DATASET argument tag:

```
rc = H.OUTPUT (dataset:"work.fromHash") ;
```

Above, the method is called assigned, as its return code is assigned to variable RC, resulting in RC=0 if the call is successful and RC≠0 otherwise. However, no useful programming action can be taken based on its value because, if the call fails, the DATA step will be stopped there and then with a run-time error. Therefore, common practice is to call the OUTPUT method unassigned:

```
H.OUTPUT (dataset:"work.fromHash") ;
```

As usual, if the output data set is written to the WORK (or USER) library, the library specification can be omitted, i.e.:

```
H.OUTPUT (dataset:"fromHash") ;
```

The calls shown above represent the most basic syntactic form of calling the OUTPUT method for two reasons:

- They result in unloading the hash table data content to the output file *as is*. That is:
  - *All* hash table items are output as observations in the logical order they are stored in the hash table.
  - *All* data portion variables of numeric and character (scalar) type end up in the output as the data set variables with exactly the same names and attributes as the corresponding PDV host variables (and in the order the latter are defined by the DEFINEDATA method). Note that non-scalar variables (such as of type hash), if present in the data portion, are ignored because they cannot be stored in a SAS data set (and a warning to that effect is written to the log).
- The argument to the DATASET argument tag is *hard-coded* as a character literal constant.

However, neither has to be the case: The functionality of the OUTPUT method is broader, and, correspondingly, its syntax is more flexible. We will discuss some of its richer features later in this section, and many examples of applying them to practical situations will be given in the parts and chapters that follow.

2.5.2 Open-Write-Close Cycle

The *Output* operation is performed strictly during the DATA step run time. It consists of three phases:

1. *Open* the data set specified with the DATASET argument tag for output access with member-level control, i.e., for writing.
2. *Write* the data portion variables to the data set, one observation per (unfiltered) item.
3. *Close* the data set when finished.

2.5.3 Open-Write-Close Cycle Encapsulation

The open-write-close cycle described above is *encapsulated* by the *Output* operation at run time. More specifically, it means that the operation is:

- Handled exclusively by the hash object, with no other DATA step I/O facilities involved. In particular, its actions are independent of the DATA step OUTPUT *statement* (implicit or explicit) and/or its timing.
- Finalized before program control moves to the executable statement following the statement containing the OUTPUT method call, and the file it has just written to is closed.

Therefore, after the operation has been successfully executed, its output data set is no longer locked for output access with member-level control. As such, the data set at this point can be:

- Reopened, read, and modified *while the DATA step keeps running*. For example, it can be viewed in the SAS viewer or used by another batch or interactive process.
- Reopened, read, and loaded into another hash table later on in the *same* DATA step (using the implicit *Insert* operation described in Chapter 3).

Because the *Output* operation cycle is run-time encapsulated, it can be performed in the same step as many times as needed to open, write (or rewrite), and close as many output data sets as program logic may dictate.

For the same reason, if the *Output* operation is successful, the output data set written by it is preserved as written in its destination library even if, later on in the DATA step, it is stopped or aborted due to a run-time error. This behavior stands in contrast with the behavior of the data sets listed in the DATA statement because they are not closed until the DATA step ceases execution.

2.5.4 Avoiding Open File Conflicts

The need to open the data set specified in the *Output* operation for writing has its own implications. Because a currently opened data set cannot be re-opened for writing, the OUTPUT method call will fail if its target data set already exists and is opened. In this event, the step will be stopped, and an error message to this effect will be written to the log. It can occur in two distinct cases:

1. The target data set already exists in the library and has been opened by another program (e.g., is being viewed in the SAS viewer or read by another program).
2. The name of the target data set is listed in the DATA statement of the same DATA step where the OUTPUT method is called. This is because all data sets listed in the DATA statement are automatically opened for writing when the step begins its execution.

The reason the open file conflict occurs in the situation #2 is that any data set listed in the DATA statement is automatically opened before the step begins its execution and locked for member-level control output access; and so the OUTPUT method called at run time cannot open it.

Therefore, the same output data set cannot be listed in the DATA statement and specified as the target for the OUTPUT method anywhere in the step. In other words, a step similar to the step schematically shown below will result in a run-time error at the time of the OUTPUT method call:

```
data ONE TWO ;
  ...
  h.output (dataset:"ONE") ; *OPEN-FILE CONFLICT;
  ...
run ;
```

However, no conflict will occur in the above step if the method call is coded, for instance, as follows:

```
h.output (dataset:"THREE") ;
```

because the *Output* operation target data set is not listed in the DATA statement. Likewise, no open file conflict of this kind is possible if the DATA statement list is _NULL_:

```
data _NULL_ ;
  ...
  h.output (dataset:"ONE") ;
  ...
run ;
```

The ability to write data to an output data set dynamically with the DATA statement data set specified as _NULL_ looked like an unusual and impressive new SAS feature at the time when the hash object was first offered.

2.5.5 Output Data Set Member Types

With the OUTPUT method, the data set specification supplied to the DATASET argument tag *cannot point to a view*, i.e., to a SAS data set of member type VIEW. It can be only a SAS data file, i.e., a SAS data set of member type DATA.

First, the method call cannot *create* a view. A call such as the following is *invalid*:

```
H.OUTPUT (dataset:"vHash/view=vHash") ; *INCORRECT!;
```

It will result in an error and corresponding error message, and the step will be stopped.

Second, the method *cannot overwrite an existing view*:

- If a view with the same name as specified to the DATASET argument tag already exists, the method will fail and the step will be also stopped with an error message.
- This behavior is consistent with the fact that a data set of member type VIEW cannot be overwritten with a data set of member type DATA and vice versa.

Bearing that in mind, a program may include a provision to check, via the dictionary tables or SAS I/O functions, whether the output data set exists and what member type it has before the name of the target data set passed to the DATASET argument tag is constructed.

2.5.6 Creating and Overwriting Output Data Set

The hash object handles the output data set specified in the DATASET argument tag differently depending on whether a data set with the same name already exists or not:

1. If it *does not* exist, a *new* data set is *created*. In this situation:

   ○ Its variable names and other attributes are inherited from the PDV host variables corresponding to the hash variables in the data portion of the table.

   ○ The variables appear in the order defined by the DEFINEDATA method, which may be *different* from the order the host variables are stored in the PDV.

2. If it *does* exist, there are two situations:

   ○ Most commonly, it is an ordinary data set that is *not* part of a generation group. In this case, it is *overwritten*. It means that from the usage standpoint (regardless of behind-the-scenes details), the existing data set is *erased* and a new data set with the same name is *created* in its stead exactly as described in #1 above.

   ○ It *is* part of a generation group. In this case, a new generation data set with the next generation number is created. Because it is a physically new file, it is treated as a data set that does not exist as described in #1.

Therefore, if in the same DATA step the OUTPUT method is called more than once with the same output data set name, each subsequent call will overwrite the data set written by the call preceding it. It can be illustrated schematically as:

```
data ... ;
  ...
  h.output (dataset:"OUT") ;
  ...
  h.output (dataset:"OUT") ; *Overwrites OUT written by call #1;
  ...
  h.output (dataset:"OUT") ; *Overwrites OUT written by call #2;
  ...
run ;
```

In this step, the first call creates a new data set Work.Out (if it does not yet exist) or overwrites it (if it already exists). The second call overwrites the like-named data set written by the first call, and the third call overwrites the data set written by the second call. Though the data in hash table H may change between the calls, the state of the data set written last reflects the most recent data the table contains.

Hence, if there is a need to save the data written by *each* call, there are two options:

1. Name the output data sets for the different calls differently - say, OUT1, OUT2, and OUT3.

2. On the first call, use the data set option GENMAX= to create a generation group with the value greater than the number of calls. For example:

   ```
   data ... ;
     ...
     h.output (dataset:"OUT(genmax=3)") ;
     ...
     h.output (dataset:"OUT") ;
     ...
   ```

```
          h.output (dataset:"OUT") ;
          ...
     run ;
```

This way, each call will write its own data to its own generation data set without overwriting the data set written by the prior call. Note that the data set option GENMAX= used in the first call is not an exception as far as using output data set options is concerned.

2.5.7 Using Output Data Set Options

Output data set options, such as KEEP=, DROP=, RENAME=, INDEX=, WHERE=, etc., can be used with the output data set in parentheses following its name. The GENMAX= option shown in the prior section is just one example.

Of particular interest is the WHERE= option because can be used to filter the data written to the output data set. For example, if the hash table had a data portion variable Runs (such as variable Runs in data set Bizarro.AtBats), the following method call would output only the items where Runs is greater than zero:

```
H.OUTPUT (dataset:"work.fromHash(WHERE=(Runs>0))") ;
```

Or, if we wanted to drop the variable from the output, we could code:

```
H.OUTPUT (dataset:"work.fromHash(DROP=Runs)") ;
```

Other output data set options can be used in the same vein and/or combined. The rules of coding them described in the SAS documentation are the same as for any data set specified as output in the DATA statement or in a SAS procedure.

2.5.8 DATASET Argument as Non-Literal Expression

Heretofore, in all examples of using the OUTPUT method, the arguments to the DATASET argument tag have been given as character literal constants, i.e., a quoted fixed string value. Just as with the DEFINEKEY and DEFINEDATA methods discussed above, the documentation describing the OUTPUT method may give an impression that using a character literal is the only option. However, this is not the case.

In actuality, the argument of the DATASET argument tag can be any valid character expression, as long as it resolves to the required data set name - if need be, together with the necessary data set options. Taking advantage of this fact can make a program using the OUTPUT method much more dynamic than using character literals alone.

In the simplest case, let us say that we want to unload a hash table H into a data set named fromHash in a library whose libname is HashOut, and we want to use the WHERE clause to filter the data on the condition Run>0. Using the DATASET argument as a character literal, we could code, as already shown above:

```
H.OUTPUT (dataset:"work.fromHash(WHERE=(Runs>0))") ;
```

Now suppose that in the program we already have a PDV character variable named arg valued as follows:

```
arg = "work.fromHash(WHERE=(Runs>0))" ;
```

before the OUTPUT method call. In this case, instead of hard-coding the DATASET argument, we can code instead:

```
data ... ;
  ...
  arg = "work.fromHash(WHERE=(Runs>0))" ;
  ...
  H.OUTPUT (dataset:arg) ;
  ...
run ;
```

The reason it can be done this way is that variable arg by itself is a character expression. The fact that, above, it is valued via an assignment statement is unimportant: It can be valued by another mechanism (such as the INPUT or RETAIN statement) or come, already properly valued, from a SAS data set.

As a more involved case, suppose that before the OUTPUT method is called, we have a number of variables representing different parts of the argument value we want to create. For example, the data set specification and the WHERE clause:

```
dsname = "work.fromHash" ;
where  = "Runs>0" ;
```

In this case, the variables can be combined into a single expression to be passed to the DATASET argument tag:

```
data ... ;
  ...
  dsname = "work.fromHash" ;
  where  = "Runs>0" ;
  ...
  H.OUTPUT (dataset: cats(dsname, "(where=(", where, "))")) ;
  ...
run ;
```

In most use cases, the components of the expression passed to the DATASET argument tag come from some kind of parameter file. This way, the output data set destination, name, filtering, etc., can be controlled dynamically based on the pre-stored control information and program logic. We will see many examples of applying this concept later in the book.

2.5.9 Output Data Order

As hash object users, we are oblivious to the order and manner in which the hash items are *physically* stored in a hash table internally. In fact, it does not matter. What really matters is the *order in which the items are accessed* during hash table operations, for this is how we use them and *logically* perceive their order in the table. This is similar to how many database systems manage their data tables.

From this *operational* standpoint, we can simply - and correctly - assume that the *logical order* in which the items are stored in the table is exactly the order in which they are written out by the *Output* operation to a data file, such as work.fromHash above. Not surprisingly, this is also precisely the order in which the hash items are accessed by the *Enumerate by Key* (*Keynumerate*) and *Enumerate All* operations, discussed later in this part of the book.

Incidentally, this is why in the realm of hash object programming the *Output* operation is a great diagnostic tool. While we cannot eyeball the hash table itself, we can always write its data content to a file, view and analyze the latter, and amend our code based on the findings.

2.5.10 Output Operation Hash Tools

- Methods: OUTPUT.
- Argument tags: DATASET.

2.6 DESCRIBE Operation

The *Describe* operation allows us to retrieve the properties of a hash table as a whole. This is done by using the tools called *hash object attributes*. Currently, two attributes are supported:

1. The NUM_ITEMS attribute. It returns the number of items currently stored in the hash table of the active instance referenced when it is called.
2. The ITEM_SIZE attribute. It returns the number of bytes the hash table *entry* occupies in computer memory.

Like the methods, the attributes are called by using the object-dot notation to reference the hash object in question. Also, just as with any method reference to the hash object name, an attribute object reference returns the information related to the table of the *active* hash object instance. Let us discuss the two attributes one at a time.

2.6.1 The NUM_ITEMS Attribute

To get the number of items stored in the table of a hash object instance referenced as H into variable N_items, we can code:

```
N_items = H.num_items ;
```

Note that in order to be used in a program, the value of the attribute does not have to be necessarily assigned to a separate variable such as N_items above. This is because H.num_items is a numeric expression and, as such, can be used in any other numeric SAS expression *directly*. For example, to make a DO loop iterate half as many times as there are items in table H, we can code:

```
do x = 1 to divide(H.num_items,2) ;
  * code inside the loop ;
end ;
```

Or, to execute some action only if the hash table is empty (i.e., has no items):

```
if H.num_items = 0 then do ;
  * action ;
end ;
```

The most valuable utility of the NUM_ITEMS attribute lies in the fact that it returns the *current* number of items in a hash table, automatically adjusted as it grows or shrinks when items are added to or removed from it. Therefore, it can be used to:

- Determine the upper index limit of an iterative DO loop used to iterate through the hash table sequentially (i.e., *enumerate* it).
- Help calculate hash table statistics that depend on the number of items in the table (for example, percentiles).
- Implement dynamic data structures, such as stacks and queues.

These uses of the NUM_ITEMS attribute will be discussed in detail and exemplified in the book later on.

2.6.2 The ITEM_SIZE Attribute

The ITEM_SIZE attribute is a *Describe* operation hash tool that returns the length of the hash table *entry* expressed in bytes. If we create an analogy between a hash table and a SAS data set, this metric roughly corresponds to the "row length" property of the SAS data set. To call the attribute and return its value into a numeric variable Entry_length, we can code:

```
Entry_length = h.item_size ;
```

Though it is difficult to think of its utility from the standpoint of dynamic programming, the attribute can be a great help in assessing the memory footprint of a future hash table. Thus, it is particularly useful in the applications where the hash object memory may be taken to the system limits, and so it is paramount to evaluate, during the program design stage, how much memory it may occupy when filled with items.

There is no hard-and-fast rule of determining the hash entry length *a priori* based on the lengths of the variables in the key and data portions, all the more true in that it varies based on the platform. However, the ITEM_SIZE attribute returns the *exact, actual* hash entry length. Thus, it can be used to form a fairly accurate idea of how much memory is needed to accommodate a hash table with a given number of items (or how many items the available memory can accommodate) on the system where it is supposed to be used.

A good use case is to determine how many items can be loaded for a given amount of memory into a table defined in a specific way. For example, imagine that we want to load data set Bizarro.At_Bats into hash table H defined with specific hash variables and, during the design phase, we need to evaluate how many items will fit in the table given 1 GB of memory. To achieve that, we would:

1. Code a DATA step with statements needed to execute the *Create* operation with the hash variables defined as we want them.
2. Obtain the value returned by the ITEM_SIZE attribute.
3. Divide 1 GB ($1024**3=2**30$ bytes) by the value and print the result in the log.

For example:

Program 2.5 Chapter 2 Number of Hash Items in Given Memory.sas

```
data _null_ ;
  dcl hash H () ;
  H.defineKey ("Player_ID", "Team_SK") ;
  H.defineData ("First_name", "Last_name", "Position_code") ;
  H.definedone () ;
```

```
  Entry_length = H.item_size ;
  N_items_in_1GB = round (2**30 / Entry_length) ;
  put (Entry_length N_items_in_1GB) (=comma16./) ;
  stop ;
  set bizarro.Player_candidates ;
run ;
```

Running the step (in this case, on the X64_7PRO platform) results in the following information printed in the SAS log:

```
Entry_length=80
N_items_in_1GB=13,421,773
```

Hence, erring on the safer side, we can reasonably expect that the table as created and defined above can accommodate about 13 million items in 1 GB of memory.

Note that the total system length of the hash variables in the key and data portions above is 43. So, if we used it to estimate hash memory usage instead of using the exact number 80 returned by the ITEM_SIZE attribute, it would be underestimated almost by the factor of 2.

2.6.3 Describe Operation Hash Tools

- Attributes: NUM_ITEMS, ITEM_SIZE.

Chapter 3: Item-Level Operations: Direct Access

3.1 Introduction

In this chapter, we will concentrate on the item-level operations performed by direct key access to a hash table without the explicit need to read the table sequentially in one form or another.

3.2 SEARCH (Pure LookUp) Operation

The *Search* operation is used only to discover whether a given key is in the hash table - and *to do nothing else*. That is why it is also dubbed as *pure look-up*. The only hash tool supporting the operation is the CHECK method. It can be called in two ways: *implicitly* without the argument tag KEY coded in or *explicitly* with the argument tag KEY coded with an argument. Let us consider these two modes separately.

3.2.1 Implicit Search: No Arguments

Suppose that H is the hash table defined with variable K in the key portion and variable D in the data portion. Correspondingly, as required by parameter type matching principle, we have host variables named K and D in the PDV.

Now suppose that we want to find out whether the current PDV value of host variable K exists in table H without affecting any data values in the PDV. In particular, we do not want any value of D in the table to affect the current value of its PDV host variable D. In other words, we need to merely search the table and do nothing else. The way to do it is to call the CHECK method designed specifically for such *"pure look-up"* purpose:

```
RC = H.CHECK() ;
```

For example, if, as SAS reads through a data set with variable K and its value in the current observation K=3, the CHECK method will return a return code of 0 (RC=0) if one or more items in the hash table has K=3. If RC=0, the key is in the table; otherwise, the method returns a *non-zero, non-missing value*. Therefore, if we want to predicate some programming actions on whether the key is in the table, we can code:

```
if RC = 0 then do ;
 <program actions for PDV value of K found the table> ;
 end ;
```

Or, if we want to create a Boolean 0/1 variable (Have_key, say) to indicate search success or failure, we can use any of the following code variations:

```
Have_key = (RC=0) ;
Have_key = not RC ;
Have_key = ifN (RC=0, 1, 0) ;
```

Since *Search* is a key-based operation, the question arises how the CHECK method call knows which key-value(s) to look for. The answer is that if it is coded *implicitly*, as above, it automatically accepts the values from the key host variables *currently in the PDV*.

3.2.2 Explicit Search: Using the KEY Argument Tag

Most of the time, calling CHECK without arguments and thus letting it infer the key to look for from the PDV host variable(s) is what is needed. However, calling it with the KEY argument tag can be much more flexible.

Suppose, for example, that we need to check if K=3 is in the table when we *do not know* the current value of host variable K and *do not want to modify it*. To call CHECK without arguments, we would need to first assign 3 to the host variable K. However, to preserve the current value of K, we would have to engage in cumbersome gymnastics of memorizing it and assigning it back:

```
 _K = K ;
 K = 3 ;
 rc = H.CHECK() ;
 K = _K ;
```

It is much simpler to achieve the same goal by using the KEY argument tag:

```
rc = H.CHECK(KEY:3) ;
```

This way, we can establish whether K=3 is in the table, leaving the current PDV value of K intact. It also results in much cleaner code and makes it more efficient by avoiding the unnecessary reassignments.

Even more flexibility in using the argument tag comes from the principle that the argument tags accept *general* SAS expressions as arguments and *not necessarily literals*. In *KEY*:3 above, 3 is a *numeric literal*, the simplest of expressions. Suppose that for some numeric variable V we want to know if table H contains a key with value V+1. Again, instead of modifying the host value K first, we can just code:

```
rc = H.CHECK(KEY:sum (V,1)) ;
```

3.2.3 Argument Tag Type Match

The *data type* of an expression assigned to an argument tag *must match* the data type it expects. For instance, the data type expected by the argument tag KEY, above, is defined by the type of hash key variable K with which the object is defined - in this case, it is numeric. Assigning a character expression to it will result in the "*Type mismatch*" run-time *error*, e.g.:

```
rc = H.CHECK(KEY:"3"); * Incorrect! Type mismatch;
```

Note that the call *will not attempt to automatically convert* "3" (a SAS character string) into 3 (a SAS number) using the BEST. format, as it would in hash-unrelated code. It will just fail. The same would happen if K were of the character type and a numeric expression were assigned to the argument tag.

Matching argument tag types is a general principle that must be borne in mind every time an argument tag (or a statement or operator parameter argument) is used.

3.2.4 Assigned CHECK Calls

When H.CHECK() is assigned to a separate variable as it was done above:

```
RC = H.CHECK() ;
```

the call, obviously, is assigned and its return code is captured in RC. Such *assigned* call generates no errors or error messages in case it fails, i.e., if the key is not found in the table. The return code can be captured because H.CHECK() is a *numeric expression*: It resolves to a numeric value, and that is the value assigned to *RC*.

But in SAS, a numeric expression can be part of a statement or any other numeric expression. In such cases, the value to which it resolves is still captured in the internal buffer and then used in whatever statement or expression it is part of. Therefore, instead of being assigned to a separate variable like RC first, H.CHECK() can be used *directly* as an expression in its own right. For example:

```
if H.CHECK() = 0 then do ;
  <program actions for PDV value of K found the table> ;
end ;
```

Likewise, it can be used directly in the Boolean expressions shown earlier (or any other numeric expressions, for that matter). For example:

```
Have_key = (H.CHECK()=0) ;
```

In all these cases, where the return code value to which the H.CHECK() expression resolves gets *captured*, it is in fact physically *assigned*, be it to a separate variable like *RC* or internally. And in all these cases, no errors or error messages are generated if the CHECK call fails- i.e., if the key is not found in the table. This is why we classify any call whose return code is captured in one way or another as *assigned*. As we will see in the next section, *unassigned* calls whose return codes are not captured behave differently.

3.2.5 Unassigned CHECK Calls

The CHECK method can be also called *unassigned* without capturing its return code either in a separate variable or internally as part of a statement or expression:

```
H.CHECK() ;
```

However, though such a *stand-alone* (or *naked*) *call* is syntactically valid, it makes no sense to use it with the CHECK method for the following reasons:

- The only information sought from the *Search* operation is whether it is a *success* (the key is in the table) or *failure* (the key is not in the table).
- The CHECK method *does nothing except* provide its return code information. Hence, without looking at it, there is no reason whatsoever to call CHECK in the first place.
- A failed unassigned call results in the error message in the log "*Key not found*".

Note that a failed unassigned call does not cause the DATA step to stop processing. However, for each failed call, it writes the error message in the log. In many real-world situations where failed CHECK calls may number in millions, this behavior is not only bothersome, especially if SAS is run in the interactive mode, but can overfill the log with useless error messages.

The takeaway is that the CHECK method should always be called assigned - either by assigning it to a separate variable or by making it part of a statement or expression. Not coincidentally, this is true for any method whose failure does not cause the DATA step to stop immediately after a failed call since further programming actions can be based on whether the method call has succeeded or failed.

3.2.6 Search Operation Hash Tools

- Methods: CHECK.
- Argument tags: KEY.

3.2.7 Search Operation Hash-PDV Interaction

- Implicit call: The key values to search for are accepted from the current PDV values of the host key variables.
- Explicit call: None.

3.3 INSERT Operation

The very purpose of a hash table is to contain keys and data to be efficiently manipulated and/or compared to values outside the table. Thus, the utility of the *Insert* operation is critical: As its name implies, it inserts items into the table. The hash object is packaged with a variety of tools and options to facilitate this operation and control its behavior in terms of handling hash items with duplicate key-values.

3.3.1 Dynamic Memory Acquisition

Before the advent of the hash object, the only way to implement a memory-resident lookup table in the DATA step was to use an array. However, since the memory needed to house an array is acquired at *compile time*, the number of its elements (i.e., its dimension) and memory footprint cannot be determined at run time. Hence, the array dimension has to be either calculated beforehand (usually at the expense of another pass through the data) or selected using the iffy "big enough" principle.

In this sense, the hash object represents a radical departure in behavior. Regardless of the tool performing the *Insert* operation, the extra memory for each item added to the table is acquired separately, at *run time*. In other words, the table grows dynamically as the items are inserted (and, as we will see later, also shrinks dynamically as they are removed).

This way, we can have a fully functional memory-resident table without the need to determine its size ahead of time. Moreover, the hash attribute NUM_ITEMS is adjusted automatically behind-the-scenes with every item inserted into or deleted from the table and can be used to return the *current* number of table items at any time.

3.3.2 Implicit INSERT

Implicit Insert occurs in two modes:

1. Calling a hash object method actuating the operation (e.g., ADD) implicitly (i.e., without the argument tags). Let us call it the "*method call mode*".
2. Inserting the items from a SAS data file by giving its name to the DATASET argument tag. Let us call it the "*argument tag mode*".

3.3.3 Implicit INSERT: Method Call Mode

This mode of the implicit *Insert* operation is typically activated by a method call like this:

```
RC = H.ADD();
```

Note the absence of content between the parentheses. Since the call is used without the available argument tags KEY and DATA, a *question* arises: Which key and data values are used for the item being added to table H? *The answer* is that, in this case, the method automatically accepts the current values of the corresponding PDV host variables. In this snippet, a (K,D) item with values (1,A) is inserted into table H:

Program 3.1 Chapter 3 Implicit Method Call Mode Insert.sas
```
data _null_ ;
  dcl hash H () ;
  H.definekey ("K") ;
  H.definedata ("D") ;
  H.definedone() ;
  K = 1 ;
  D = "A" ;
  rc = H.ADD() ;
run ;
```

If there is a need to add more (K,D) items, the statements in bold can be repeated with other (K,D) values either explicitly or in a loop. For example, to load (K,D) pairs with values (1,A), (2,B), (3,C) consecutively, we could replace the lines in bold above with the following:

```
  do K = 1 to length ("ABC") ;
    D = char ("ABC", K) ;
    rc = H.ADD() ;
  end ;
```

The implicit *Insert* operation is also convenient when a hash table is loaded from a data file (i.e., a SAS data file or an external file via the INPUT statement). This is because in this case, new host variable values are placed into the PDV automatically every time the next record is read in. Suppose we want to create a hash table from Bizarro.Player_candidates (for example, intending to search for Position_code by Player_ID):

Program 3.2 Chapter 3 Implicit Method Call Mode Insert from a File.sas

```
data _null_ ;
  dcl hash H () ;
  H.definekey ("Player_ID") ;
  H.definedata ("Player_ID", "Position_code") ;
  H.definedone() ;
  do until (lr) ;
     set bizarro.Player_candidates end = lr ;
     rc = H.ADD() ;
  end ;
  H.output (dataset: "Players") ; *Check content of H;
  stop ;
run ;
```

Every time a new record is read in, the pair of values (Player_ID, Position_code) from it is moved to the PDV, and from there they are implicitly consumed by the ADD method call. (The only purpose of the OUTPUT call above is *diagnostic*, i.e., to check the hash table content.)

3.3.4 Implicit INSERT: Methods Other Than ADD

The ADD method is not the only one available to actuate the implicit *Insert* operation. The two other methods, REF and REPLACE, can also be called implicitly to insert items into a hash table. As far as implicit calls are concerned, their actions are similar to ADD in the sense that they also infer the key and data values to be inserted from the current values of the corresponding PDV host variables. In other respects, their actions differ from those of ADD:

- If duplicate-key items are allowed in the table, the ADD method inserts a new item unconditionally, regardless of whether the key-value it accepts is already in the table or not.
- By contrast, the REF method inserts a new item *only* if the key-value it accepts is not already present in the table, irrespective of whether duplicate-key items are allowed or not.
- The ability of the REPLACE method to insert an item is a side effect of its primary function of updating the data portion variables. Namely, if the key is not in the table, there is nothing to update; hence, a new item is inserted. Other details of the REPLACE method's behavior will be discussed later on in the appropriate sections.

3.3.5 Implicit INSERT: Argument Tag Mode

This mode of the implicit *Insert* operation is related to loading a table from a SAS data file, whose name is specified as a *character expression* assigned to the argument tag DATASET. It can be done regardless of whether the DECLARE|DCL statement or the _NEW_ operator is used to instantiate the table. Suppose, for example, that we intend to load the table from the Bizarro.Player_Candidates data set. If the declaration and instantiation are combined, in order to implicitly load the values from the data set into the table, we can code:

```
dcl hash H (DATASET: "bizarro.Player_candidates") ;
```

Or, in the case when the declaration and instantiation are separated:

```
dcl hash H ;
H = _new_ hash (DATASET: "bizarro.Player_candidates") ;
```

This insertion mode is obviously *implicit* because no method to trigger the *Insert* operation is called explicitly. However, what occurs with this input mode behind the scenes is *exactly identical* to the implicit

Insert operation done by reading the file one record at a time and calling the ADD method for each item to be inserted. Hence, the snippet below populates table H with the same content as the ADD method used above to read the file in the DO loop:

Program 3.3 Chapter 3 Implicit Argument Tag Mode Insert from a File.sas

```
data _null_ ;
  dcl hash H (dataset: "bizarro.Player_candidates") ;
  H.definekey ("Player_ID") ;
  H.definedata ("Player_ID", "Position_code") ;
  H.definedone () ;
  H.output (dataset: "Players") ; *Check content of H;
  stop ;
  set bizarro.Player_candidates (keep = Player_ID Position_code) ;
run ;
```

Note that above, the SET statement serves to place the host variables into the PDV at compile time. Without it (or another valid means of parameter type matching), the DEFINEDONE method call will fail. Coding SET after STOP ensures that it reads *no actual data* from bizarro.Player_candidates at run time, letting the compiler read its *descriptor* at compile time.

Which mode of the implicit *Insert* operation to select when the key and data values to be inserted come from a data file is dictated by both convenience and utility. Their relative advantages and disadvantages by feature are presented below:

Table 3.1 Implicit INSERT Method Call Mode vs Argument Tag Call Mode

Feature	Method Call Mode	Argument Tag Mode
Lines of code	More	Fewer
File name as a run-time expression	No	Yes
Input file data set options	Yes	Yes
Input subsetting	IF and WHERE	WHERE only
Using SET, MERGE, UPDATE, MODIFY, INPUT statements for input	All	SET only (implicit)
Input BY-group processing if needed	Yes	No
Creating PDV host variables	Automatic	Needs extra statement

3.3.6 Explicit INSERT

The implicit *Insert* operation relies on the current values of the PDV host variables for its input. However, there are situations when we want to load a key or data value which is the result of an *expression*. To achieve that using *Insert* via an implicit method call, the expression must first be assigned to the PDV host variable in question.

For example, suppose that before loading the pair (Player_ID, Position_code) into table H, we need to increment Player_ID up by 1. Using an implicit ADD call, we could code:

```
Player_ID = sum (Player_ID, 1) ;
rc = H.ADD() ;
```

Such an approach, while certainly doable, presents a problem: The current PDV value of Player_ID has thus been *altered*, while we may want to *preserve it*. Doing so while keeping *Insert* implicit involves more code and unsightly variable reassignment gymnastics. For instance:

```
_Player_ID = Player_ID ;
Player_ID = sum (Player_ID, 1) ;
rc = H.ADD() ;
Player_ID = _Player_ID ;
drop _Player_ID ;
```

Moreover, if the DATASET argument tag is used for implicit *Insert*, the subterfuges shown above cannot be used, for the table can be loaded only with the values as they come from the input file.

Thankfully, all hash methods supporting the *Insert* operation are furnished with the *argument tags* KEY and DATA, both of which accept *SAS expressions as arguments* (just as any other hash argument tag, for that matter). Therefore, instead of resorting to the kludge used above, we can simply code:

```
rc = H.ADD(KEY:sum(Player_ID, 1), DATA:Position_code) ;
```

Such an *explicit Insert operation* - that is, calling a hash method with its valued argument tags - provides a wide range of run-time flexibility. This is because the expression assigned to an argument tag can be *any valid SAS expression*. For example, it can include references to array items, formats, informats, functions, or be a mere literal, such as 1 or "A".

The only limitation imposed on the expression is that it *must be of the data type* the argument tag expects (in the specific position where it is listed. See the next section). In this respect, it is important to remember that:

- The data type the KEY or DATA argument tag expects is determined by the data type of the host variable of the hash variable for which the argument tag is used.
- Non-scalar data types, such as hash and hash iterator, cannot be used with the KEY argument tag because the key portion can contain only scalar variables.
- Hash object method calls *do not* provide for automatic data type conversions. Hence, if the expression is not of the expected data type, the method call will fail.

3.3.7 Explicit INSERT Rules

Implicit and explicit method calls cannot be mixed. In the example above, coding the DATA argument tag *cannot be omitted* with the assumption that the value of Position_code will be inferred from the PDV host variable, as it happens in an implicit call. Leaving it out will cause the method to fail.

This behavior corresponds to a general rule for *explicit method calls*: The number of argument tags and the data types of the arguments assigned to them *must coincide exactly* with the number and data types of the key and data portion hash variables defined to the hash object instance.

As an example, suppose that we have a hash table X defined with the following hash variables and in the following order:

1. Key portion: KN (numeric), KC (character).
2. Data portion: DN (numeric), DC (character).

Then an exlicit call to the ADD method must be of the form:

```
X.ADD(KEY:   <numeric    expression for KN>
     ,KEY:   <character expression for KC>
     ,DATA:  <numeric    expression for DN>
     ,DATA:  <character expression for DC>
     )
```

The KEY and DATA argument tags in the list do not necessarily have to be grouped together as above. However, their relative sequence *must mirror* the relative sequence in which the key and data portion variables, respectively, are defined. In other words, in our case, the expression for KN *must* precede that for KC; and the expression for DN *must* precede that for DC. Thus, this form of the ADD call is also correct:

```
X.ADD(DATA:  <numeric    expression for DN>
     ,KEY:   <numeric    expression for KN>
     ,DATA:  <character expression for DC>
     ,KEY:   <character expression for KC>
     )
```

It is also important to remember that *none of these argument tag assignments can be omitted* without failing the call. Any attempt to do so will result in an error message to the effect that the number of the key or data variables is incorrect.

From our geek perspective, it might be regrettable that a method cannot be called part-explicitly and part-implicitly. In other words, it would be nifty to be able to value only the argument tags we want and imply the rest of the inserted values from the host variables. However, c'est la vie. Hence, if the hash entry contains many variables and we want to alter just a few before insertion, it may be less fuzzy to use an implicit call after assigning the needed values to their PDV host variables accordingly (and add some variable reassignment or renaming code to preserve the PDV values if needed).

3.3.8 Implicit vs Explicit INSERT

As we have seen, both styles have their respective advantages. We will see numerous examples of using both further in the book. Choosing one over the other is dictated by the programmatic situation and, to a certain degree, by personal preferences. But whatever the choice, together they provide all the tools needed to load a hash table from any imaginable data source.

3.3.9 Unique Key and Duplicate Key INSERT

The *Insert* operation, regardless of the tools used, always begins behind the scenes with the *Search* operation to determine whether the key for the item to be inserted is already in the table. After that, the actions of the *Insert* operation depend on the hash object definition specifications and the tool used to perform the operation.

The *Insert* operation can be performed in two principal ways geared toward different data processing tasks:

1. *Unique Insert*. Duplicate key items are prohibited. This is the default behavior, which occurs if the MULTIDATA argument tag (a) is not used at all or (b) is used with an argument other than "YES"|"Y".

2. *Duplicate Insert*. Duplicate key items are allowed. This mode occurs if the MULTIDATA argument tag is coded in the statement creating the hash object instance and values as "YES"|"Y".

Unique Insert is useful when we want the uniqueness of the keys loaded into a hash table to be automatically enforced by the hash object without having to code for it explicitly. For example, if we want to unduplicate the keys on input or produce aggregates for each unique key, letting the object reject duplicate-key items all by itself is the obvious choice.

Duplicate Insert is unavoidable in many practical situations when we need to load a hash table with duplicate-key items and have the capability to manipulate them within any same-key item group. For example, we may want to *enumerate* the group of items with a given key to extract their data. Or we may need to delete or update *some items* in the group *selectively*. The hash object offers a set of tools designed to handle items with duplicate keys in a number of ways.

Let us now consider the unique and duplicate *Insert* operation modes separately.

3.3.10 Unique INSERT

When the MULTIDATA argument tag is not specified or specified but valued with anything but "Y"|"YES", we let the hash object handle the duplicate rejection process according to its internal rules. In this case, questions arise:

1. Which input instance of a duplicate key does the *Insert* operation keep in the table?
2. How does the hash object respond and what, if any, errors does it generate if an attempt to insert a duplicate key item is made?

To illustrate more tangibly what these questions mean, suppose that table H is defined as:

```
dcl hash H   () ;
H.definekey ("K") ;
H.definedata("D") ;
H.definedone() ;
```

and that our table H input consists of the following key and data value pairs:

K	D
A	1
A	2
A	3

If duplicate-key items are filtered out overtly using BY-group programming logic, any of the three input pairs above can be selected. But if we do not code MUTIDATA:"Y" and thus let the hash object handle duplicate-key input all by itself, we will see that it has its own rules with respect to (a) which input occurrence is selected and (b) how it reacts to an attempt to insert a duplicate-key item in terms of error generation and messaging.

First of all, when the hash object is left to its own devices in the unique *Insert* mode, it chooses either the *first* or the *last* duplicate-key input occurrence. In other words, with our sample three-record input above, either the (A,1) or (A,3) value pair is selected depending on the particular tool used in the *Insert* operation. Let us see how different tools handle the situation.

- The ADD method:
  - The *first* duplicate-key input occurrence is kept.
  - If ADD is called *unassigned*, an error message "*ERROR: Duplicate key*" is written to the log for *every* duplicate-key occurrence rejected (which can be quite pesky). However, automatic

variable _ERROR_ is *not* set to 1, the step is *not* stopped, and the table load process proceeds till completion.

- o If ADD is called *assigned*, no errors or error messages are generated.
- The REF method:
  - o The *first* duplicate-key input occurrence is kept.
  - o The REF method internal logic is "*If the key is not in the table, insert; else ignore*".
  - o No errors or error messages are generated regardless of whether REF is called assigned or unassigned.
  - o In this sense, calling REF is exactly equivalent to calling ADD assigned.
- The REPLACE method:
  - o The *last* duplicate-key input occurrence is kept.
  - o Due to the way REPLACE works, this is actually not a result of *rejecting* duplicate input keys. Rather, it is a result of *overwriting* the data portion values with new ones every time REPLACE is called with a key already present in the table.
  - o REPLACE inserts an item only once, when the key it accepts is not yet in the table. Every ensuing call with the same key-value repeatedly overwrites the data values already stored. So at the end of the process it appears as if the last duplicate-key input occurrence was inserted.
  - o No errors or error messages are generated regardless of whether REPLACE is called assigned or unassigned.
- Input from a data file specified to the DATASET argument tag:
  - o If the DUPLICATE argument tag is not also specified, the *first* duplicate-key input occurrence is kept. No errors or error messages are generated.
  - o If the DUPLICATE argument tag is coded with the "REPLACE"|"R" argument, the *last* duplicate-key occurrence is kept. No errors or error messages are generated in this case, either.
  - o If the DUPLICATE argument tag is coded with "ERROR"|"E" argument, an attempt to insert a duplicate-item causes the DATA step to be immediately terminated. No further statements are executed, and an error message is written to the log.

3.3.11 Duplicate INSERT

In this mode, i.e., if MULTIDATA:"Y" is specified, the *Insert* operation is *always successful* regardless of the tool - provided, of course, that it is used correctly. The latter means, for example, using correct syntax and the number of the argument tags and the expected data types for their arguments. Therefore, in the duplicate *Insert* mode, we need not be concerned with any errors or warnings when an attempt is made to insert a duplicate-key item. Nor is there any difference, in this regard, between assigned and unassigned calls. (In fact, it may be better to leave them unassigned since it helps reduce unnecessary code clutter.) Also, in this mode, there is no reason nor need to use the DUPLICATE argument tag: Since duplicates are permitted, it has no effect.

However, with respect to the duplicate *Insert* mode not all tools are created equal. Moreover, their effects depend on whether SAS 9.4 or earlier is used. Again, we can look at the similarities and differences all at once:

Table 3.2 Duplicate-Key INSERT Operation Actions

		Action	
Method	**Duplicate Keys**	**Before SAS 9.4**	**As of SAS 9.4**
ADD	Yes	Insert	Insert
	No	Insert	Insert
REPLACE	Yes	Insert	*Update all items with the key*
	No	Insert	Insert
REF	Yes	*Ignore the key*	*Ignore the key*
	No	Insert	Insert

To illuminate the differences of the REPLACE method duplicate-key behavior before and as of SAS 9.4, suppose that a hash table is already loaded with the following items for key K="A":

K	D
A	1
A	2
A	3

Now if we call the REPLACE method with the (K,D) values of (A,4) running SAS 9.4, all the items whose K="A" will be updated with the value of D=4:

K	D
A	**4**
A	**4**
A	**4**

If we are running a version before SAS 9.4, the items will *not* be updated. Instead, an item with (K,D)=(A,4) will be *inserted* in the table:

K	D
A	1
A	2
A	3
A	**4**

3.3.12 Insertion Order

From the programming perspective, the *logical order* of the items in a hash table is the only order that matters. As already mentioned above, this is the order in which they are written out by the OUTPUT method or accessed by the *Enumerate* operation. This way, we know how to determine the logical order of the hash table items in the table *after* they have been inserted.

But it raises an interesting question: Can we predict the order in which the hash items *will be* inserted before it has actually happened? The answer is "yes" and "no", or rather "it depends". Still, there are rules:

- If the table is not explicitly sorted in ascending or descending key-value order via the argument tag ORDERED, the relative order in which the *same-key item groups* follow each other is *random*. In particular, it means that if the table is operated in the *unique-key mode*, i.e., with 1 item per same-key group, the order of its individual items in the table is *random*.

- Otherwise, the *same-key groups* follow each other in key-value order specified by the argument of the ORDERED argument tag.
- Within *each same-key item group*, the relative sequence order is *exactly the same* in which the items *are received from input*. This is true regardless of whether the table is ORDERED or not.

These rules can be easily verified by running the code snippet below. In it, the argument of ORDERED is varied from "N" (undefined order) to "A" (ascending) to "D" (descending); and the argument of MULTIDATA is varied from "Y" to "N". Each time, table H is re-created and reloaded, and its content is output to a separate data set named using the six combinations of the above values:

Program 3.4 Chapter 3 Insertion Order Demo.sas

```
data _null_ ;
  do dupes = "Y", "N" ;
    do order = "N", "A", "D" ;
      dcl hash H   (multidata: dupes, ordered: order) ;
      H.definekey  ("K") ;
      H.definedata ("K", "D") ;
      H.definedone () ;
      do K = 2, 3, 1 ;
        do D = "A", "B", "C" ;
          rc = h.add() ;
        end ;
      end ;
      h.output (dataset:catx ("_", "Hash", dupes, order)) ;
    end ;
  end ;
run ;
```

As a result, we have 6 output data sets mirroring the content of 6 hash tables created with 2 MULTIDATA and 3 ORDERED argument tag choices, automatically named as follows:

Table 3.3 Data Set Names for Different MULTIDATA and ORDERED Arguments

	Ordered:"N"	Ordered:"A"	Ordered:"D"
Multidata:"Y"	Hash_Y_N	Hash_Y_A	Hash_Y_D
Multidata:"N"	Hash_N_N	Hash_N_A	Hash_N_D

Of note in this example is our use of expressions to create the various alternatives and to name the output data sets to indicate to which combinations they correspond. We need not modify hard-coded values (or use macro language logic) to generate the alternatives. In the table below, the contents of these data sets (i.e., also the content of the corresponding hash tables) are shown side by side:

Table 3.4 Hash Table Order for Different MULTIDATA and ORDER Arguments

	Input Order		Ordered: "N"		Ordered: "A"		Ordered: "D"	
	K	**D**	**K**	**D**	**K**	**D**	**K**	**D**
Multidata:"Y"	2	A	2	A	1	A	3	A
	2	B	2	B	1	B	3	B
	2	C	2	C	1	C	3	C
	3	A	1	A	2	A	2	A

		3	B		1	B		2	B		2	B
		3	C		1	C		2	C		2	C
		1	A		3	A		3	A		1	A
		1	B		3	B		3	B		1	B
		1	C		3	C		3	C		1	C
Multidata:"N"		2	A		2	A		1	A		3	A
		3	A		1	A		2	A		2	A
		1	A		3	A		3	A		1	A

For MULTIDATA:"Y", i.e., when duplicate-key items are allowed, it shows that if table H is left unordered, the item groups with K=(2,1,3) follow each other in some random order, different from the input order in which the keys are received, i.e., K=(2,3,1). However, *within each same-key group*, the items follow each other exactly in the *order they are received from input*, i.e., D=(A,B,C). It also shows that ordering the table via the ORDERED argument tag causes the same-key items groups as a whole to be permuted into required order by key. The sequence of items themselves within each same-key group remains intact and always mirrors their input sequence regardless of the table order.

For MULTIDATA:"N" (equivalent to omitting it altogether), the duplicate-key items are rejected. Note that the relative order of the same-key groups by table key is the same as before, except that now each group contains one item only. This is the item D="A" received from the input first because the ADD method was used. Using the REPLACE method would result in keeping the last input item, D="C", instead.

The randomizing default behavior of the *Insert* operation is due to the fact that the hash function used behind the scenes to place the keys in the table is random by nature. Though it can be, at a certain angle, viewed as unwelcome, it has useful applications we will see in this book later on.

3.3.13 Insert Operation Hash Tools

- Methods: ADD, REPLACE, REF.
- Arguments Tags: DATASET, DUPLICATE, ORDERED.

3.3.14 INSERT Operation Hash-PDV Interaction

- Implicit Insert: The key portion and data portion values are populated from their PDV host variables counterparts.
- Explicit Insert: None.

3.4 DELETE ALL Operation

The *Delete All* operation is just the opposite of *Insert*:

- If a given key-value is in the table, it deletes *all the items* with this key-value. In other words, it eliminates the entire group of items sharing the key.
- If the table is constrained to unique keys only, every same-key item group contains just one item, and so this is the item that gets deleted. Otherwise, if duplicate-key items are allowed, the whole group of items with this key is deleted, whether it includes a single item or more.
- For every item deleted from the table, memory occupied by it is released and the value of the NUM_ITEMS attribute is automatically decremented by 1.

The operation can facilitate data processing in many ways. To mention just a couple:

- We can eliminate the items whose keys are deemed by programming logic "already used" and no longer needed for the purposes of other operations, such as *Search*, *Retrieve*, or *Enumerate*. Also, the operation comes in handy when hash tables are used to join data and some items need to be removed prior to performing the join.
- Thanks to its dynamic nature, *Delete All* coupled with *Insert* can be used to implement run-time dynamic structures, such as stacks, queues, etc. Examples of constructing them are offered in the book later on.

Let us now look at the implementation of the *Delete All* operation using hash object tools.

3.4.1 DELETE ALL Implementation

The *Delete All* operation is supported by the REMOVE method. In the snippet below, table H is populated with 3 items with the following (K,D) key-data pairs: (1,A), (2,B), (2,C). Note that the last 2 items share the same key-value, K=2. After the table is loaded, the REMOVE call deletes both items with K=2 (the current PDV value of K), so that only the pair (1,A) remains in the table:

Program 3.5 Chapter 3 Removing an Item.sas

```
data _null_ ;
  dcl hash H  (multidata:"Y") ;
  H.definekey ("K") ;
  H.definedata("D") ;
  H.definedone() ;
  do K = 1, 2, 2 ;
    q + 1 ;
    D = char ("ABC", q) ;
    rc = H.add() ;
  end ;
  rc = H.REMOVE() ; *implicit/assigned call
run ;
```

The key and data content that the ADD method above loads in table H is as follows:

K	D
1	A
2	B
2	C

After the REMOVE method is used, the table H content looks as follows:

K	D
1	A

Note that at the time of the REMOVE call, the value of key host variable is K=2. Since the call is *implicit* (no argument tags are used), this is the key value accepted by the method. And since the key is in the table, the call is successful, so the entire group of items with K=2 gets deleted. In this case, any of the following alternative REMOVE calls would achieve the same result with no errors:

```
H.REMOVE() ;                    * implicit, unassigned ;
rc = H.REMOVE(key: 2) ;         * explicit,   assigned ;
H.REMOVE(key: sum(_N_,1)) ; * explicit, unassigned ;
```

Let us make a few germane observations:

- Since the calls are successful (the key is in the table), they can be left unassigned.
- The last line works because at the time of call _N_=1.
- In fact, any *expression* assigned to the KEY argument tag in an *explicit call* will work here as long as it is of the numeric data type (because K is compiled as numeric) and resolves in 2.
- If MULTIDATA:"Y" were not coded, the only change would be that, for K=2, only the (K,D)=(2,B) item would be inserted in table H due to the unique key constraint. Therefore, only this item would be deleted.

A *different picture* emerges if the method is given a *key that is not present in the table*. For example:

```
K = 3 ; rc = H.REMOVE() ; * assigned call - no error ;
K = 5 ; H.REMOVE() ;        * unassigned call - error! ;
rc = H.REMOVE(key:7) ;      * assigned call - no error ;
H.REMOVE(key:7) ;           * unassigned call - error! ;
```

Again, a couple of notes are due:

- Since none of the keys used above is in the table, all these calls *fail*, and so no items are deleted.
- If a failing call is left *unassigned*, the step will stop processing with an error message.

Hence, a general *safety rule: If a method can fail, it should never be called unassigned.* Either assign it to a variable to hold its return code or include it in another statement - for example, a conditional clause. (The REMOVE method is one of the methods that *can* fail - in contrast to such methods as REPLACE or REF, which never fail due to their design.)

3.4.2 DELETE ALL and Item Locking

There exists a scenario under which an item group (including a single-item group) cannot be deleted by calling the REMOVE method, even if the key-value that it is called with is in the table. Namely, it happens when the item group targeted for deletion is locked by a *hash iterator* linked to the table. Since this phenomenon is associated with the *Enumerate All* operation, the issue and ways to deal with it will be discussed in detail in the next chapter.

3.4.3 DELETE ALL Operation Hash Tools

- Methods: REMOVE.
- Argument tags: KEY.

3.4.4 DELETE ALL Operation Hash-PDV Interaction

- An implicit REMOVE call accepts the current key values from the PDV host key variables.
- None. If the operation is successful, the corresponding item or items are merely deleted from the table. Neither PDV host variables nor hash variables change their values.

3.5 RETRIEVE Operation

The *Retrieve* operation *extracts the values of the data portion variables* from the hash table *into their corresponding host variables* in the PDV. *Retrieve* is quite an important part of hash object programming since it is an integral component of such data processing tasks as data aggregation, joining tables, and a host of others.

In fact, in many cases *Retrieve* is done behind the scenes even when it is not performed *directly*. This is because it is indirectly invoked during the *Enumerate* operation. We ought to be cognizant of it to ensure that the values of PDV host variables get overwritten only when this is part of the intent and programming logic. We will leave these behind-the-scenes actions for the section where the *Enumerate* operation is reviewed later on. In this section, we will discuss the *direct Retrieve* operation only.

3.5.1 Direct RETRIEVE

Direct *Retrieve* is invoked by calling the FIND method. It triggers the following sequence of actions:

- Given a value of the table key, search the table for it.
- If the key *is* in the table: Extract the values of *all data portion* hash variables for the item with this key into the respective PDV host variables, overwriting their values. Generate a *zero* return code to indicate that the call is *successful*.
- If the key *is not* in the table: Generate a *non-zero* return code to indicate that the call has *failed*. If the call is *unassigned*, write an error message in the log to the effect that the key is not found.

Note that if the key is not found, the values of the PDV host variables remain unchanged- i.e., the same as before the method call.

To illustrate these *Retrieve* actions in the SAS language, we can use the same table H as in the example above, populated with the (K,D) tuples (1,A), (2,B), (2, C) in the following snippet:

Program 3.6 Chapter 3 Direct Explicit Assigned Retrieve.sas

```
data _null_ ;
  dcl hash H  (multidata:"Y") ;
  H.definekey ("K") ;
  H.definedata("D") ;
  H.definedone() ;
  do K = 1, 2, 2 ;
    q + 1 ;
    D = char ("ABC", q) ;
    H.add() ;
  end ;
  D = "X" ;              *pre-call value of PDV host variable D;
  RC = H.FIND(KEY:1) ;
  put D= RC= ;
run ;
```

3.5.2 Successful Direct RETRIEVE

The step, as shown above, prints the following in the SAS log:

```
D=A  RC=0
```

This demonstrates the effect of the *successful* direct *Retrieve* operation: Since K=1 is found in the table, RC=0 and the value of hash variable D="A" from the hash item with K=1 overwrites the original value "X" of PDV host variable D.

If, instead of using the FIND call with key-value 1, we key it with 2, the operation will be also successful because K=2 is in the table as well. For example, any of the following calls will work:

```
RC = H.FIND() ;             *Implicit, K=2 accepted from PDV;
RC = H.FIND(KEY: 2) ;       *Explicit, 2 is a numeric literal;
RC = H.FIND(KEY: _N_+1) ; *Explicit, _N_+1 resolves to 2;
```

In both cases, the PUT statement prints in the log:

```
D=B RC=0
```

Evidently, the original PDV value D="X" is overwritten with "B" - that is, with the data value from the item (2,B). However, it raises a *question*: The table has another item with K=2, (2,C), so why is (2,B) chosen? The *answer* is that if a group of same-key items has more than one item, a call to FIND with this key *always points to the logically first item in the group*. To recap, the logically first item within any same-key item group is the one inserted first, and in this case, it is (2,B).

Direct *Retrieve*, via the FIND method, can perform the operation *only for the logically first item* in a same-key group. In order to get to the rest of the items in the group and retrieve their data, the group has to be *enumerated*, which involves hash tools other than FIND. They will be reviewed in the section devoted to the *Enumerate* operation.

3.5.3 Unsuccessful Direct RETRIEVE

Suppose that the FIND call used in the step above is replaced with one of the following:

```
K = 5 ; RC = H.FIND() ;
RC = H.FIND(KEY: 5) ;
```

In both cases, the step prints:

```
D=X RC=160038
```

Since K=5 is not in the table, the operation is unsuccessful. Hence, RC is not zero and there is no hash variable D value available to overwrite its host counterpart in the PDV, so the PDV value D="X" remains intact.

Note that all the sample FIND calls above are issued *assigned*. Since the method can fail when the key-value accepted by it is not in the table and, generally speaking, we do not know ahead of time whether the key is in the table or not, calling it unassigned may result in an error. So, as with all methods that can potentially fail for the same reason (e.g., CHECK or REMOVE) it is a prudent practice to *always call FIND assigned*, lest the DATA step generate an "*ERROR: key not found*" message along with the unpleasant note "*The SAS System stopped processing this step because of errors*".

3.5.4 Implicit vs Explicit FIND Calls

We have already discussed the relative merits of *implicit* and *explicit* calls, and the same considerations apply to the FIND method. To recap:

- The choice is dictated by a number of factors, such as the number of component hash variables in the key if it is composite, whether or not we want to preserve the current values of the PDV host key variables, etc.
- Both styles work, and often opting for one against the other is a matter of programming convenience and code brevity.
- The FIND method, if called explicitly, requires only the argument tag KEY. This is because the *Retrieve* operation requires nothing but a key-value in order to extract the data (as opposed to, for example, the Update All operation).
- If the table key is composite, as many KEY argument tags must be listed as there are constituent key variables in the key; and expressions assigned to them must be of the same data types as the corresponding key variables.

3.5.5 RETRIEVE Operation Hash Tools

- Methods: FIND.
- Argument tags: KEY, MULTIDATA.

3.5.6 RETRIEVE Operation Hash-PDV Interaction

- The data flow is from the hash table data portion variables to their PDV host variables.
- If the operation is successful, the values of the former overwrite the values of the latter.

3.6 UPDATE ALL Operation

A successful *Retrieve* operation overwrites the values of the PDV host variables with the values of the corresponding data portion hash variables. A successful *Update All* operation, in a sense, does the opposite: It overwrites the values of the data portion hash variables with the values of the corresponding PDV host variables (or values supplied by the program). The operation is critical for any dynamic data exchange between the PDV and the hash table. It is indispensable for a number of data processing tasks, notably, for data aggregation.

The *Update All* operation works according to the following pattern:

- *If the key it accepts is in the table*, the data portion variables of *all* the items with this key are *updated*, i.e., overwritten, with the data values accepted by the operation.
- *Otherwise*, a new item with the key and data values accepted by the operation is inserted into the table.

Therefore, the *Update All* operation is always successful, regardless of whether the key it accepts is in the table or not or whether the argument tag MULTIDATA:"Y" is in effect:

- If the key is in the table:
  - If MULTIDATA:"Y" *is not* in effect, the data portion variables in the single hash item with this key are updated.
  - If MULTIDATA:"Y" *is* in effect, the data portion variables in *all* the hash items with this key are updated.
- If the key *is not* in the table, a new item with this key is inserted in the table.

The *Selective Update* operation is possible only if MULTIDATA:"Y" is specified. In this case, we can update the data portion variables in a *particular item* within a group of items with the same key.

3.6.1 UPDATE ALL Implementation

The hash object tool implementing this operation is the REPLACE method. To demonstrate its mechanics, let us turn to basically the same simple DATA step we have already used for illustrative purposes, with some wrinkles:

Program 3.7 Chapter 3 Explicit Unassigned Update All.sas

```
data _null_ ;
  dcl hash H (multidata:"Y") ;
  H.definekey ("K") ;
  H.definedata ("D") ;
  H.definedone() ;
  do K = 1, 2, 2 ;
    q + 1 ;
    D = char ("ABC", q) ;
    rc = H.add() ;
  end ;
  H.REPLACE (KEY:1, DATA:"X") ; ❶
  H.REPLACE (KEY:2, DATA:"Y") ; ❷
  H.REPLACE (KEY:3, DATA:"Z") ; ❸
run ;
```

The evolution of the hash table H content after each successive REPLACE call is shown below, left to right:

Table 3.5 Results of UPDATE ALL Operation

	As Loaded		After update by REPLACE call					
			❶		❷		❸	
	K	D	K	D	K	D	K	D
	1	A	1	*X*	1	X	1	X
Multidata:"Y"	2	B	2	B	2	*Y*	2	Y
	2	C	2	C	2	*Y*	2	Y
							3	*Z*
	1	A	1	*X*	1	X	1	X
Multidata:"N"	2	B	2	B	2	*Y*	2	*Y*
							3	*Z*

If the argument tag MULTIDATA:"N" is used (or the argument tag is omitted), the second input item with K=2 (2,C) is auto-rejected by the *Insert* operation invoked via the ADD method call. However, regardless of whether a same-key item group contains one item or more, the *Update All* operation invoked via the REPLACE method behaves in exactly the same manner: If the key is in the table, the data portion hash variables are overwritten with the values accepted by the call for *all* items in the group; and if it is not in the table, a new item with these values is inserted.

3.6.2 Assigned vs Unassigned REPLACE Calls

As far as this distinction is concerned, "*vs*" is a misnomer here because all REPLACE calls are successful by the nature of its design: The method can fail neither if the key it accepts is found in the table, nor if it is not found. Hence, there is no reason to call REPLACE assigned, as in, for instance:

```
RC = REPLACE(KEY:3, DATA:"Z") ;
```

In any case, the result of the method call will be RC=0. So, we can always call REPLACE *unassigned* - and probably *should* - based on the principle that any piece of coding not caused by programming necessity and included *just in case* is to some degree misleading.

3.6.3 Implicit vs Explicit REPLACE Calls

All the considerations concerning relative merits of *implicit* and *explicit* REPLACE calls are precisely the same as those spelled out for the *Insert* operation and the ADD method. Their calling style is a matter of programmatic convenience based on (a) code brevity and (b) whether or not it is important to keep the values of the host variables intact.

3.6.4 Selective UPDATE Operation Note

The *Update All* operation is one of two update operations that can be performed on a hash table. *Update All* updates all the data portion variables within a same-key item group with one set of updating values. By contrast, the *Selective Update* operation can update specific items in the group selected according to the required programming logic (including all of them if it so dictates). Moreover, it can update different items with different sets of updating values. However, it can work only in conjunction with the *Keynumerate* operation because the latter is the only way to latch onto a particular item to be updated - a situation similar to the *Selective Delete* operation. Hence, we will postpone reviewing *Selective Update* till the time when we get to describe the *Keynumerate* operation.

3.6.5 UPDATE ALL Operation Hash Tools

- Methods: REPLACE.
- Argument tags: KEY, DATA.

3.6.6 UPDATE ALL Operation Hash-PDV Interaction

- *Implicit* calls: The data portion hash variables are overwritten with the values of the PDV host variables if the key-value supplied with the call is in the table.
- *Explicit* calls: PDV variable values (not necessarily of host variables) may be used in the expressions supplied to the argument tags KEY and DATA.

3.7 ORDER Operation

The *Order* operation causes the *same-key item groups* to be *logically ordered* within the table according to their key-values into a *specified sequence*. After the operation is complete:

1. If the *Output* operation is used to write the content of the table to a SAS data set, it accesses the table one item at a time in their logical order. Hence, the *physical sequence* of the output file records will repeat the *logical sequence* of the items in the table.

2. Likewise, the *Enumerate* operation also accesses the table in its *logical sequence*. Thus, if, for example, the table is ordered *ascending* and enumerated *forward*, the item group with the lowest key-value is accessed first, and so on until the group with the highest key-value is accessed last.

3.7.1 ORDER Operation Invocation

Technically, invoking the Order operation is exceedingly simple: It is done via the ORDERED argument tag at the time when a hash object instance is created - that is, either in the compound DECLARE (DCL) statement or the _NEW_ operator statement. For instance, for table H to be ordered ascending, both approaches below work:

```
dcl hash H(ORDERED:"A") ;
```

Or:

```
dcl hash H ;
H = _NEW_ hash (ORDERED:"A") ;
```

The order is determined by the value of the argument given to the argument tag ORDERED. The acceptable variants of that value are listed in the table below:

Table 3.6 Arguments for ORDERED Argument Tag

ORDERED Argument	Resulting Table Order
"ASCENDING" | "A" |"YES" | "Y"	Ascending
"DESCENDING" | "D"	Descending
"NO" | "N"	Undefined (random)
Not coded	Undefined (random)

3.7.2 ORDERED Argument Tag Plasticity

Just like with any other argument tag in the hash tools arsenal, the argument of ORDERED is, generally speaking, an expression. In this case, it is required to be of the character type. It affords the flexibility to have a table ordered into a specific sequence based on a condition or control information. For example, in the step below, variable *Orders* represents a primitive "control table" telling which hash table instance gets which sorting order according to the position of the corresponding character. The DO loop creates 3 instances for hash object H, of which instance 1 is ordered ascending, instance 2 - descending, and instance 3 - internally:

Program 3.8 Chapter 3 ORDERED Argument Tag as Expression.sas

```
data _null_ ;
  retain Orders "ADN" ;
  dcl hash H ;
  do i = 1 to 3 ;
    H = _NEW_ hash (ORDERED: char (Orders, i)) ;
  end ;
run ;
```

This kind of functional plasticity may seem a bit geeky at first. However, it illustrates the important concept of the hash object's dynamic nature. Also, it can be very useful if we intend to store hash object instance identifiers in another hash table and want the respective instances structured differently according to control (driver) information.

3.7.3 Hash Items vs Hash Item Groups

Note that when we are talking about the sequence of items within a hash table, it is not in terms of the *individual items* but in terms of the *same-key item groups*. The general principle for the sequence of the items in a hash table can be tersely formulated as follows:

- Regardless of whether or not the ORDERED argument tag has been used and/or with which argument, when items with duplicate key-values are added to the table:
  - The items with the same key-value are always grouped together.
  - Within each same-key item group, the sequence of its items is the same as that in which they have been added.

Now let us take a more detailed look:

- When duplicate-key items are allowed in the table via MULTIDATA:"Y", the *items with the same key-value are always logically grouped together*.
- The relative sequence of the items within every same-key group with more than one item is *always exactly the same as received from input* during the *Insert* operation.
- Both statements above are true regardless of the argument given to the ORDERED argument tag ("A", "D", or "N"). The latter affects only the positions of the *same-key item groups as whole units* relative to one another. The items with the same key-values always stick together regardless of the sequence of the groups to which they belong. The sequence of the items *within each same-key group always remains as inserted*.
- If "A" or "D" is specified, the *Order* operation *permutes same-key groups as whole units* to force them into ascending or descending by their key-values. The sequence of items within each group is not affected, and it is preserved as inserted.
- If "N" is specified (or if ORDERED is not coded at all), the *Order* operation is still performed, albeit *implicitly*. The only difference from "A" or "D" is that now the sequence of different same-key groups relative to one another is internal and, in fact, random. The sequence of items within each group is not affected, and it is preserved as inserted.
- If a group of items sharing the same key contains just one item, it is merely a partial case: It is still a group, and relative to the rest of the groups it behaves exactly as described above.
- If MULTIDATA:"Y" is *not in effect*, it forces each same-key group to have a single item with its unique key-value. Or it may happen that each item comes from input with its own unique key. In these situations, it *appears* that the items themselves, rather than their same-key groups, are ordered accordingly. However, it is just a mirage created by the fact that in such a case each group contains a single, unique-key item.

To inculcate this essential concept better using an example, let us run the following step, varying the value of variable *Order* from "N" to "A" to "D" in a loop. For each value, table H is created anew, populated using the ADD calls the same way, and its content is output to a data set named using the current value of *Order*.

Program 3.9 Chapter 3 Duplicate-Key Table Ordered 3-Way.sas

```
data _null_ ;
  do order = "N", "A", "D" ;
    dcl hash H   (multidata:"Y", ORDERED:Order) ;
    H.definekey  ("K") ;
    H.definedata ("K", "D") ;
    H.definedone () ;
    K = 1 ; D = "A" ; H.add() ;
    K = 2 ; D = "B" ; H.add() ;
    K = 2 ; D = "C" ; H.add() ;
    K = 3 ; D = "D" ; H.add() ;
    K = 3 ; D = "E" ; H.add() ;
    K = 3 ; D = "F" ; H.add() ;
    H.output (dataset: catx ("_", "Hash", Order)) ;
  end ;
run ;
```

The step generates an output data set for each of the three different *Order* values: Hash_N, Hash_A, and Hash_D. Each mirrors the content of its own correspondingly ordered hash table. In the table below, their contents are presented side by side along with the original input order, with the same-key item groups shaded identically:

Table 3.7 Same-Key Item Groups and Hash Items Relative Order

Input Order		Ordered: "N"		Ordered:"Y"		Ordered:"D"	
K	**D**	**K**	**D**	**K**	**D**	**K**	**D**
1	A	2	B	1	A	3	D
2	B	2	C	2	B	3	E
2	C	1	A	2	C	3	F
3	D	3	D	3	D	2	B
3	E	3	E	3	E	2	C
3	F	3	F	3	F	1	A

Now, if we omit MULTIDATA:"Y" (or recode it as MULTIDATA:"N") and repeat the procedure, only the first item with a given input key-value is kept, and so we will get the following picture:

Table 3.8 Hash Items Relative Order for Unique Keys

Input Order		Ordered: "N"		Ordered:"Y"		Ordered:"D"	
K	**D**	**K**	**D**	**K**	**D**	**K**	**D**
1	A	2	B	1	A	3	D
2	B	1	A	2	B	2	B
3	D	3	D	3	D	1	A

This simple data experiment clearly confirms the principal points made above:

- With any specified sequence, the *Order* operation permutes the same-key groups as whole units.
- The input sequence of the items within each same-key item group is always preserved.
- These principles hold true regardless of the number of items in any same-key group.

3.7.4 OUTPUT Operation Effects

As we have seen, the *Output* operation, regardless of the specified order (ascending, descending, or undefined/random) reshapes the input data sequence by forcing the same-key items to bunch up into same-key groups. There are several takeaways from this fact:

- If the goal of having the table ordered is to enable BY-group processing of the data set created from the table by the *Output* operation, specifying ascending or descending order is extraneous. Even if the order is left (or specified) internal, the same-key items are still grouped together, so the output data set can be safely processed using the BY statement with the NOTSORTED option.

- By the same token, it is unnecessary to explicitly order the table into ascending or descending order if the only goal of a task at hand is to locate a group of (one or more) items with a given key and process or use the data within the group. Within this scope, the position of the group in the table relative to other same-key groups is irrelevant. Typical tasks of this nature include a Slowly Changing Dimensions Type 2 (SCD2) table lookup or harvesting same-key items for one-to-many or many-to-many joins.

- Under other circumstances, such as processing *partially sorted data*, forcing the hash table into ascending or descending order may be crucial. Usually, it occurs when a file is processed one BY group at a time and for every BY group, a number of hash tables, each sorted by a different key, need to be populated, post-processed, and reinitialized for the BY group to be processed next.

The result of the *Order* operation with ORDERED:"A" or ORDERED:"D" is known formally as *stable sorting*. It means that within every same-key group, the relative input sequence of items is preserved, while the groups themselves are permuted into the specified order according to their key-values. This is precisely analogous to the action of the SORT procedure when its EQUALS option is in effect.

The result of the implicit *Order* operation with ORDERED:"N" (or when the argument tag is omitted) can be called, in the same vein, *stable grouping*: Though the same-key groups, relative to each other, are disordered, items with the same key always abut each other, and within each same-key group, the relative input sequence of items is preserved.

3.7.5 General Hash Table Order Principle

The principal takeaway from this section can be condensed as follows:

1. The items in any hash table are always grouped by the table key.
2. The sequence of the same-key groups relative to each other is determined by the specified (or implied) order.
3. Within each same-key group, the relative position of the items is the same as in input.

3.7.6 Ordering by Composite Keys

If the hash table key is composite, i.e., comprises more than one hash variable, it is always processed by the hash object as if it were a single concatenated key. That is why the *Order* operation can order the table only by its composite key as a whole either ascending or descending. It cannot order it ascending by one key hash variable and descending - by another. The only way this kind of effect can be achieved is by using certain programming subterfuges, such as multiplying a numeric key variable by -1, but the purpose and value of doing so are questionable.

3.7.7 Setting the SORTEDBY= Option

When a hash table sorted by the *Order* operation is written out to a SAS data file, the SORTEDBY= option on this file *is not set*. Therefore, if it is desirable to indicate, for the sake of downstream data processing, that the file is already sorted, the option has to be spelled out explicitly. For example, if in the DATA step shown above the argument tag ORDERED is valued as "A", we may include SORTEDBY= in the DATASET argument tag specification for the OUTPUT method call:

```
H.output (dataset: cats ("Hash_", Order, "(SORTEDBY=K)") ;
```

In this case, the character expression assigned to DATASET would resolve to:

```
Hash_A(SORTEDBY=K)
```

This way, if an attempt to sort the file should be made, the SORT procedure will recognize the flag and skip sorting. Also, the SORTEDBY=K flag may cause the SQL optimizer to take the fact that the file is already sorted into account and use it to construct a more efficient query.

3.7.8 ORDER Operation Hash Tools

- Methods: OUTPUT.
- Argument tags: ORDERED, MULTIDATA.

3.7.9 ORDER Operation Hash-PDV Interaction

- None. All required sorting is done internally by the hash object.

Chapter 4: Item-Level Operations: Enumeration

4.1 Introduction

Under a number of real-life programming scenarios a hash table, in one way or another, needs to be processed sequentially one item at a time- i.e., *enumerated*. The item-level operations discussed thus far do not provide for this capability: They process either a single item or all items in a same-key item group at once. However, the hash object includes special tools designed to enumerate a hash table in a number of variations. In this chapter, we are discussing the hash table operations based on these tools. In the chapters that follow they will be further exemplified in a variety of ways.

4.2 Enumeration: Basics and Classification

The SAS hash object supports two enumeration operations on its tables: *KeyNumerate* (aka *Key Enumerate*) and *Enumerate All*. In the next two sections we will discuss them separately in detail. However, before getting to it, let us briefly dwell on what "*enumeration*" means in general and what it means specifically as applied to a hash object table.

4.2.1 Enumeration as a Process

Generally speaking, to *enumerate* means to *list items in a collection sequentially in their intrinsic order*. There are a number of reasons to enumerate. For example, the goal may be just to determine the number of items in the collection or to find out how many of them carry information, in which case it is sufficient merely to list the items without retrieving the data stored in them. On the other hand, it may be desirable to extract the data in order to process it or store elsewhere. Enumeration is an intrinsic part of SAS programming. Just a few examples of enumeration include (you can doubtless think of more):

- Reading a SAS data file sequentially.
- Scanning through an array.
- Scanning through a character string one character at a time.

Note that in all these cases, enumeration can be related either to the whole data collection or just part of it. Also, the items be simply listed to discover certain properties of the data collection, or the data they carry can be extracted into the PDV as well.

4.2.2 Enumerating a Hash Table

Just like in the case of a general data collection, enumerating a hash table means accessing its items sequentially. However, in terms of a SAS hash object, enumerating the table always means not merely listing the items, but also retrieving the values of the data portion variables into their respective PDV host counterparts. In other words, when a hash table is enumerated, the *Retrieve* operation is *always* performed latently on each item it visits.

In terms of the hash object, there are two different ways to enumerate a hash table corresponding to two different enumeration operations. Let us look at them separately.

4.2.3 KeyNumerate (Key Enumerate) Operation

This operation works by taking the following course of actions:

- *Given a key-value*, find the group of items sharing this key-value.
- If an item group with this key-value exists in the table, access the items in *this, and this group only,* sequentially, starting from the *logically first* item in this group.

Also, the *KeyNumerate* operation has certain limitations:

- It requires a key-value in order to point to a specific same-key item group on which to work.
- It cannot cross the boundaries of a given same-key item group. Only the items in this group can be enumerated. To enumerate another item group, another key-value must be used.
- It is supported only if the MULTIDATA:"Y" argument tag is set.

4.2.4 Enumerate All Operation

The *Enumerate All* operation works according to the following three alternative modes:

1. Start from the logically *first* item in the table and access one or more items, one at a time, in the direction of the table logical order, i.e., *forward*.
2. Start from the logically *last* item in the table and access one or more items, one at a time, against the direction of the table logical order, i.e., backward.
3. Start with an item with a given key-value and access one or more items, one at a time, either forward or backward. The items visited can be *any* of the items in the table, regardless of the same-key item group or *any* subset of the items in the group thereof.

The *Enumerate All* operation eliminates the limitations of the *Keynumerate* operation since it is not constrained to a single same-key item group:

- In modes 1 and 2, it does not require a key in order to operate at all.
- In mode 3, it needs a key-value only to locate the starting point. Thereafter, no key is required for it in order to work.
- It can access items across all same-key item groups, i.e., the entire table.
- It operates in the same exact way regardless of whether MULTIDATA:"Y" is set or not.
- However, in order to work, it requires an auxiliary structure, the *hash iterator object*, linked to the instance of the hash table in question.

4.2.5 Template DATA Step

We will discuss the *KeyNumerate* and *Enumerate All* operations and the hash tools they are implemented with in the two following sections separately. However, we will be using the same sample hash table to illustrate the concepts in both. Consider the following *template* DATA step:

Program 4.1 Chapter 4 Template Data Step.sas

```
data _null_ ;
  dcl hash H (multidata:"Y", ordered:"N") ;
  H.definekey ("K") ;
  H.definedata ("D") ;
  H.definedone () ;
  do K = 1, 2, 2, 3, 3, 3 ;
    q + 1 ;
    D = char ("ABCDEF", q) ;
    H.ADD() ;
  end ;
  h.output(dataset: "AsLoaded") ;
  /*...Insert demo code snippets below this line...*/
  stop ;
run ;
```

Note that ORDERED:"N" in the table H declaration indicates that it uses its default internal ordering rather than explicit "A" (ascending) or "D" (descending). Also, MULTIDATA:"Y" is intentionally set to allow items with duplicate key-values. After all input value pairs (K,D) have been inserted in the table via the ADD method, the content of the loaded table H is as follows:

K	D
2	B
2	C
1	A
3	D
3	E
3	F

- To test-run the code snippets presented in the next two sections, all we need is to paste the snippet in question into the template step above after the comment line, just before the STOP statement.
- Note, however, that for each snippet, the programs supplied with the book present the full template step with the relevant snippet included.
- The call to the OUTPUT method is included in the step for diagnostic purposes, so that the data portion of the table could be viewed as reflected in the content of data set work.AsLoaded.
- The reason for setting ORDERED:"N" and MULTIDATA:"Y" is that it makes the ensuing discussion more generic and applies equally to the hash table with these particular argument tag values or any other values.

4.3 KEYNUMERATE Operation

Suppose that, given table H above, we need to list all the items with K=3 and retrieve the values of D (i.e., "D", "E", "F") from each item into the PDV host variable D. Such need, for example, arises when hash tables are used to perform a one-to-many or many-to-many table join and in a number of other use cases as well.

Doing so is a two-step process:

1. Point at the *logically first* item of the K=3 same-item group- i.e., the item with D="D" and retrieve the value of hash variable D into PDV host variable D.
2. Step down the list one item at a time. Each time, *retrieve* the value of D. If there are no more items in the list, terminate.

4.3.1 KeyNumerate Operation Mechanics

It is important to get a good grip on what is happening within the key group on which the *Keynumerate* operating is working in order to apply the operation's programming logic cognizantly and avoid befuddling surprises. In the SAS language, the two-step plan outlined above can be implemented as shown in this snippet (inserted in the template step above for testing):

Program 4.2 Chapter 4 Keynumerate Mechanics Snippet.sas

```
call missing (K, D, RC) ;        ❶
put "Before Enumeration:" ;
put +3 (RC K D) (=) ;
put "During Enumeration:" ;
RC = H.FIND(KEY:3) ;             ❷
do while (RC = 0) ;              ❸
   put +3 (RC K D) (=) ;
   RC = H.FIND_NEXT() ;          ❹
end ;
put "After Enumeration:" ;
put +3 (RC K D) (=) ;
```

Running this snippet within the template step prints the following information in the SAS log:

```
Before Enumeration:
   RC=. K=. D=
During Enumeration:
   RC=0 K=. D=D
   RC=0 K=. D=E
   RC=0 K=. D=F
After Enumeration:
   RC=160038 K=. D=F  ❺
```

❶ Before the *KeyEnumerate* operation is invoked, K, D, and RC in the PDV are set to nulls just to see how their values will change.

❷ `H.FIND(KEY:3)` call points to the group of items with K=3. Another way to phrase it is that the call *sets* the list of items with K=3. Setting the list places the enumeration pointer at the logically first item in the group. FIND executes the direct *Retrieve* operation, extracting the hash value D="D" from the first item in the group into its PDV host variable. Thus, the missing value for D is overwritten with "D". Since the call is explicit, there is no need to assign K=3 in the PDV.

❸ A DO loop is launched to call **H.FIND_NEXT()** repeatedly until such time when the method call fails with RC≠0.

❹ **H.FIND_NEXT()** is called in the loop. Its first iteration moves the pointer from the first item (where the FIND call placed it) to the second, and every next iteration moves it one more item down the list. In each iteration, FIND_NEXT retrieves the hash value of D from the corresponding item into host variable D. This way, D="D" is first overwritten with "E", and then with "F", which is what the log records indicate.

❺ At this point, there are no more items in the group left to enumerate. Thus, when FIND_NEXT is called once again, it fails returning RC=160038. Since RC≠0, the loop terminates at the next iteration at the top of the loop because of the RC=0 condition in the WHILE clause. This action moves the pointer outside the K=3 item group, which *unsets* the item list and terminates the operation.

A few more points deserve to be emphasized:

- Since K is not specified in the data portion of table H, neither the FIND nor FIND_NEXT method retrieves its value from the table into its host variable, and thus K=. in the PDV remains unchanged during the whole process.

- After the item list has been *unset*, the only way to enumerate the item group anew is to set the item list again by calling FIND again.

- Incidentally, in addition to FIND, both the CHECK and REF method calls can set an item group for enumeration. In other words, they also place the enumeration pointer on the first item in the group and enable further enumeration using FIND_NEXT. However, unlike FIND or FIND_NEXT, neither CHECK nor REF retrieves the data portion values from the first item due to their design. Thus, they can be useful in terms of *Keynumerate* only in a hypothetical situation where the data has to be retrieved from all items in an item group except the logically first. Whether there can be a realistic use case for such a functionality is moot. However, the capability does exist, and we would like you to be aware of it, even if it is not much more than a curiosity.

4.3.2 FIND_NEXT: Implicit vs Explicit

Note that the FIND_NEXT call is *implicit*- i.e., it does not have the KEY argument tag specified. However, the method is provided with the option of calling it explicitly, so why not do it? Actually, there is a reason behind it:

- The FIND_NEXT method always works only with a key-value group set by the initial call to the FIND method (or, as an option, CHECK or REF). Without such a priming call, it would not work at all because the item list for it to operate upon would not be set.

- Hence, calling FIND_NEXT *explicitly* is pointless: If, instead of the implicit call, we coded the method explicitly with *any value* of the argument tag KEY, it would change nothing since the arguments provided for the KEY argument tag are ignored. For example, the calls H.FIND_NEXT(KEY:0), H.FIND_NEXT(KEY:3), or H.FIND_NEXT(KEY:999) would result in the same outcome as H.FIND_NEXT().

4.3.3 Other KeyNumerate Coding Styles

The *Keynumerate* code section can be written more concisely by dropping the FIND call before the DO loop and instead including it in the loop's index specification:

Program 4.3 Chapter 4 Keynumerate Loop Style2 Snippet.sas

```
do RC = H.FIND(KEY:3) BY 0 while (RC = 0) ;
  put +3 (RC K D) (=) ;
  RC = H.FIND_NEXT() ;
end ;
```

By including FIND in the loop, above, we call it automatically on the first iteration, and the RC value the call returns is evaluated at the top of the loop. If the key-value provided with the FIND call is not in the table, there is no same-key group to enumerate, so the loop stops right there and then. Otherwise, it proceeds to calling FIND_NEXT repeatedly in the same manner as before. The BY 0 clause is a subterfuge used to keep the loop iterating: Without it, the loop would terminate at the first iteration according to the DO loop index rules. Above, we want the loop to be controlled not by its index but by the WHILE condition and terminate when RC≠0.

Yet another, even more concise, style is based on calling FIND conditionally and using the UNTIL condition instead of WHILE:

Program 4.4 Chapter 4 Keynumerate Loop Style3 Snippet.sas

```
if H.FIND(KEY:3) = 0 then do until (H.FIND_NEXT() ne 0) ;
  put +3 (RC K D) (=) ;
end ;
```

This style is more convenient in the sense that it does not require including the FIND_NEXT call in the body of the DO loop, much in the same vein as the DO_OVER technique discussed below.

4.3.4 Version 9.4 Add-On: DO_OVER

The styles of invoking *KeyNumerate* shown above work in SAS 9.2 and later. Since the advent SAS 9.4, the operation can be activated without calling the combination of FIND and FIND_NEXT, but by calling the DO_OVER method instead:

Program 4.5 Chapter 4 Keynumerate Loop DO_OVER Snippet.sas

```
do while (H.DO_OVER(KEY:3) = 0) ;
  put +3 (K D) (=) ;
end ;
```

On the first call, DO_OVER accepts a key-value from the PDV or explicitly and sets the pointer to the first item of the group with the key. Then it retrieves the data portion values and moves the pointer to the second item. On the subsequent calls, it proceeds down the list one item at a time. For each item on which it works, it performs the indirect *Retrieve* operation, snatching the data portion values from the table and overwriting the host variables with them. When it is called for the last item in the list, it moves the pointer outside the group, thus unsetting it. This causes the next call to fail and the loop to stop because of a non-zero return code. Thus, effectively, the DO_OVER method:

- Acts as the FIND method on the first call.
- Acts as the FIND_NEXT method on the ensuing calls.
- Is called unassigned in the WHILE clause and thus there is no need for RC=.

If the DO_OVER method, as a tool for keyed enumeration is so convenient and concise that a question arises: Why not call it alone rather than use other methods described above for the same purpose? The answer is three-fold:

- We have found out that while DO_OVER works without a glitch when we merely need to harvest the hash data from an entire same-key group (as in the snippet above), it has issues with the *Selective Delete* and *Selective Update* operations (Sections 4.3.10-4.3.14). At the same time, the theoretically equivalent combination of FIND and FIND_NEXT works as expected under those scenarios as well.
- As we will see below, combining it with a priming call to CHECK or REF adds a degree of flexibility not offered by calling DO_OVER alone.
- SAS version independence: The FIND and FIND_NEXT combination works in SAS 9.2 through SAS 9.4.

4.3.5 Forward and Backward, In and Out

When a given key is in the table and the list of the items with this key is set, we can use the techniques shown above to *Keynumerate* one, more, or all items in the group *forward*. Is it possible to do it backward? The answer is "yes, but to a degree", and here is why:

- There exists a hash tool, the FIND_PREV method, designed to *Keynumerate* backward. Its name speaks for itself.
- Unfortunately, it cannot be used *starting with the bottom* of the list. This is because any method setting the item list (e.g., FIND) always points to the *logically first* item.
- Therefore, to scan backward starting with a given item, we need *to scan forward* first in order to reach it.
- In doing so, we must be careful not to let the pointer escape the item group when it dwells on the last and first item. Such an attempt will unset the item list, thus precluding any further enumeration, forward or backward, unless the list is reset anew.

To make these points more eminent, let us recall the content of table H created in our template DATA step (note that the column Item# below is *not* a variable in the table - it is included just to indicate the logical order of the items):

Item#	K	D
1	2	B
2	2	C
3	1	A
4	3	D
5	3	E
6	3	F

Suppose that we want to enumerate the item group with K=3 *backward* from item #6 to item #4 and then *forward* to item #6. To accomplish it, we can proceed in this manner:

- Call FIND with K=3 to set the list and latch onto item #4.
- Call FIND_NEXT twice to reach item #6.
- Call FIND_PREV twice, enumerating items #5 and #4 backward in succession.
- Call FIND_NEXT twice, enumerating items #5 and #6.

The problem with this fine plan is that we have based our actions on knowing, ahead of time, the item group content and thus knowing exactly when to stop calling FIND_NEXT and FIND_PREV to get what we want without unsetting the item list. Calling either method one more time than indicated above would unset the list, pushing the enumeration pointer out of either the bottom or the top of the item group. Therefore, in order to be able to code dynamically, we need a means to tell whether the pointer currently dwells on the last or first item of the group without knowing the number of items in it beforehand.

4.3.6 Staying within the Item List (Keeping It Set)

In real-life situations, we usually do not know ahead of time how many items are in any same-key group, nor what the values of D for each item are. Therefore, if we need to stay within the item list during the Keynumerate operation without unsetting the list, we need to be able to call FIND_NEXT and FIND_PREV only if the next or previous item, respectively, still exists.

The auxiliary methods HAS_NEXT and HAS_PREV, respectively, are designed to provide the means of knowing it. By calling them, we can predict whether to call FIND_NEXT or FIND_PREV another time in order to stay within the item list without unsetting it. This snippet causes the *Keynumerate* operation to walk forward, then backward, and then again forward within the K=3 item group without ever moving the enumeration pointer out of it:

Program 4.6 Chapter 4 Keeping Item List Set Snippet.sas

```
put "Forward:" ;
RC = H.FIND(KEY:3) ;
do while (1) ;
  put +3 RC= D= ;
  RC = H.HAS_NEXT(RESULT:NEXT) ;
  if not NEXT then LEAVE ;
  RC = H.FIND_NEXT() ;
end ;
put "Backward:" ;
do while (1) ;
  RC = H.HAS_PREV(RESULT:PREV) ;
  if not PREV then LEAVE ;
  RC = H.FIND_PREV() ;
  put +3 RC= D= ;
end ;
put "Forward:" ;
do while (1) ;
  RC = H.HAS_NEXT(RESULT:NEXT) ;
  if not NEXT then LEAVE ;
  RC = H.FIND_NEXT() ;
  put +3 RC= D= ;
end ;
```

Note that the methods HAS_NEXT and HAS_PREV are supplied with the argument tag RESULT. It accepts the *unquoted* name of a numeric variable (in the snippet above, NEXT and PREV), into which the result of the call is to be placed. We will dwell on this distinctive feature in Section 4.3.7. Running the template step with this snippet inserted into it prints the following in the SAS log:

```
Forward:
    RC=0  D=D
    RC=0  D=E
    RC=0  D=F
Backward:
    RC=0  D=E
    RC=0  D=D
Forward:
    RC=0  D=E
    RC=0  D=F
```

This diagnostic log output confirms that, by keeping the item list set (or preventing it from being unset), we can make the operation scan the item group to and fro as intended.

4.3.7 HAS_NEXT and HAS_PREV Peculiarities

It should be noted that these methods, compared to other hash methods, possess a few peculiar traits:

- Judging from their return codes, both methods are always successful. That is, they generate a zero return code, even if no data item list is set. This is odd since the way the methods work makes any sense only within the concept of a group of same-key items whose list is set by calling FIND, CHECK, REF, or DO_OVER. By contrast, the FIND_NEXT and FIND_PREV methods always return a non-zero code if they are called when no item list is set, which is par for the course.

- As already noted, the methods are packed with the argument tag RESULT. It accepts the *unquoted* name of a *numeric variable*, to which the call returns the result (not the return code!) it generates. This is in sharp contrast with other methods whose argument tags accept variable names only as *character expressions*.

- If the variable name given to RESULT already exists in the PDV and it is numeric, it is fine; but if it is character, it creates a data type conflict. This is because, unlike the other methods, HAS_NEXT and HAS_PREV actually look to see if the variable with the name specified to the RESULT argument tag already exists in the PDV. If it does not, they create a numeric variable with this name in the PDV. But if it does and it has a different data type, a conflict ensues and the step is terminated. Unlike its behavior in many other cases, with the hash object SAS does not provide for implicit character-to-numeric conversion (along with the corresponding notes in the log).

- The fact that these two methods create PDV variables on their own accord sets them apart from the rest of the methods, which accept variable names as character expressions only if they already exist in the PDV. By contrast, with these two methods, the RESULT argument tag expects not a character expression resolving to a valid SAS variable name, but an *unquoted* SAS variable name itself.

- The values returned into the variable (whose name is assigned to the RESULT argument tag such as variables *Next* and *Prev* above) follow standard SAS conventions for logical expressions. That is, if the next or previous item exists, the respective method returns 1 (i.e., logical True); otherwise, they return 0 (logical False). This is the opposite of the *return codes* all method calls generate, for which RC=0 (i.e., logical False) means success and non-missing RC≠0 (logical True) means failure.

4.3.8 Harvesting Hash Items

So far, we have concentrated on a single same-key group (with K=3) for the purpose of explanation and illustration of a number of principles. However, in the real data management world, we are usually given *a set* of key-values and need to perform an action on all same-key groups whose keys are found in the table.

In most cases, the action means collecting - or "*harvesting*" - the values from the data portion hash variables for each key-value in the set. And most often, its key-values come from a file - as it happens, for instance, when the *Keynumerate* operation is used to combine data sets (discussed in Chapter 6).

In the sample step below, a set of key-values (including both the key-values present and not present in our hash table) is coded into an array. For each array element, whose value matches the key of an item group in hash table H, the group is enumerated, so that the value of D is retrieved from every item in the group (i.e., *harvested*), and a record is added to output data set work.Harvest. (Compared to the template step, in the step below the DATA statement list is not _NULL_, so instead of just showing the code snippet, the step is presented in full.)

Program 4.7 Chapter 4 Harvesting Items via Implicit FIND and FIND_NEXT.sas

```
data harvest (keep = K D) ;
  dcl hash H    (multidata:"Y", ordered:"N") ;
  H.definekey   ("K") ;
  H.definedata  ("D") ;
  H.definedone () ;
  do K = 1, 2, 2, 3, 3, 3 ;
    q + 1 ;
    D = char ("ABCDEF", q) ;
    H.add() ;
  end ;         ❶
  array keySet [5] _temporary_ (0 1 5 7 3) ;   ❷
  K = . ;       ❸
  do i = 1 to dim (keySet) ;    ❹
    K = keySet[i] ;    ❺
    do RC = H.FIND() by 0 while (RC = 0) ;    ❻
      output ;
      RC = H.FIND_NEXT() ;    ❼
    end ;
  end ;
 stop ;
run ;
```

Note that in this step:

- Table H is loaded with the same content as before.
- The name of the output data set is defined in the DATA statement, rather than via the OUTPUT method argument tag DATASET.

● Key variable K is *not* defined in the data portion, as shown below.

Key portion	Data portion
K	**D**
2	B
2	C
1	A
3	D
3	E
3	F

The fact that K is not in the data portion in addition to the key portion has its important implications, as we will see below. Let us take a look at some details of the code snippet above.

❶ At this point, table H has been loaded with the content shown above.

❷ Temporary array keySet contains 5 items values at compile time. They contain some (but not all) key-values present in table H (1, 3) and also key-values not loaded in the table (0, 5, 7).

❸ PDV host variable K is set to missing just to observe what happens to its value; from the standpoint of harvesting, it is irrelevant.

❹ The key-values to be searched for and harvested from table H are extracted from the array in a loop one at a time.

❺ With the intent to use implicit method calls, the value of the current key-value from the array is assigned to the PDV host variable K, overwriting its value.

❻ The FIND method is called *implicitly* with the current key-value of K as a key.

If the value of K *is not present* in the table, FIND returns RC≠0, and the DO loop terminates without executing any of the statements inside it because of the RC=0 condition in its WHILE clause. This causes program control to go back to the outer DO loop and get the next element from the array (if there are any left).

If the value of K *is present in the table*, the method call (a) sets the item list for enumeration, (b) retrieves the value of hash variable D into the PDV host variable D for the first item in the group, and (c) returns RC=0, which causes the DO loop to continue its current iteration and execute the statements inside the loop. This way, after the successful FIND call, the current PDV value pair (K,D) is written to the output data set as a record, and then program control is passed to the FIND_NEXT call.

❼ If the FIND call was successful, FIND_NEXT is called. If there is more than one item in the group, it retrieves the value of hash variable D from the second item (and third, and so on) and sets RC=0. If the second item does not exist, FIND_NEXT returns RC≠0. In both cases, after the method call, program control is passed to the top of the DO WHILE loop. If RC≠0, the loop terminates at the top. Otherwise, the current PDV value pair *(K,D)* is written to the output file as a record. Then FIND_NEXT is called again to enumerate the next item. The process continues in this manner for every iteration of the loop until there are no more items for FIND_NEXT to retrieve and it returns RC≠0, thus terminating the loop. At this point, program control is passed to the outer loop to get the next key-value from the array.

Since the only array key-values matching the keys in table H are *keySet[2]*=1 and *keySet[5]*=3, only the items from the item groups with K=(1,3) are harvested. As a result, running this step generates output data set work.Harvest with the following content:

Obs	K	D
1	1	A
2	3	D
3	3	E
4	3	F

Instead of using the FIND and FIND_NEXT pair, the harvesting DO loop can be rewritten more simply and tersely by using the DO_OVER method:

Program 4.8 Chapter 4 Harvesting Items via Implicit DO_OVER Snippet.sas

```
do i = 1 to dim (keySet) ;
  K = keySet[i] ;
  do while (H.DO_OVER() = 0) ;
     output ;
  end ;
end ;
```

Both of the *implicit call* techniques work, and the OUTPUT statement writes out the expected values of K and D. It means that at the time the statement is executed, both PDV host variables have the correct values. However, because K is not defined in the data portion, the DO_OVER calls extract only the values of hash variable D into its PDV host variable D, while PDV host variable K gets the proper values because they are assigned to it via the array references. Below we will see how to get both K and D retrieved from the table in the absence of the assignment statement.

4.3.9 Harvesting Hash Items via Explicit Calls

Instead of assigning the needed key value to PDV host variable K and making *implicit* calls, the FIND and DO_OVER methods can also be called *explicitly* by using the array reference directly as an expression with the argument tag KEY. For example, in the case of using DO_OVER, we could code:

Program 4.9 Chapter 4 Harvesting Items via Explicit DO_OVER Snippet.sas

```
K = . ;
do i = 1 to dim (keySet)
  do while (H.DO_OVER(KEY:keySet[i]) = 0) ;
    output ;
  end ;
end ;
```

Note that above, the assignment statement `K=keySet[i]` is omitted, and the array reference is essentially moved from it to the argument tag. However, if we rerun the template step with this little change, the output we get will look as follows:

Obs	K	D
1	.	A
2	.	D
3	.	E
4	.	F

Clearly, it is not what we want. So, why is it that the *implicit calling* techniques shown above do work and the one with the explicit call does not? The reason is that with both *implicit calling* techniques we get correct output values of K merely because the expected values are assigned to the PDV host variable K from the array - and not because the calls retrieve the values of hash variable K into its PDV host variable K. Moreover, the latter *cannot happen* as long as K is not part of the data portion. The obvious remedy is to add K to the data portion:

```
H.definedata ("K", "D") ;
```

The content of table H loaded with this definition in place will now look like so:

Key portion	Data portion	
K	K	D
2	2	B
2	2	C
1	1	A
3	3	D
3	3	E
3	3	F

With the hash table entry defined this way, the current hash value of K is retrieved into the PDV host variable K alongside with D with every successful DO_OVER call, and so now the step will generate correct output. Incidentally, with K now added to the data portion, the variant with the FIND and FIND_NEXT pair can be recoded with an explicit call to FIND:

Program 4.10 Chapter 4 Harvesting Items via Explicit FIND and FIND_NEXT Snippet.sas

```
do i = 1 to dim (keySet) ;
  do RC = H.FIND(KEY:keySet[i]) by 0 while (RC = 0) ;
    output ;
    RC = H.FIND_NEXT() ;
  end ;
end ;
```

As explained earlier, the FIND_NEXT call can be left implicit, since even if an expression is assigned to its argument tag KEY, it is disregarded because the FIND_NEXT method operates only within the confines of the item group set by calling the FIND method.

The behavior observed above vis-a-vis the absence or presence of key portion variables in the data portion adheres to the general principle that any *Retrieve* operation, either direct or indirect, extracts the values *only* from the hash variables defined in the data portion. You will quite often see hash table definitions which include the key variables also included in the data portion for exactly this reason.

4.3.10 Selective DELETE and UPDATE Operations

Let us recall that the *Delete All* operation - activated by the REMOVE method - *obliterates the entire group* of hash items with a given key-value if the key is in the table. However, often it is desirable to delete, not all the items in the same-key item group, but only specific items based on certain conditions.

Likewise, the *Update All* operation (in SAS 9.4 and later) - activated by the REPLACE method - updates the data portion hash variables for all items with a given key-value if it is in the table. But yet again, by the same token as with *Delete*, we may need to update not all the items in a same-key item group, but only certain items in the group that meet specific criteria, including those based on the data portion values of the items targeted for deletion.

For same-key item groups containing but a single item - for example, if the uniqueness of the table key is enforced programmatically - *Delete All* and *Update All* can be used for conditional deletion and update just fine. The reason is that in this case there is no need to enumerate multiple items. Otherwise, to select items for deletion or update from a multiple-item same-key group, we need the capability to scroll through the group in order to decide which items to act upon.

This capability is provided by the *Selective Delete* and *Selective Update* operations. They are built on the *Keynumerate* operation as a backbone. The reason is two-fold. On the one hand, in order to delete or update a specific item in a multi-item same-key group, the item has to be reached and latched onto first. On the other hand, the hash methods REMOVEDUP and REPLACEDUP designed for *Selective Delete* and *Update*, respectively, work only if the item list for the group is set and thus can be enumerated. We will discuss these two operations one at a time.

4.3.11 Selective DELETE: Single Item

To illustrate the operation, we will use, as a guinea pig, the same template step and hash table H as we have already used in this section. Imagine that in the item group with K=3 we need to delete the item whose data-value D="E", i.e., item #5. Again, please note that Item# is not in the table and is shown here just to indicate the logical order:

Item#	K	D	Action
1	2	B	
2	2	C	
3	1	A	
4	3	D	
5	3	E	Delete this item
6	3	F	

In the plug-in snippet below, the item group with K=3 is enumerated beginning with table item #4, and for every item encountered, the data-value of hash variable D is retrieved into host variable D. When the latter becomes "E", program control dwells on the item to be deleted, which is then done by calling the REMOVEDUP method. The next enumerating loop is used only to check the content of the item group *K*=3 after the deletion.

Program 4.11 Chapter 4 Selective Delete Single Item Snippet.sas

```
put "Enumerate to delete item with D=E:" ;
do RC_enum = H.FIND(KEY:3) by 0 while (RC_enum = 0) ;
  if D in ("E") then RC_del = H.REMOVEDUP() ;
  RC_enum = H.FIND_NEXT() ;
end ;
put +3 RC_enum= RC_del= D= /;
put "Enumerate again to check result:" ;
do RC_enum = H.FIND(KEY:3) by 0 while (RC_enum = 0) ;
  put +3 RC_enum= D= ;
  RC_enum = H.FIND_NEXT() ;
end ;
```

Running the template step with this snippet prints the following in the SAS log:

```
Enumerate to delete item with D=E:
   RC_enum=160038 RC_del=0 D=E
Enumerate again to check result:
   RC_enum=0 D=D
   RC_enum=0 D=F
```

It confirms that the item with D="E" has indeed been deleted. So, in this particular case - when the deletion criterion is specific to *a single item* - everything *appears* to work right. Indeed, we wanted to delete the item (K,D)=(3,E), and that is what happened.

However, the log tells us that, after the first loop, we have D="E" from item #5 in the PDV. It suggests that RC_enum=160038 is triggered not by not having any more items to enumerate (in which case we would have D="F" from item #6) but by the fact that after REMOVEDUP successfully removed item #5, FIND_NEXT failed to enumerate item #6, generating RC≠0 and stopping the loop. In turn, it means that if, after deleting item #5, we also needed to delete item #6, we could not do it in the same loop.

Unfortunately, this is indeed the case. So, if more than one item satisfies the deletion criterion, things become more complicated. Let us look how the problem can be addressed.

4.3.12 Selective Delete: Multiple Items

Suppose that now we would like to delete both the item where D="E" *and* the next item where D="F", like so:

Item#	K	D	Action
1	2	B	
2	2	C	
3	1	A	
4	3	D	
5	3	E	Delete this item
6	3	F	Delete this item

It would be a natural inclination to merely alter the IF statement accordingly to:

```
        if D in ("E", "F") then RC_del = H.REMOVEDUP() ;
```

However, if we rerun the step with this change, we will quickly discover that what the step prints in the log is *exactly the same* as it printed before. In other words, while the item with D="E" has been deleted, the *item with* D="F" *has NOT been deleted*! We have already hinted at the reason why it happens in the previous subsection. However, it is important enough to add more gory details to the explanation:

- If we look at the log attentively, we will see that after the first enumerating loop terminated, the value of D is still "E". But if after the deletion of item #5 the enumeration process continued unabated, the last item visited would be item #6 and so the value of D after the loop ended would be "F". Since the value of D is still "E", that means the first loop had terminated before enumerating the remaining values. The loop terminated because RC_enum was not 0. The REMOVEDUP method caused the FIND_NEXT method call to fail.
- It fails because a successful REMOVEDUP call not only removes the item *but also unsets the item list*, and this causes the next FIND_NEXT call trying to progress to the next item to fail. This is

why - and not because it ran out of the items in the set item list - FIND_NEXT returned RC_enum=160038 and caused the loop to end, having never visited item #6.

The conclusion is that during a single pass of *Keynumerate* it is possible to delete *only the item* from a same-key group that meets the deletion criterion. In other words, in order to delete more than one item, we need to enumerate the group as many times as there are items to be deleted, resetting the item list every time.

Unfortunately, this fact complicates selective *Delete* programming, for not only do we have to enumerate repeatedly, but in addition we need to check to see whether the process has progressed to the very last item in the group. Yet fortunately, it is not difficult to do by making use of the HAS_NEXT method:

Program 4.12 Chapter 4 Selective Delete Multiple Items Snippet.sas

```
put "Keynumerate to delete items with D in ('E','F'):" ;
do ENUM_iter = 1 by 1 until (not MORE_ITEMS) ;       ❶
  do RC_enum = H.FIND(KEY:3) by 0 while (RC_enum = 0) ;       ❷
    H.HAS_NEXT(RESULT:MORE_ITEMS) ;       ❸
    if D in ("E", "F") then H.REMOVEDUP() ;       ❹
    RC_enum = H.FIND_NEXT() ;       ❺
  end ;
  put +3 ENUM_iter= MORE_ITEMS= ;
end ;
put "Keynumerate again to check result:" ;
do RC_enum = H.FIND(KEY:3) by 0 while (RC_enum = 0) ;       ❻
  put +3 RC_enum= D= ;
  RC_enum = H.FIND_NEXT() ;
end ;
```

By plugging the snippet into the template step and running it, we will have the following result in the SAS log:

```
Enumerate to delete items with D=(E,F):
   ENUM_iter=1 MORE_ITEMS=1
   ENUM_iter=2 MORE_ITEMS=0
Enumerate again to check result:
   RC_enum=0 D=D
```

Now, both required items have been deleted - at the expense of performing the *Keynumerate* operation twice, as indicated by ENUM_iter=2. Let us take a closer look at the details of the code snippet above:

❶   Organize a DO loop to make as many *Keynumerate* passes through the K=3 item group as necessary until all required items are deleted.

The ENUM_iter variable counts the iterations of the loop for diagnostic and display purposes.

Boolean variable MORE_ITEMS tells this DO loop when to stop by indicating that in its final iteration, no more items are left to enumerate, i.e., the last item in the group has been processed.

The UNTIL clause causes the loop to execute the statements inside it at least once. Hence, there is no need to initialize MORE_ITEMS beforehand - it will be set inside the loop to 1 when the iteration process hits the bottom item; and to 0 for each item prior to that. The only way the UNTIL condition evaluates true is if we are pointing to the last item in the group; and the only way that can happen is if there were no more items to remove.

❷ *Keynumerate* the item group with K=3 by using an explicit call to FIND to set the item list and enumerate the first item in the group. Variable RC_enum controls the termination of the loop. It is set to 0 on the first successful call to FIND. Thereafter, it is set to RC_enum≠0 by the FIND_NEXT call either after a successful REMOVEDUP call or having reached the bottom of the item list and thus failing to locate the next item to enumerate. Either way, the loop is terminated, and program control is passed back to the outer DO loop.

❸ Call HAS_NEXT to detect whether the item, on which the enumeration pointer currently dwells, is the last in the item group. If so, MORE_ITEMS is set to 0; else it is set to 1. If it is set to 0, it will make the outer DO loop terminate as soon as program control leaves the inner loop, due to the UNTIL clause (whose condition is checked at the bottom of the outer loop).

❹ If the value retrieved into the PDV host variable D in the prior FIND or FIND_NEXT call is "E" or "F", delete the item.

❺ *Keynumerate* the next item in the item list by calling FIND_NEXT. It returns RC_enum≠0 if (a) the prior item has been successfully deleted, or (b) the bottom item has already been enumerated, so there are no more items to fetch, or (c) both.
Otherwise, RC_enum is set to 0.
Either way, program control is passed to the top of the inner loop. If RC_enum=0, the loop goes into the next iteration. Otherwise, it terminates, and program control is passed back to the bottom of the outer loop. If at this point MORE_ITEMS=0, the outer loop terminates, thus making the current *Keynumerate* pass final.

❻ The item group with K=3 goes through another *Keynumerate* pass, merely to show in the SAS log which items are left in the item group after the deletions.

The same result, as far as *Selective Delete* is concerned, can be achieved more tersely code-wise by using the DO_OVER method instead of the tandem of FIND and FIND_NEXT in the first (nested) loop:

```
do until (not MORE_ITEMS) ;
  do while (H.DO_OVER(KEY:3) = 0) ;
     RC_next = H.HAS_NEXT(RESULT:MORE_ITEMS) ;
     If D in ("E", "F") then RC_del = H.REMOVEDUP() ;
  end ;
end ;
```

However, regardless of the methods activating the *Keynumerate* operation, the same principle still stands: Since a successful call to REPLACEDUP unsets the item list, *Keynumerate* must be performed repeatedly if more than one item satisfies the deletion criterion. The nested loop shown in the two last snippets works regardless of how many items are to be deleted; and since usually the latter is unknown ahead of time, we recommend using this coding pattern. If it so happens that there is only a single item satisfying the deletion criterion, the worst case scenario is enumerating the item group twice. In general, this approach will enumerate the group one time more than that the number of items to be deleted.

4.3.13 Selective UPDATE

Just like selective *Delete*, selective *Update* can work only in tandem with the *Enumerate* operation. But, unlike selective *Delete*, the operation works in a straightforward manner regardless of the number of items updated within a same-key item group. Suppose, for example, that we need to perform the following updates:

Item#	K	D	Action
1	2	B	
2	2	C	
3	1	A	
4	3	D	Update "D" with "X"
5	3	E	
6	3	F	Update "F" with "Z"

The scheme to attain the goal is fairly plain:

1. Enumerate the item group where K=3.
2. For each item visited in the group in the process, whose data-value of D meets the criterion, update the value of hash variable D by calling the REPLACEDUP method.

The code snippet below does exactly that. The second loop is used only to display the data content of the updated item group in the SAS log:

Program 4.13 Chapter 4 Selective Update Snippet.sas

```
put "Enumerate to update D=(D,F) with D=(X,Z):" ;
do RC_enum = H.FIND(KEY:3) by 0 while (RC_enum = 0) ;
  D = translate (D, "XZ", "DF") ;
  RC_updt = H.REPLACEDUP() ;
  RC_enum = H.FIND_NEXT() ;
end ;
put +3 RC_updt= ;
put "Enumerate again to check result:" ;
do RC_enum = H.FIND(KEY:3) by 0 while (RC_enum = 0) ;
  put +3 RC_enum= D= ;
  RC_enum = H.FIND_NEXT() ;
end ;
```

Running the template step with this snippet plugged in prints in the SAS log:

```
Enumerate to update D=(D,F) with D=(X,Z):
   RC_updt=0
Enumerate again to check result:
   RC_enum=0  D=X
   RC_enum=0  D=E
   RC_enum=0  D=Z
```

So, all the necessary selective updates can be done in a single pass over the same-key item group because a successful REPLACEDUP call, unlike REMOVEDUP, *does not unset the item list*, and thus the enumerating process continues unabated until after the logically last item in the group is reached.

Note that in the snippet above, REPLACEDUP is called *implicitly*. It means that it accepts its values from the PDV host variables. To let it accept the desired values, the assignment statement preceding the call:

```
D = translate (D, "XZ", "DF") ;
```

changes the value of PDV host variable D as required by the updating logic before the method call.

However, under certain scenarios, alerting the host variable values may be undesirable. To avoid it, REPLACEDUP can be called *explicitly* instead. In addition, the first DO loop (the one actually doing selective *Update*) can be coded more concisely by replacing the tandem of FIND and FIND_NEXT with the DO_OVER method:

Program 4.14 Chapter 4 Selective Update Explicit REPLACEDUP Snippet.sas

```
do while (H.DO_OVER(KEY:3) = 0) ;
  RC_updt = H.REPLACEDUP(DATA: translate (D, "XZ", "DF")) ;
end ;
```

4.3.14 Selective DELETE vs Selective UPDATE

In addition to the obvious distinction between two operations with regard to their self-explanatory purposes and results, let us underscore the difference they have on the *Keynumerate* operation, without the aid of which neither can succeed:

- The *Selective Delete* operation is a bit tricky. This is because once an item is removed from a same-key group, it *unsets* the item list and thus precludes enumerating this group further till the end of the list. Therefore, in order to ensure that all the items meeting the deletion criteria are deleted, the item list has to be reset and the enumeration process repeated until we have discovered that the item on which the enumerating pointer dwells is now the logically last in the item group.
- The selective *Update* operation presents no such complications. All the items in the requisite same-key group fitting the updating criteria can be updated accordingly in a single *Keynumerate* pass through the item group.

4.3.15 KeyNumerate Operation Hash Tools

- Methods: FIND, CHECK, REF, FIND_NEXT, FIND_PREV, DO_OVER, HAS_NEXT, HAS_PREV, REMOVEDUP, REPLACEDUP.
- Argument tags: MULTIDATA, ORDERED, RESULT.

4.3.16 KeyNumerate Operation Hash-PDV Interaction

- The *Keynumerate* operation performs the indirect *Retrieve* operation on every hash item it visits. To wit, it extracts the values from the *data portion* variables into the corresponding PDV host variables, overwriting their values with those from the table. Hence, it directs data movement from the hash table to the PDV.
- The *Selective Delete* and *Selective Update* operations rely on the *Keynumerate* operation. They work in the opposite directions by, respectively, (a) deleting specific hash table items and (b) updating their data portion values with the values of their PDV host variables or expressions specified to the argument tags used in explicit method calls.

4.4 ENUMERATE ALL Operation

The *KeyNumerate* (Enumerate by Key) operation discussed in Section 4.2.4 is predicated on *using a key-value* to access the table and enumerates the items in the item group with this value if the key is in the table. However, often we need to enumerate one or more table items:

1. Without any prior knowledge and regardless of their key-values, merely starting from the *logical beginning* or *logical end* of the table.
2. From the logically first item of the item group with a given key across not only this group but also across other item groups whose keys we do not know beforehand.

This important hash table functionality (in a way similar to reading a SAS data file forward and backward using the SET statement option POINT=) is delivered by the *Enumerate All* operation. It is supported by a structure known as the *hash iterator object* and the methods specifically designed to control its actions against the hash table to which it is linked.

4.4.1 The Hash Iterator Object

Before discussing how the hash iterator object can be used practically to support the *Enumerate All* operation, we need to point out a few fundamental facts about its nature. To materialize an iterator and make it operable, we need to perform a number of actions described below.

1. Declare a hash iterator object:

```
DECLARE hiter IH ;
```

This *compile time* statement does the following:

-   Declares a hash iterator object named IH.
-   Defines a PDV variable IH of *hash iterator* type.

The statement must conform to a set of rules:

-   The "DECLARE" keyword can be abbreviated as "DCL". Only these two spellings are accepted.
-   "Hiter" is also a keyword, so it must also be spelled exactly.
-   The statement must appear *before* variable IH is referenced.
-   It can appear with the same iterator name (in this case, IH) only once. The reason is that a non-scalar variable, once defined, cannot be defined again.

2. Create a new hash iterator object instance:

```
IH = _new_ hiter("H") ;
```

This *run-time* assignment statement does the following:

-   Creates a new instance of iterator object IH.
-   Generates a distinct value of type hash iterator to identify the iterator instance.
-   Assigns the value to PDV variable IH.
-   Links it to the *active* instance of hash object H named in the statement.

This statement also must satisfy a number of rules:

- "_NEW_" and "hiter" are keywords, and so they cannot be misspelled.
- Argument "H" is a reference to a PDV variable of type hash whose current value identifies an active instance of hash object H. This is the value the statement uses to link the newly created hash iterator instance to a specific instance of hash object H.
- Hence, by the time this statement has program control, the active hash object instance identified by the value of variable H must already exist; and IH must have been defined in the PDV as a non-scalar variable of type hash iterator.

If everything above is done according to the rules:

- The newly created instance of hash iterator IH is linked to a specific instance of hash object H.
- The current PDV value of variable IH identifies the *active* hash iterator instance. That is, unless and until this value is changed, any hash iterator method referencing IH will use that iterator instance with the hash object instance (i.e., its hash table) to which it is linked.
- Once a hash iterator instance is linked to a hash object instance, it remains permanently paired with the latter, *and only* with the latter.
- Whenever IH is referenced, the program uses the active iterator instance, i.e., the one identified by the current PDV value of IH.
- This active hash iterator instance operates only against the hash object instance to which it has been linked - irrespective of whether the latter is currently active or not.
- An iterator instance cannot be linked to more than one hash object instance - otherwise the iterator would not know which hash table to enumerate.
- However, more than one iterator instance can be linked to the same hash object instance, in which case any of them can be used to enumerate the same hash table as a matter of choice.

Just as with the hash object, the DECLARE statement and _NEW_ operator statements can be fused into a single compound DECLARE|DCL statement:

```
declare hiter IH("H") ;
```

Also similarly, the compound statement works partially at compile time and run time:

- At compile time, the actions related to the stand-alone DECLARE statement are performed.
- At run time, the actions related to the stand-alone _NEW_ operator statement are performed.

One significant difference between the two separate statements and compound statement is that if more than one instance of hash iterator IH needs to be created at different points in the program, the stand-alone _NEW_ operator statement can be explicitly repeated after the object is declared:

```
dcl hiter IH ;
...
IH = _new_ hiter("H") ;
...
IH = _new_ hiter("H") ;
```

Each _NEW_ operator call creates a new hash iterator instance. (The ellipses above denote program logic that may need to be included between the statements.) However, the compound DECLARE statement referencing the same variable IH cannot be repeated for the same reason the stand-alone DECLARE statement cannot: It would be an attempt to define a non-scalar variable more than once. Therefore, to use the compound statement in order to create more than one hash iterator object instance, it has to be enclosed in a loop. For example:

```
do i = 1, 2 ;
   ...
   dcl hiter IH("H") ;
   ...
end ;
```

Alternatively, the program can be constructed to let the implied DATA step loop to do the work. For example, if the intent is to create a new iterator instance only in its first and second iteration, one might code:

```
data ... ;
  ...
  if _n_ in (1, 2) then dcl hiter IH("H") ;
  ...
run ;
```

In both cases, the compiler sees the DECLARE statement referencing IH only once, whereas its *run-time* component is executed twice, each time creating a new hash iterator instance.

4.4.2 Creating and Linking the Iterator Object

Now that we know the rules, let us consider an example. The DATA step below is a program template for the rest of the chapter. The code snippets presented throughout can be merely pasted in the space marked "*Demo code snippet using iterator* IH". (Note that for every snippet, the sample programs supplied with the book contain the complete template step including the snippet in question.)

Program 4.15 Chapter 4 Create and Link Hash Iterator.sas

```
data _null_ ;
  dcl hash H (multidata:"Y", ordered:"N") ;
  H.definekey ("K") ;
  H.definedata ("D", "K") ;
  H.definedone () ;
  do K = 1, 2, 2, 3, 3, 3 ;
     q + 1 ;
     D = char ("ABCDEF", q) ;
     H.add() ;
  end ;
  DECLARE HITER IH ;
  IH = _NEW_ HITER ("H") ;
  /*Demo code snippet using iterator IH */
  stop ;
run ;
```

We can make a few relevant notes from the outset:

- The iterator object instance is *linked* to the hash object *instance* by specifying the name of the latter as the character literal constant "H".
- However, it can be any valid *character expression* as long as it resolves to the name of the hash object we need to link - in this case, object H.
- The iterator instance is created *after* the hash object is instantiated but *before* the iterator is to be used.

The statement with the _NEW_ operator can be repeated or used in a loop to create multiple instances of hash iterator object IH, each with its own distinct identifier. This may be necessary when the tables of multiple hash object instances need to be enumerated independently. This functionality will be exemplified in (the more advanced) Chapter 9.

Most of the time, though, a single iterator object instance linked to a single hash object instance is used. In this case, it is simpler to use the compound DECLARE described above:

```
declare hiter IH ("H") ;
```

This is all it takes to create an instance of a hash iterator for the *Enumerate All* operation. Its different actions are activated by the respective methods supplied with the hash iterator object. In particular, they support the following actions:

- **Direct access**, i.e., non-sequential access from any item in the table or from outside it:
    a. To the *logically first* hash table item.
    b. To the *logically last* hash table item.
    c. To the *key-item*, i.e., the *logically first item in the item group* with a given key-value.
- **Sequential access**, i.e., one item at a time:
    a. To all items, starting from any item, to the end of the table in the direction of the logical order, i.e., *Enumerate forward*.
    b. To all items, starting from any item, to the beginning of the table in the opposite direction, i.e., *Enumerate backward*.
    c. Note that while it rarely makes sense to *Enumerate forward* from the last item or *backward* from the first item, nothing precludes being able to do this.
- **Data retrieval**:
    a. For *every* item accessed in any manner described above, retrieve the values of its data portion variables into their corresponding PDV host variables.

Before we proceed with the description of these *Enumerate All* actions, let us recall that the content of hash table H created and loaded by the template step above, shown in its *logical order* indicated by Item# below (which is *not* a hash variable in the table), looks as follows:

Item#	Key portion	Data portion	
	K	D	K
1	2	B	2
2	2	C	2
3	1	A	1
4	3	D	3
5	3	E	3
6	3	F	3

4.4.3 Hash Iterator Pointer

The *hash iterator pointer* is a concept convenient for helping visualize the enumeration process. It is somewhat similar to that of the file pointer used to describe the mechanics of reading a file.

One can imagine the iterator pointer as a cursor that at any given point in time dwells in one of the two general domains:

1. On a particular hash item *inside* the table.
2. *Outside* the table, i.e., on no hash item at all.

If no enumeration action has been taken yet, the pointer can be thought of as currently located *outside* the table. In this state, the *Enumerate All* operation can begin an enumeration process by moving the pointer *into* the table *directly* to one of the following locations inside it:

- The logically first item. In our sample table H above, it is item #1.
- The logically last item. In our table H, it is item #6.
- The key-set item. In our table H, for the item group with K=3, it is item #4.

As a result, the pointer will be located inside the table dwelling on a certain item. Let us call it *the pointer item*. After the pointer has been moved into the table, an enumeration action that follows can move the iterator pointer from the pointer item:

- To the item immediately *following* the pointer item - if the pointer item is not the *last* table item.
- To the item immediately *preceding* the pointer item - if the pointer item is not *first* table item.
- Outside the table - if the pointer item is the first or the last table item (or both in case there is only one item in the table).

The item to which the pointer has moved now becomes the pointer item itself. It is said that the pointer item is *locked by the hash iterator*. What it really means is that while the pointer dwells on the item, it *cannot be removed*. (In fact, as we will see in more detail later, all the items in the item group with the same key as the pointer item are also locked.) As opposed to that, when the pointer is located outside the table, no pointer item exists and so no item is locked. Note that the iterator pointer can be moved out of the table only if the current item is first or the last - it cannot be done *directly* from any other pointer item in the middle of the table.

Overall, the current position of the iterator pointer indicates what kind of enumerating actions can or cannot be performed next. Thus, the iterator pointer concept is useful from the standpoint of thinking about an enumeration process visually. Also, it appears to be handy semantically. There is no scalar or non-scalar field accessible to our DATA step program that represents the value of the pointer. It is just a semantic construct to aid in the understanding of the *Enumerate All* operation facilities.

4.4.4 Direct Iterator Access: First Item

The hash iterator object provides a means to access the *logically first* item *regardless of its key* directly. The latter means that the iterator pointer can be moved to the first item regardless of its current location (either inside or outside the table) and without the need to visit any other items. Let us look at how the hash object tools facilitate this functionality and bear in mind that everything said about direct access to the logically first item pertains, almost to a tee, to direct access to the *logically last* item as well.

The primary hash iterator tool that activates this action is the FIRST method, its name speaking for itself. It can be called both when the iterator pointer (a) is originally located outside the table or (b) already dwells on a pointer item. Let us first consider the case when it is located outside the table at the time of call. Consider this code snippet:

Program 4.16 Chapter 4 Access First Item Snippet.sas

```
call missing (K, D) ;
put K= D= ;
RC = IH.FIRST() ;
put K= D= RC= ;
```

Because no enumerating action had been taken yet, the iterator pointer is outside the table at the time of the method call. Here the host variables K and D are preset to missing values only for the sake of illustrating the principle.

Before running the template step with this snippet, let us recall that in it, key variable K was added to the data poriton on purpose:

```
H.definedata ("D", "K") ;
```

Now if we insert the snippet into the template step above and run it, it prints in the SAS log:

```
K=. D=
K=2 D=B RC=0
```

As we see, as a result of the successful (RC=0) method call, the originally missing values of PDV host variables K and D have been replaced with the values of their corresponding hash variables from the logically first item. (The reason why K has not remained missing is that it was defined in the data portion in addition to the key portion as shown above.) In other words, the method call accesses the logically first item and performs the data retrieval.

As a result of the FIRST method call, the pointer has been moved to the logically first item *inside* the table. Let us call the method once again:

Program 4.17 Chapter 4 Access First Item Twice Snippet.sas

```
call missing (K, D) ;
put K= D= ;
RC = IH.FIRST() ;
put K= D= RC= ;
RC = IH.FIRST() ;
put K= D= RC= ;
```

Rerunning the template step with this snippet inserted produces the following in the SAS log:

```
K=. D=
K=2 D=B RC=0
K=2 D=B RC=0
```

In other words, calling the FIRST method results in the same outcome irrespective of whether the call pointer was outside the table or already dwelled on the first item. If your power of inductive reasoning makes you suspect that calling FIRST always results in the same outcome regardless of the pointer's pre-call location, your intuition does not deceive you: This is indeed the case.

Let us draw some conclusions and add a few notes:

- Calling the FIRST iterator method accesses the logically first item in the table directly.
- It moves the iterator pointer to the first table item, no matter where the pointer is located before the call, whether it is on any other item or outside the table.
- Therefore, calling FIRST repeatedly always results in the same outcome (provided that, between the calls, the first item has not been deleted or updated).
- It extracts *only the values of the data portion*, but not of the key portion, variables into their PDV host variables; in this case, into variables D and K. The value of K is retrieved because it is also included in the data portion.
- The values of the key PDV host variables before the call are irrelevant. The FIRST method does not use them; or, for that matter, it *needs no key at all* to do its job. It merely moves the interator pointer directly to the first item to access it.
- Calling the FIRST method is always successful, *except when the table is empty* - i.e., contains no items. Hence, if the table is not empty, it can be used *unassigned*.
- The method has no argument tags. Thus, it is always called *implicitly*.

4.4.5 Direct Iterator Access: Last Item

Accessing the *logically last* directly is ideologically identical to that of accessing the logically first item directly. The only difference is that instead of the FIRST method, the LAST method must be used.

Still keeping variable K in the data portion, let us replace the last snippet above with this:

Program 4.18 Chapter 4 Access Last Item Twice Snippet.sas

```
call missing (K, D) ;
put K= D= ;
RC = IH.LAST() ;
put K= D= RC= ;
RC = IH.LAST() ;
put K= D= RC= ;
```

and rerun the template step. It will print the following in the SAS log:

```
K=. D=
K=3 D=F RC=0
K=3 D=F RC=0
```

This agrees with the (K,D)=(3,F) values of the logically last item. The picture observed here logically matches to a tee the picture after the two successive FIRST method. The only difference is that instead of the logically first item, the iterator pointer accesses the logically last item, and so the different data portion values are retrieved into the PDV.

Other than that, everything else said with regard accessing the first item directly using the FIRST method and the details of its activity applies to accessing the last item as well.

4.4.6 Direct Iterator Access: Key-Item

This component of the *Enumerate All* operation stands apart from the rest because it is the only one which *requires a key-value* to work. This is because this action, unlike the rest of *Enumerate All* operation actions, is used to access a specific *key-item* based on its key-value. It can work either stand-alone to retrieve the data just from the key item or used to start enumerating other items in the table forward or backward sequentially.

It may seem that this kind of functionality is already provided by the *Keynumerate* operation. Indeed, we can use the latter to point at the item group with a given key and enumerate the items within the group. However, with *Keynumerate*, such access is restricted to that particular item group, whose boundaries the operation cannot cross without unsetting its item list. *Enumerate All* is free of such constraints: After a key-item has been latched onto, it can use it as a starting point to enumerate any number of items in the table below or above the key-item crossing any item group.

Let us look at our sample table H and imagine that we want to access, via the hash iterator, an item with K=3 directly, regardless of where the iteration pointer currently dwells. This purpose is served by the SETCUR method. However, a question immediately arises. In our table H, there are three items in the item group with K=3, as shown in this subset:

Item#	K	D
4	3	D
5	3	E
6	3	F

So, which one of these items are we going to get as the *key-item*? The way the SETCUR method works resolves the ambiguity by making the choice for us. Let us insert this code snippet in the template step and run it:

Program 4.19 Chapter 4 Access Key-item Snippet.sas

```
call missing (K, D) ;
put K= "IS NOT in the table:" @ ;
RC = IH.SETCUR() ;
put +1 (K D RC) (=) ;

put "K=3 EXISTS in the table:" @ ;
RC = IH.SETCUR(KEY:3) ;
put +1 (K D RC) (=) ;
```

In response, we get this printout in the SAS log:

```
K=. IS NOT in the table: K=. D=  RC=160038
K=3 EXISTS in the table: K=3 D=D RC=0
```

This output demonstrates the following:

- The first SETCUR call is implicit. Hence, it accepts the current PDV host value of K, which at this point is missing. Since there is no such key-value in the table, the call fails with RC≠0 and no data retrieval occurs, so the host variable values remain intact.
- The second SETCUR call is explicit and is given the key-value of 3. While the table has an item group with 3 items with this key, the method sets the iterator pointer at the *logically first* item in the group (item #4). Then it retrieves its data-values, overwriting the values of the PDV host variables - in this case, both D and K because both are included in the data portion.

Note that the result of calling SETCUR does not depend on how the table is ordered. Ordering the table by its key, as we already know, does not change the relative sequence of the items *within* the same-key item groups, which always remains as originally inserted.

4.4.7 Sequential Access

Direct access to the first item, last item, and key-item are important components of the *Enumerate All* operation, for they can access items from which the whole table can be enumerated. But sequential access to table items is the backbone of the operation because it makes it possible to scroll through a hash table one item at a time using a previously accessed item as the point of departure. Thus, by enumerating sequentially, we can cover the entire table or any part of it thereof.

To enumerate the table sequentially in its logical order and in reverse, the *Enumerate All* operation works in essentially the same way. The only difference between the two modes is the direction in which the items are enumerated and the methods called to activate them.

There are a number of conceivable ways an iterative enumeration process can be organized:

- Enumerate forward starting from the logically first item in the table.
- Enumerate backward starting from the logically last item in the table.
- Enumerate forward or backward starting from a key-item.
- Enumerate forward or backward starting from the current pointer item.

Before any enumeration action has been taken yet, the iterator pointer dwells outside the table. So, to start enumerating sequentially, it must be moved into the table. As we already know, the effect can be achieved by priming the process by accessing the first item, the last item, and a key-item directly. Let us proceed in this order.

4.4.8 Enumerating from the End Points

The end points are the logically first and last items. Starting from the logically first item can be done by calling the FIRST method. After that, the next item can be enumerated by calling the NEXT method. This is what the snippet below does:

Program 4.20 Chapter 4 Enumerate from First Item Snippet.sas

```
call missing (K, D) ;
RC = IH.FIRST() ;
put RC= K= D= ;
RC = IH.NEXT() ;
put RC= K= D= ;
```

Inserting it in the template step (with K still kept in the data portion) and running it prints the following in the SAS log:

```
RC=0 K=2 D=B
RC=0 K=2 D=C
```

If we want to enumerate more items going forward, calling the NEXT method in this manner, one call at a time, is rather impractical. The obvious solution is to enclose the NEXT method call in a DO loop. For example, to enumerate all items in the table forward, we can code:

Program 4.21 Chapter 4 Enumerate from First Item Forward Snippet.sas

```
call missing (K, D) ;
do RC = IH.FIRST() by 0 while (RC = 0) ;
  put RC= K= D= ;
  RC = IH.NEXT() ;
```

```
     end ;
     put RC= ;
```

(As a side note, the *by*-specification is included to ensure that the DO loop will iterate more than once. This is because when only the *from*-specification is present, it stops after the first iteration. The *by*-specification is set to 0 to prevent the RC value set in the from-specification from being incremented.) Running the template step with this excerpt inserted, we get the following output:

```
RC=0  K=2  D=B
RC=0  K=2  D=C
RC=0  K=1  D=A
RC=0  K=3  D=D
RC=0  K=3  D=E
RC=0  K=3  D=F
RC=160038
```

It corresponds exactly with the content and logical order of the table. A few points can be emphasized here:

- The return code of the FIRST method call is assigned to the FROM expression of the DO loop, RC being used as the loop index. The call enumerates the logically first item. Hence, by the time the PUT statement is executed, the values of the pair (K,D)=(2,B) are already in the PDV host variables, and so this is what the PUT statement writes.
- Since the TO expression is absent from the loop, **BY 0** is coded in to ensure that the loop does not stop after the first iteration and the subsequent *RC* modifications occur only inside the body of the loop.
- The WHILE condition is what causes the loop to continue and to terminate. This dovetails with the BY 0 construct and the absence of the TO construct.
- After the fifth NEXT call has enumerated the logically last item, it is called again. This time, there are no more items left, and so the call fails with $RC \neq 0$, thus terminating the loop. At the same time, it moves the iterator pointer out of the table.

Enumerating sequentially backward is completely analogous to enumerating forward. The only differences are:

- *Enumerate Backward* begins with the logically last item.
- The primary move of the iterator pointer into the table is done using the LAST method.
- Stepping backward through the table is done using the PREV method.

Correspondingly, to enumerate the whole table backward, the last snippet has to be changed to:

Program 4.22 Chapter 4 Enumerate from Last Item Backward Snippet.sas

```
call missing (K, D) ;
do RC = IH.LAST() by 0 while (RC = 0) ;
  put RC= K= D= ;
  RC = IH.PREV() ;
end ;
put RC= ;
```

Running it as part of the template DATA step prints the following in the SAS log:

```
RC=0  K=3  D=F
RC=0  K=3  D=E
```

```
RC=0  K=3  D=D
RC=0  K=1  D=A
RC=0  K=2  D=C
RC=0  K=2  D=B
RC=160038
```

In this case, the output sequence totally agrees with listing the items in the sequence opposite to the logical order of the table, which is what the intent of enumerating backward is in the first place. Note that in this case, just like with the NEXT method before, the final call to PREV fails and, by doing so, terminates the loop and moves the iterator pointer out of the table.

4.4.9 Iterator Priming Using NEXT and PREV

The enumerating loops shown above can be coded more simply, and here is why. When the iterator pointer dwells outside the table - *and only outside the table*:

- Calling the NEXT method has the same effect as calling the FIRST method.
- Calling the PREV method has the same effect as calling the LAST method.

Therefore, in the *Enumerate Forward* loop shown above, FIRST can be recoded as NEXT, without any change in the outcome:

Program 4.23 Chapter 4 Iterator Priming Next Method Snippet.sas

```
do RC = IH.NEXT() by 0 while (RC = 0) ;
  put RC= K= D= ;
  RC = IH.NEXT() ;
end ;
```

But then there no longer exists any reason for coding NEXT twice, and so the loop can be further reduced to the following form:

Program 4.24 Chapter 4 Iterator Priming Next Method Terse Snippet.sas

```
do while (IH.NEXT() = 0) ;
  put K= D= ;
end ;
```

Likewise, the backward-enumerating loop can be recoded as:

Program 4.25 Chapter 4 Iterator Priming Prev Method Terse Snippet.sas

```
do while (IH.PREV() = 0) ;
  put K= D= ;
end ;
```

Again, neither recoded loop changes the result of respective forward and backward enumeration.

4.4.10 FIRST/LAST vs NEXT/PREV

Coding this way being rather concise and elegant, it raises a question: Why not ditch the FIRST and LAST methods altogether and use the NEXT and PREV methods at all times? There are a number of reasons and meaningful distinctions.

Namely, because the FIRST and LAST methods are *direct access* methods, the following are true:

- They move the iterator pointer to their respective end points *unconditionally*.
- That is, they do it regardless of where the pointer currently dwells, whether on some pointer item inside the table or outside it.
- The result of calling either method repeatedly is always exactly the same.
- They fail only if the table has no items. Otherwise, they always succeed.
- Neither method can move the pointer out of the table (and this unlock the respective end point item).

By contrast, because the NEXT and PREV methods are sequential access methods, the following are true:

- They move the iterator pointer to the logically first and last items, respectively, only when before the call, the pointer is not inside the table.
- Hence, these methods are good for setting the end points only: (a) if no enumeration action has been taken yet; and (b) after either method has moved the iterator out of the table as a result of being called past the respective end point.
- Calling these methods repeatedly results in enumerating the next or previous item and moving the pointer one item further in the corresponding direction.
- In addition to failing when the table is empty, these methods also fail if there are no more items to enumerate forward or backward, respectively. In this case, they unlock the end point items by moving the iterator out of the table.

In other words, compared to the NEXT and PREV methods, the FIRST and LAST methods possess a number of distinct features that are valuable under specific programming scenarios. In fact, combining any or all of these methods in the manner dictated by a specific task is what makes the *Enumerate All* operation powerful and flexible.

4.4.11 Keeping the Iterator in the Table

The advantage of ending the loop using RC≠0 from the extra call to NEXT or PREV past their respective end point items is that it makes the iterator pointer leave the table, and so neither item is any longer locked.

In certain situations, however, it is useful to keep the iterator pointer confined inside the table. For example, imagine that we need to enumerate our sample table H repeatedly, without re-enumerating the end points, in the following manner:

- Looping forward, from item #1 through item #6.
- Looping backward from item #5 through item #1.

Stopping the forward loop with the failing call to the NEXT method is ill-suited for the task. Indeed, since it moves the pointer out of the table, it has to be moved into the table again to enumerate backward. But it means revisiting item #6, which we want to avoid. Thus, to perform the task as outlined, another way of terminating the loop has to be found.

One such way is offered by using the NUM_ITEMS attribute. It tells us the number of items in the table, at any given run-time point, *a priori*. So, the goal can be attained by coding the two successive DO loops. The entire step, rather than the snippet, is shown since its DATA statement differs from that in the template step:

Program 4.26 Chapter 4 Keeping Iterator in Table.sas

```
data Forward (keep=ItemNo K D)
     Backward(keep=ItemNo K D)
  ;
 dcl hash H    (multidata:"Y", ordered:"N") ;
 H.definekey  ("K") ;
 H.definedata ("D", "K") ;
 H.definedone () ;
 do K = 1, 2, 2, 3, 3, 3 ;
   q + 1 ;
   D = char ("ABCDEF", q) ;
   H.add() ;
 end ;     ❶
 dcl hiter IH ("H") ;

 do ItemNo = 1 to H.Num_Items ;    ❷

   RC = IH.NEXT() ;     ❸
   output Forward ;
 end ;

 /* end of forward loop */    ❹

 do ItemNo = H.Num_Items - 1 by -1 to 1 ;    ❺

   RC = IH.PREV() ;    ❻
   output Backward ;
 end ;
run ;
```

❶ At this point, table H has been loaded with the same content as before:

Item#	Key portion	Data portion	
	K	**D**	**K**
1	2	B	2
2	2	C	2
3	1	A	1
4	3	D	3
5	3	E	3
6	3	F	3

❷ Make this DO loop execute exactly as many times as there are items in the table.

❸ Call the NEXT method. The first call moves the iterator pointer to the first item in the table. Each ensuing call enumerates the next item in the table order and retrieves the values of hash variables (K,D) into their PDV host variables (K,D). The OUTPUT statement writes these PDV values, along with that of *ItemNo*, to the output data set *Forward*.

❹ At this point, the iterator pointer dwells on the last item.

❺ Make this DO loop execute one fewer time than the number of items in the table because we now want to enumerate items #5 through #1 backward.

❻ Call the PREV method. The first call moves the pointer from item #6 to item #5. Each ensuing call enumerates the next item in the reverse table order and retrieves the values of hash variables (K,D) into their PDV host variables (K,D). The OUTPUT statement writes these PDV values, along with that of *ItemNo*, to the output data set *Backward*.

Running the template step generates two output test data sets *Forward* and *Backward*. Their contents are shown in the matrix below side by side:

Forward			Backward		
K	**D**	**Item#**	**K**	**D**	**Item#**
2	B	1	3	E	5
2	C	2	3	D	4
1	A	3	1	A	3
3	D	4	2	C	2
3	E	5	2	B	1
3	F	6			

The output shows that the enumeration process works as intended. In this manner, merely by using a single variable like ItemNo to track the location of the iterator pointer, we can walk forward and backward through the table without ever leaving its boundaries.

4.4.12 Enumerating Sequentially from a Key-Item

As we already know, accessing the key-item directly using the SETCUR method with a given key-value moves the iterator pointer, regardless of where it is currently located, to the key-item, which is the logically first item of the same-key item group with the given key-value. Hence, by doing it first, it is possible to enumerate beginning from the key-item, rather than from either end point of the table. The ability to do so has a number of useful applications and performance advantages. This is because the ability to enumerate directly from the key-item in either direction makes it unnecessary to iterate toward it from either end point of the table in order to reach it.

For example, suppose that we have our table H sorted using the argument tag ORDERED:"A" instead of ORDERED:"N", so its content now looks as follows:

Item#	K	D
1	1	A
2	2	B
3	2	C
4	3	D
5	3	E
6	3	F

Suppose further that we want to discover the key and data values of the 2 items with K<3 immediately preceding the item group with K=3. Of course, we can enumerate from the first item of the table forward until we hit K=3 and then enumerate two steps backward. Or, we can enumerate from the end of the table backward till we hit K<3 and then take one more step backward. The problem with both of these propositions is that since we do not know *a priori* where the item group with K=3 is located in the table, we do not know which path - from the beginning of from the end - is shorter or longer. Moreover, by guessing either approach, we may end up scanning almost the entire table, while in fact we only need to enumerate but two items.

The ability to access the key-item with K=3 *directly* rids us of this dilemma because we can use SETCUR to latch onto the logically first item in the item group with K=3 at once and then step sequentially only two items backward. This is what the following snippet does:

Program 4.27 Chapter 4 Enumerating from Key-Item Snippet.sas

```
RC = IH.SETCUR(KEY:3) ;
do count = 1 to 2 while (RC = 0) ;
  RC = IH.PREV() ;
  if RC = 0 then put K= D= RC= ;
end ;
```

Running this snippet as part of the template step with ORDERED:"Y" results in the following text in the SAS log:

```
K=2  D=C  RC=0
K=2  D=B  RC=0
```

It shows that the code works as intended. The RC=0 condition in the WHILE clause is needed just as a precaution to stop the loop in case there is only one item with K<3 (meaning that it is the first item in the table). To check it for robustness, we can change **KEY:3** to **KEY:2**, which means that now we would be seeking two items with K<2, while there is just one item like that in the table with K=1, and it is the first in the table. Running the step with this change would produce:

```
K=1  D=A  RC=0
```

In other words, the PREV method enumerates the only item available to it, and stops the loop by generating RC≠0 when there are no more items in the table for it to access.

4.4.13 Harvesting Same-Key Items from a Key-Item

Previously, we have shown how the data can be "harvested" from all the items in a same-key item group using the *KeyNumerate* operation. The availability of the hash iterator object and its ability to enumerate from a key-item offers an alternative.

For example, suppose that we want to collect the data from all the items where K=2. As we know, if we call the SETCUR with K=2, it will point to the logically first item in the item group with K=2- i.e., to item #1 in the case of our table H. Also recall that no matter how the table is ordered, the items sharing the same key-value are always grouped together. Hence, starting with the key-item, we can simply enumerate *forward* until either (a) a different key is encountered or (b) we have reached the end of the table. Translating it into the SAS language, we can code:

Program 4.28 Chapter 4 Harvesting Same Key Items from Key Item Snippet.sas

```
call missing (K, D) ;
_K = 2 ;
do RC = IH.SETCUR(KEY:_K) by 0 while (RC = 0 and K = _K) ;
  put K= D= RC= ;
  RC = IH.NEXT() ;
```

```
  end ;
  put RC= ;
```

Running the template step with this snippet included, we get in the SAS log:

```
  K=2  D=B  RC=0
  K=2  D=C  RC=0
  RC=0
```

For the item group with *K*=2, the loop is terminated because a different key, K=1, is encountered. Since at this point the last item in the table has not been reached, we still have RC=0.

The group with K=3 represents the other situation where the last item in the group is at the same time the last item in the table. So, if we recode _K=2 above as _K=3 and rerun, the step will print in the SAS log:

```
  K=3  D=D  RC=0
  K=3  D=D  RC=0
  K=3  D=F  RC=0
  RC=160038
```

This time, the loop is stopped not because a different key-value has been encountered but due to the NEXT method call's attempt to enumerate past the last item in the table, thus setting RC≠0.

4.4.14 The Hash Iterator and Item Locking

Before we introduce the issue, let us again refresh in the mind the image of our sample table H created and populated in the template DATA step (with ORDERED:"N"):

Item#	K	D
1	2	B
2	2	C
3	1	A
4	3	D
5	3	E
6	3	F

Next, let us consider the following test code snippet:

Program 4.29 Chapter 4 Iterator Locking Snippet.sas

```
    _K = 2 ; * <--Key-value to delete ;    ❶
    RC = IH.FIRST() ;       ❷
*   RC = IH.NEXT() ;        ❸
*   RC = IH.NEXT() ;
    put "Item:" +3 K= D= RC= ;     ❺
    RC = H.REMOVE(KEY: _K) ;    ❻
    put "Remove:" +1 RC= ;
    RC = H.CHECK(KEY: _K) ;     ❹
    put "Check:" +2 RC= ;
```

Before running the snippet as part of the template step, let us review what this code does:

❶ An auxiliary variable _K is assigned the key-value for the items intended for deletion. By varying it, we can see how it affects the outcome.

❷ Call the FIRST method to set the iterator pointer at item #1. Its key is K=2. Note that this has nothing to do with the fact that we assigned _K=2.

❸ For the time being, both NEXT calls are inactivated. By activating one or both of them, we can observe how moving the pointer forward to items #2 and #3 affects the result.

❹ The CHECK method is called to see whether the *Delete* operation has succeeded. If it has, the call will return RC≠0.

❺ The PUT statements are used to write test information to the SAS log.

❻ The REMOVE method call is an attempt to perform the *Delete All* operation against the item group with K=2. It should be successful, both items #1 and #2 would be purged, and the call would return RC=0.

Now, it we insert the snippet in the template step and run it, we will discover the following error messages in the SAS log:

```
Item:   K=2 D=B RC=0
ERROR: Cannot remove record from a hash object which has been locked by an
       iterator at line <X> column <Y>.
ERROR: DATA STEP Component Object failure. Aborted during the EXECUTION phase.
```

Note that the PUT statements following the REMOVE call print nothing. These statements are never executed because the step generates an error and is terminated altogether immediately as a result. The error message makes it clear that the issue is in the locking of hash items.

4.4.15 Locking and Unlocking

This simple test illustrates the phenomenon of the *iterator item locking*. It makes it pretty obvious that the item pointed to by the iterator pointer is *locked* by the iterator and thus cannot be deleted from the table. In turn, it raises the question: Does this cause any *other items to be locked* as well; and if yes, then which ones?

To find out, let us first change the _K=2 to _K=3 and rerun the step. The intent is to see if the pointer dwelling on item #1 with K=2 prevents items with another key from being deleted. This time, the step runs normally and reports the following in the log:

```
Item:    K=2 D=B RC=0
Remove:  RC=0
Check:   RC=160038
```

In other words, the item group with K=3 was successfully deleted, even though the first item with K=2 is locked as a result of the FIRST call. Thus, having an item locked in one same-key item group does not affect the ability to delete an item group with a different key.

Now let us restore _K=2 and uncomment the first NEXT call:

```
  _K = 2 ;
  RC = IH.FIRST() ;
  RC = IH.NEXT() ;
* RC = IH.NEXT() ;
```

The NEXT call causes the pointer to move to item #2, whose key is also K=2. Running the template step this way brings back the error message and step termination. Logically, it adds up because now item #2 is locked, while the REMOVE call aims to delete both item #2 and #1, both having K=2.

From what we have seen above for _K=3, it is logical to expect that if we move the pointer out of the item group with K=2 to the next item #3, we should be able to delete the item group with K=2 successfully since the iterator will no longer lock either item in the K=2 item group. This is indeed the case: If we now activate, in addition to the first, the second call to NEXT, the latter is expected to move the pointer to item #3 with K=1:

```
  _K = 2 ;
  RC = IH.FIRST() ;
  RC = IH.NEXT() ;
  RC = IH.NEXT() ;
```

So, with this change, the step runs normally and prints the following in the SAS log:

```
  Item:    K=1 D=A RC=0
  Remove:  RC=0
  Check:   RC=160038
```

Expectedly, the CHECK call fails with *RC≠0* because the items with K=2 have been deleted and so the key is no longer in the table.

In this case, the second NEXT call moves the pointer out of the K=2 item group to the *immediately adjacent* item group where K=1. Does it matter from the standpoint of unlocking the K=2 item group; or will moving the pointer *anywhere out* of the group unlock it? The answer to this question is: It *does not matter where the pointer is moved* as long as it is moved out of the group, including out of the table altogether. For example, we can replace the pair of the NEXT calls above with *either* of the following two calls:

```
  RC = IH.PREV() ;
```

Or:

```
  RC = IH.SETCUR(KEY:3) ;
```

Because the FIRST call makes the first table item the pointer item, the PREV method has no item to fetch, and so the call will move the pointer *out of the table*. Calling SETCUR as shown above moves the pointer directly to the first item in the K=3 item group. Either way, the pointer is moved out of the K=2 item group. So, if the template step is now rerun with the call to PREV or call to SETCUR as shown above instead of the two NEXT calls, it still runs normally and the K=2 item group is deleted successfully.

Note that when SETCUR method is used to move the pointer out of a same-key item group where it currently dwells, it must be called with a key-value different from the one identifying the group but still present in the table. Calling it with a key-value absent from the table will *not* move the pointer out of its current position.

4.4.16 Locking Same-Key Item Groups

The examples presented above leave one final aspect unclear. Since in all of them the *Delete All* operation attempts to delete all items from a same-key item group, the removal is impossible regardless of which item in the group is locked. However, in addition to *Delete All*, we also have the *Selective Delete* operation to delete individual items from a same-key item group rather than the entire group.

So, a germane question arises: Does the hash iterator lock the *individual item* on which the iterator pointer dwells; or does it lock the *entire same-item group* to which the pointer item belongs? If we could delete one item from a same-key group while the iterator pointer dwells on another, the answer would be "*individual point item*"; otherwise, it would be "*entire same-item group*". The question is easy to answer using the item group with K=2 (items #1 and #2) from our sample table H as a guinea pig. Let us insert this code snippet in the template step:

Program 4.30 Chapter 4 Locking Same Item Key Group Snippet.sas

```
RC = IH.FIRST() ;          ❶
RC = H.FIND(KEY:2) ;       ❷
RC = H.FIND_NEXT(KEY:2) ;  ❷
RC = H.REMOVEDUP(KEY:2) ;  ❸
put RC= ;
```

The intent of this test case is to:

❶ Call the FIRST method to set the iterator pointer on item #1, the first item in the K=2 group.

❷ Call the FIND and then FIND_NEXT methods to move the *Keynumerate pointer* to item #2. Note that these methods are *not iterator object methods* and thus do not affect the position of the *iterator pointer*, which after these calls keeps dwelling on item #1.

❸ Call the REMOVEDUP method in an attempt to remove item #2, *presumably* not locked because the iterator pointer still dwells on item #1.

Running the template step with this snippet results in the same log error message "*locked by the iterator*" we have seen earlier. The unescapable conclusion is that when a pointer item is locked by the iterator:

- *All the items* in the same-key item group to which the pointer item belongs are locked.
- In other words, iterator locking occurs, not at the level of an individual point item, but at the level of its entire same-key item group.

4.4.17 Locking the Entire Hash Table

The verbiage in the "*locked by the iterator*" SAS log error message suggests that the entire table is locked when the iterator pointer dwells inside it. However, it is more intimidating than necessary, for as we have established earlier, it is true only in the situation when the table contains a single same-key item group - in other words, when all items in the table have the same key-value. Otherwise, only the same-key item group containing the pointer item is locked; and the items from the rest of the groups can be deleted successfully, whether unconditionally or selectively. To unlock an item group, all you need to do is to move the iterator pointer out of it, no matter where.

The impression that "the entire table is locked by the iterator" is most often created when an attempt to perform the *Clear* operation is made- i.e., to delete every single item from the table. Since the CLEAR method aims to purge every item, it does not matter on which particular item the iterator pointer dwells since the item is inevitably a target for deletion. Therefore, the only way to make the CLEAR method succeed when a hash table has an iterator object linked to it is to *ensure that the iterator pointer is out of the table*.

Often, this state occurs naturally as a result of the enumerating logic. For example, if an enumerating loop is terminated with the NEXT or PREV method called beyond the last or first table item, respectively, the iterator pointer is moved outside the table as a result, and so no item in the table is locked any longer.

Otherwise, if the iterator pointer, for one reason or another, still dwells inside the table when it has to be cleared, a surefire method to guarantee that the CLEAR method succeeds is to *force* the iterator pointer out of the table. Either pair of the statements below:

```
RC = H.FIRST() ;
RC = H.PREV() ;
```

Or:

```
RC = H.LAST() ;
RC = H.FIRST() ;
```

is equally good for the purpose. The first moves the pointer directly from any location to the first item and then forces it out of the table with the failing call past the end point. The second does the same in the opposite direction. After either of these statement pairs has been executed, the CLEAR method will always succeed since no item in the table will be locked by the iterator.

4.4.18 ENUMERATE ALL Operation Hash Tools

- Statements: DECLARE (DCL) HITER.
- Operators: _NEW_.
- Methods: FIRST, LAST, NEXT, PREV, SETCUR.
- Argument Tags: KEY.

4.4.19 Hash-PDV Interaction

Any iterator method, if it succeeds, performs the indirect *Retrieve* operation by extracting the values of the data portion hash variables and overwriting their PDV host variable counterparts with them.

Part Two—The WHAT and the WHY of the SAS Hash Object

The goal of Part Two is to provide a rich set of examples that address typical business intelligence type questions. It uses the story line of a Proof of Concept project to describe to both business and IT users WHY they should consider the use of the hash object and WHAT it can be used for.

Part Two contains four chapters:

1. Chapter 5 provides a description of a set of transactional data that was generated as the source data for this Proof of Concept.
2. Chapter 6 expands upon the HOW discussion in Part One by presenting simple examples of questions that can be answered using the SAS hash object.
3. Chapter 7 describes using the SAS hash object to take that transactional data as input to create a data mart/warehouse that can be used to produce the summary data and metrics to be reported on.
4. Chapter 8 provides a rich set of example programs that address typical data aggregation and reporting requirements.

Part Two is targeted to both our business and IT users. We suggested that as they read through the examples that they refer back to the more detailed HOW discussion in Part One as needed.

There are also valid reasons to begin your exploration of this book by starting directly with Part Two (especially if you already have some experience with the SAS hash object).

Chapter 5: Bizarro Ball Sample Data

5.1 Introduction

As mentioned previously, we have been approached by the headquarters office for Bizarro Ball about their interest in using analytics to improve the quality of the game. Specifically, the business users are interested in using the detailed data about all the events in a game to answer what we would call business intelligence questions like:

- How well is each batter performing overall?
- How well is each pitcher performing?
- Are there differences due to external factors like when and where the games were played?
- And so on.

The results data they want to use describe what happened for:

- Each pitch.
- Each plate appearance (which is often referred to as an *at bat*).
- Each runner who gets on base.

The business users have a number of different metrics they are interested in calculating using this data and have asked for a Proof of Concept (PoC) for how to analyze such data to calculate those metrics as well as some new metrics they believe might be useful. Given that there is no source of data for the PoC, we agreed to write sample SAS programs to generate data to approximate the kind of data they would need to collect on an ongoing basis.

The rest of this chapter describes the data sets we generated. That data will be used as the source data for the examples in the rest of the book. You can access the programs, the sample data, and the blog entries that are used in this book. Note that the blog entries include additional examples (including the programs that created the sample data). These are available from the author page at http://support.sas.com/authors. Select either "Paul Dorfman" or "Don Henderson." Then look for the cover thumbnail of this book, and select "Blog Entries." Those programs also make extensive use of the SAS hash object; however, that sample code is not directly related to the point of the Proof of Concept, and so they are not described here.

5.2 Sample Data Descriptions

We created a number of data sets to simulate the Bizarro Ball data. The data sets are not normalized- e.g., the AtBats data contains all the available fields about a player such as their name and team. The data corresponds to what a person watching the game would record or write down. The data sets contain:

- 2,880 games (the Games data set).
- Approximately 875,000 pitches (the Pitches data set).
- Almost 290,000 at bats/plate appearances (the AtBats data set).
- Roughly 78,000 runs scored (the Runs data set).
- And more.

The following subsections describe seven data sets that are the primary source data for the examples in the rest of the book. These data sets reflect the basic data needed for Bizarro Ball that is also needed for our Proof of Concept. Should the business users decide to move forward and formally collect all the relevant data from Bizarro Ball games, that data would likely contains many more fields; for the purposes of our Proof of Concept we focused on creating a minimal set of fields needed to highlight how the SAS hash object could be used to answer questions about the data.

Each of the following subsections includes:

- A brief description of the data set.
- A description of how some of the fields in each data set can be used.
- A listing of the fields in each data set.
- A sample of the data (the first five rows).

Also please note the following about the sample data:

- A number of field names have a suffix of _SK which stands for **S**urrogate **K**ey. That is a standard convention in many data tables. For example, the Teams data set could be uniquely identified by its name; however this can create data access issues if a team changes it name. A standard approach is to generate an arbitrary value and use that as the key to uniquely identify a data row.
- The Game_SK field is constructed from a number of source fields. Its display value is 32 characters that have no meaning and so the listings below only show the first four characters following by ellipses (. . .) in order to designate that only part of the value is displayed.

5.2.1 AtBats

A plate appearance (PA) in Bizarro Ball, like baseball, corresponds to each time a batter faces an opposing pitcher in a game. There is one observation in the AtBats data set for each such occurrence. A PA counts as an at bat (AB) based on what the results are of the batter facing the pitcher. For example, if the batter draws a walk (4 pitches out of the strike zone), that PA does not count as an AB. In response to a request from the business users, we agreed to call this data set AtBats as that is the term most commonly used to describe these events. The distinction between an AB and a PA is used in calculating some of the metrics we will generate later in this book.

The AB_Number field is a sequential counter for each team within a game that uniquely identifies what the result of each plate appearance was. It provides a one-to-many link to the Pitches data set. Every observation in the AtBats data set has at least one row in the Pitches data set. It also provides a one-to-many (including 0) link to the Runs data set. For example, there is an observation in the Runs data set for each run scored as the result of a plate appearance. If no runs are scored as the result of the plate appearance, then there are no rows in the Runs data set for that AB_Number.

Many of the fields in this data set will be used as class variables in later chapters. Other fields are aggregated to calculate the various metrics the business users are interested in. For example, the fields whose names begin with Is_A_ are Boolean (missing/0 or 1) fields which specify how they should be counted when calculating a number of the metrics:

- Is_An AB: Designates if the plate appearance contributes a count of 1 to the number of at bats. Aggregating this field, for example, will provide the denominator for the calculation of a batter's batting average (BA).
- Is_An_Out: The batter made an out. This field can be used to aggregate how many innings a pitcher pitched (3 outs is an inning).
- Is_A_Hit: The batter got a hit (a single, double, triple, or home run). Aggregating this field, for example, will provide the numerator for the calculation of the batter's batting average (BA).
- Is_An_OnBase: The batter reached base, regardless of how (e.g., a hit or a walk). Aggregating this field, for example, will provide the numerator for the on base percentage metric (OBP).

Since each observation in the data set contributes a count of 1 to the number of plate appearances, we did not create a specific field for that calculation; the number of rows can be counted/calculated.

Output 5.1 AtBats Contents

Name	Label	Type	Length
Game_SK	Game Surrogate Key	char	16
Date	Game Date	num	8
Time	Game Time	num	8
League	League	num	8
Away_SK	Away Team Surrogate Key	num	8
Home_SK	Home Team Surrogate Key	num	8
Team_SK	Team Surrogate Key	num	8
Batter_ID	Batter ID	num	8
First_Name	Batter First Name	char	12
Last_Name	Batter Last Name	char	12
Position_Code	Batter Position	char	3
Inning	Inning	num	8
Top_Bot	Which Half Inning	char	1
Bats	Bats L, R, or Switch	char	1
Throws	Throws L or R	char	1
AB_Number	At Bat Number in Game	num	8
Result	Result of the At Bat	char	16

Name	Label	Type	Length
Direction	Hit Direction	num	8
Distance	Hit Distance	num	8
Outs	Number of Outs	num	8
Balls	Number of Balls	num	8
Strikes	Number of Strikes	num	8
onFirst	ID of Runner on First	num	8
onSecond	ID of Runner on Second	num	8
onThird	ID of Runner on Third	num	8
onBase	Number of Men on Base at Beginning of AB	num	8
Left_On_Base	Number of Men Left on Base at End of AB	num	8
Runs	Runs Scored	num	8
Is_An_AB	Counts as an AB	num	8
Is_An_Out	Is an Out	num	8
Is_A_Hit	Is a Hit	num	8
Is_An_OnBase	Counts as an On Base	num	8
Bases	Number of Bases for the Hit	num	8
Number_of_Pitches	Number of Pitches This AB	num	8

Output 5.2 First 5 Observations in AtBats

Obs	Game Surrogate Key	Game Date	Game Time	League	Away Team Surrogate Key	Home Team Surrogate Key	Team Surrogate Key	Batter ID	Batter First Name	Batter Last Name	Batter Position
1	0D58 . . .	2017-03-20	7:00 PM	2	165	228	165	16923	Willie	Garcia	CF
2	0D58 . . .	2017-03-20	7:00 PM	2	165	228	165	21791	Terry	Martinez	C
3	0D58 . . .	2017-03-20	7:00 PM	2	165	228	165	58751	Dylan	Martin	2B
4	0D58 . . .	2017-03-20	7:00 PM	2	165	228	228	31019	Anthony	Price	2B
5	0D58 . . .	2017-03-20	7:00 PM	2	165	228	228	10760	Jeremy	Gray	LF

Obs	Inning	Which Half Inning	Bats L, R, or Switch	Throws L or R	At Bat Number in Game	Result of the At Bat	Hit Direction	Hit Distance	Number of Outs	Number of Balls	Number of Strikes	ID of Runner on First
1	1	T	R	R	1	Out	5	193	0	0	1	.
2	1	T	R	R	2	Strikeout	.	.	1	0	3	.
3	1	T	R	R	3	Out	17	126	2	2	1	.
4	1	B	L	R	1	Out	5	77	0	0	1	.
5	1	B	L	R	2	Strikeout	.	.	1	2	3	.

Obs	ID of Runner on Second	ID of Runner on Third	Number of Men on Base at Beginning of AB	Number of Men Left on Base at End of AB	Runs Scored	Counts as an AB	Is an Out	Is a Hit	Counts as an On Base	Number of Bases for the Hit	Number of Pitches This AB
1	.	.	0	0	0	1	1	0	0	0	2
2	.	.	0	0	0	1	1	.	0	.	4
3	.	.	0	0	0	1	1	0	0	0	4
4	.	.	0	0	0	1	1	0	0	0	2
5	.	.	0	0	0	1	1	.	0	.	5

5.2.2 Games

The Games data set is the schedule of games and contains information on which teams play each other on what date.

The fields in this data set are used as class variables in many of the aggregates.

Output 5.3 Games Contents

Name	Label	Type	Length
Game_SK	Game Surrogate Key	char	16
Date	Game Date	num	8
Time	Game Time	num	8
Year	Year	num	8
Month	Month	num	8
DayOfWeek	Day of the Week	num	8
League	League	num	8
Home_SK	Home Team Surrogate Key	num	8
Away_SK	Away Team Surrogate Key	num	8

Output 5.4 First 5 Observations in Games

Obs	Game Surrogate Key	Game Date	Game Time	Year	Month	Day of the Week	League	Home Team Surrogate Key	Away Team Surrogate Key
1	C3F0 . . .	2017-03-20	7:00 PM	2017	3	2	1	317	342
2	3149 . . .	2017-03-21	7:00 PM	2017	3	3	1	317	342
3	DF10 . . .	2017-03-22	7:00 PM	2017	3	4	1	317	342
4	F948 . . .	2017-07-03	7:00 PM	2017	7	2	1	342	317
5	66FC . . .	2017-07-04	7:00 PM	2017	7	3	1	342	317

5.2.3 Leagues

The Leagues data set contains just two rows – one for each of the two Bizarro Ball leagues. Its primary use case is to provide a label for aggregate metrics produced at or rolled up to the league level. The business users have indicated they are interested in comparing results between the two leagues.

Output 5.5 Leagues Contents

Name	Label	Type	Length
League_SK	League Surrogate Key	num	8
League	League	char	7

Output 5.6 All Observations in Leagues

Obs	League Surrogate Key	League
1	1	Eastern
2	2	Western

5.2.4 Pitches

The Pitches data set contains one row for every pitch in every game. The value of the Result field on the last Pitch for an AtBat is the same as the value of the Result field in the AtBats data set for each AtBat. The AB_Number field can be used to link the Pitches and AtBats data and can be used when generating metrics that require data from both the AtBats and Pitches data (e.g., summarize the performance of pitchers vs. specific batters and vice versa).

As for the AtBats data set, many of the fields here will be used as class variables. The fields Is_A_Ball and Is_A_Strike are also Booleans (missing/0 or 1) and can be aggregated to evaluate pitch distribution. The fields Balls (how many pitches that are balls have been thrown in the current plate appearance) and Strikes (how many pitches that are strikes have been thrown in the current plate appearance) can be used as filter criteria as well as class variables. The variable Strikes is used as a filter variable in the sample case study in Chapter 13.

Output 5.7 Pitches Contents

Name	Label	Type	Length
Game_SK	Game Surrogate Key	char	16
Date	Game Date	num	8
Team_SK	Team Surrogate Key	num	8
Pitcher_ID	Pitcher_ID	num	8
Pitcher_First_Name	Pitcher_First_Name	char	12
Pitcher_Last_Name	Pitcher_Last_Name	char	12
Pitcher_Bats	Bats L, R, or Switch	char	1
Pitcher_Throws	Throws L or R	char	1
Pitcher_Type	Starter or Reliever	char	3
Inning	Inning	num	8
Top_Bot	Which Half Inning	char	1
Result	Result of the At Bat	char	16

Name	Label	Type	Length
AB_Number	At Bat Number in Game	num	8
Outs	Number of Outs	num	8
Balls	Number of Balls	num	8
Strikes	Number of Strikes	num	8
Pitch_Number	Pitch Number in the AB	num	8
Is_A_Ball	Pitch is a Ball	num	8
Is_A_Strike	Pitch is Strike	num	8
onBase	Number of Men on Base at Beginning of AB	num	8

Output 5.8 First 5 Observations in Pitches

Obs	Game Surrogate Key	Game Date	Team Surrogate Key	Pitcher ID	Pitcher First Name	Pitcher Last Name	Bats L, R, or Switch	Throws L or R	Starter or Reliever	Inning
1	0D58 . . .	2017-03-20	165	13014	Louis	Watson	R	R	SP	1
2	0D58 . . .	2017-03-20	165	13014	Louis	Watson	R	R	SP	1
3	0D58 . . .	2017-03-20	165	13014	Louis	Watson	R	R	SP	1
4	0D58 . . .	2017-03-20	165	13014	Louis	Watson	R	R	SP	1
5	0D58 . . .	2017-03-20	165	13014	Louis	Watson	R	R	SP	1

Obs	Which Half Inning	Result of the At Bat	At Bat Number in Game	Number of Outs	Number of Balls	Number of Strikes	Pitch Number in the AB	Pitch is a Ball	Pitch is Strike	Number of Men on Base at Beginning of AB
1	T	Swinging Strike	1	0	0	1	1	.	1	0
2	T	Out	1	0	0	1	2	.	1	0
3	T	Called Strike	2	1	0	1	1	.	1	0
4	T	Foul	2	1	0	2	2	.	1	0
5	T	Foul	2	1	0	2	3	.	.	0

5.2.5 Player_Candidates

The Player_Candidates data set contains 10,000 rows which resulted from a Cartesian product of 100 unique first names and 100 unique last names. This data set is the pool of available players for the 32 Bizarro Ball teams (25 players per team). A total of 32*25 observations in this data set were randomly assigned to specific positions (e.g., the Position_Code field designates if the player is a starting pitcher, a first baseman, and so on) and teams (the Team_SK field). The fields in this data set are descriptive in nature and will primarily be used as descriptive labels or classification fields.

Output 5.9 Player_Candidates Contents

Name	Label	Type	Length
Player_ID	Player ID	num	8
Team_SK	Team Surrogate Key	num	8
First_Name	First Name	char	12
Last_Name	Last Name	char	12
Position_Code	Batter Position	char	3
Bats	Bats L, R, or Switch	char	1
Throws	Throws L or R	char	1

Output 5.10 First 5 Observations in Player_Candidates

Obs	Player ID	Team Surrogate Key	First Name	Last Name	Batter Position	Bats L, R, or Switch	Throws L or R
1	11611	269	Henry	Young	UT	R	R
2	28519	269	Joseph	Davis	UT	S	L
3	39778	161	Lawrence	Diaz	UT	S	R
4	13977	161	Matthew	Howard	UT	R	R
5	26952	165	Wayne	Carter	UT	L	R

5.2.6 Runs

The Runs data set contains information on each scored run – both the ID of the batter as well as the ID of the runner. The AB_Number field can be used to retrieve data values needed for various metrics from the AtBats and Pitches data sets. For example, we might want to rank the importance of what kinds of hits result in the most runs scored. Alternatively, what is the distribution of runs scored based on the number of outs. Aggregating the count of rows (i.e., runs scored) using Batter_ID as a classification variable calculates the Runs Batted In (aka RBIs) metric; using Runner_ID calculates the Runs_Scored metric for the player.

Output 5.11 Runs Contents

Name	Label	Type	Length
Game_SK	Game Surrogate Key	char	16
Date	Game Date	num	8
Batter_ID	Batter ID	num	8
Inning	Inning	num	8
Top_Bot	Which Half Inning	char	1
AB_Number	At Bat Number in Game	num	8
Runner_ID	ID of Runner Who Scored	num	8

Output 5.12 First 5 Observations in Runs

Obs	Game Surrogate Key	Game Date	Batter ID	Inning	Which Half Inning	At Bat Number in Game	ID of Runner Who Scored
1	0D58 . . .	2017-03-20	16923	3	T	10	13014
2	0D58 . . .	2017-03-20	16923	3	T	10	16923
3	0D58 . . .	2017-03-20	24000	4	B	13	10760
4	0D58 . . .	2017-03-20	24000	4	B	13	14241
5	0D58 . . .	2017-03-20	58751	5	T	21	16923

5.2.7 Teams

The Teams data sets contains 32 observations with the league and team name for each team. In real life, it would contain many more fields such as Owner, City, Manager, and so on. This data set is used to provide classification variables and labels.

Output 5.13 Teams Contents

Name	Label	Type	Length
Team_SK	Team Surrogate Key	num	8
Team_Name	Team Name	char	16
League_SK	League Surrogate Key	num	8

Output 5.14 First 5 Observations in Teams

Obs	Team Surrogate Key	Team Name	League SK
1	317	Mountaineers	1
2	342	Gladiators	1
3	132	Vikings	1
4	251	Hurricanes	1
5	259	Grizzlies	1

5.3 Summary

These Bizarro Ball data sets will be used for the examples in the next few chapters. As mentioned in the Introduction, they were designed to be transactional in nature (i.e., what would be collected during games). In Chapter 7 this data will be used to create a rudimentary data warehouse/mart that can be more easily used for business intelligence type questions.

Chapter 6: Data Tasks Using Hash Table Operations

6.1 Introduction

In Chapters 2-4, we have discussed how the hash object tools can be used to perform hash table operations. In this chapter, in response to a request from the IT users at Bizarro Ball headquarters, we will discuss how the operations can be applied to address a number of common data processing tasks using terminology that they are familiar with. In later chapters in this part we will refocus our efforts on the requirements (and terminology) of the business users.

In particular, we will apply hash table operations to:

- Subset, unduplicate, combine, split, and order data.
- Implement or supplant dynamic data structures, such as arrays, stacks, queues, etc.
- Use hash tables as dynamic storage media to discover certain data properties.

The examples and code snippets in this chapter are kept at the complexity level needed to demonstrate the principle. However, they are complete enough to serve as practical templates for more sophisticated examples found in the next chapters, as well as in the programs used to create the sample data.

Our challenge is to first convince the IT users that the underlying technology and functionality can address their concerns while at the same time meeting all of the business users' requirements. In the following sections we will lead with terminology that the IT users are most comfortable with and also describe the functionality that the business users are comfortable with. Anyone who has dealt with describing technology solutions knows that IT and business users quite often don't speak the same language, and so we have to serve as translators.

6.2 Subsetting Data

Subsetting is an act of selecting a subset from one data collection (let us call it *A*) based on values from another data collection (let us call it *B*). "Data collection" can be pretty much anything: a flat file, an array, a set of macro variables, etc. Most commonly, the collection *A* is represented by a SAS data file or its logical equivalent. The collection whose values are used to do the subsetting (*B*) can take various forms depending on its size and other circumstances. However, it can be assumed without a loss of generality that the values of *B* are also stored in a SAS data file since the file can be used to generate subsetting code using any lookup technique in the SAS arsenal.

For the benefit of our business users, subsetting data might include something like the example in the case study in Chapter 12 where our subset is defined using the data in the Pitches data set (all the data where the first two pitches are strikes) – data collection B; and data collection A is the Runs data set. In other words, we want to create a subset of the Runs data corresponding to just those AtBats where the first two pitches were strikes. But before we can get to that example, we need the buy-in from the IT users. Thus, this section will present both a number of alternative subsetting requirements as well as some wrinkles.

6.2.1 Two Principal Methods of Subsetting

There are two principally distinct methods of subsetting, regardless of a particular implementation:

- **Sorting and merging**:
  - Order both data collections *A* and *B* the same way by the key to be used for the subsetting.
  - Execute the sequential matching algorithm (exemplified in SAS by the MERGE statement) based on ordered runs through each.
  - Select records in *A* for matching key-values found in *B*.
  - The method is especially appealing when *A* and *B* are already intrinsically sorted.
  - However, if they are not, and forced sorting is required, it may prove to be extremely computationally expensive, particularly in the today's world of big data.
- **Table lookup**:
  - Pre-store the keys from *B* in some kind of a lookup table.
  - For every record read from *A*, search the table for the key-value coming from it.
  - Select the record from *A* based on whether there is a key match.

- ○ The method does not require sorting either *A* or *B*.
- ○ However, in order for the method to be effective, the lookup table has to be supported by a reasonably fast search algorithm that scales well as the number *N* of its distinct key-values grows.
- ○ Practically speaking, well-scaling algorithms include those running in *O(log(N))* time (e.g., the binary search) or, ideally, in *O(1)* time (as hashing algorithms).

The hash object supports its tables in terms of both searching speed and scaling. Let us now consider examples of subsetting a data file using a hash table.

6.2.2 Simple Data File Subsetting via a Hash Table

Suppose that we are interested to know the distinct players from file Bizarro.Player_candidates who have:

- played for team Huskies (Team_SK=193)
- been at bat
- hit one or more triples

Below is a visual sample for a few players from the file (picked to both satisfy and fail the requirements above). In terms of our subsetting process described above, this file is going to play the role of data collection *A*.

Output 6.1 Bizarro.Player_candidates Sample

Player_ID	Team_SK	First_Name	Last_Name	Position_Code	Bats	Throws
43400	193	Andrew	Anderson	UT	L	L
55438	193	Bobby	Miller	UT	R	L
52320	193	Jose	James	RP	R	R
37661	193	Bryan	Rivera	SP	S	R

The information about the results at bat for all games in the season is stored in file Dw.AtBats, which does not include the player candidates who have not batted. In our subsetting scheme, it is going to play the role of data collection *B*. Here is a sample from it for the players listed in Output 6.1:

Output 6.2 Dw.AtBats Sample

Batter_ID	Result
55438	Out
37661	Strikeout
37661	Walk
43400	Single
55438	Triple
37661	Out
55438	Triple
43400	Walk
43400	Double
37661	Triple

Note that this sample does not contain the player with Player_ID=52320 because he is not in the file (meaning that he has not been at bat). Also, though all records with Result="Triple" for these players are included in the sample, only a few records with other results are included, merely for illustrative purposes.

If we used SQL, the following query could be used to achieve the subsetting goal:

Program 6.1 Chapter 6 Simple Subsetting via SQL Subquery.sas

```
proc sql ;
  create table Triples as
  select * from bizarro.Player_candidates
  where  Player_ID in
        (select Batter_ID from Dw.AtBats where Result = "Triple")
  and    Team_SK in (193) ;
quit ;
```

To get the same result using a hash table, we need three hash table *operations*:

1. *Create* to declare, define, and instantiate a hash object.
2. *Insert* to load its hash table with the key-values of Batter_ID from Dw.AtBats for which Result="Triple".
3. *Search* to find which Player_ID key-values coming from Bizarro.Player_candidates are in the hash table.

In the snippet below, the DATASET argument tag is used to perform the indirect *Insert* operation:

Program 6.2 Chapter 6 Simple Subsetting via Hash Table.sas

```
data Triple ;
  if _n_ = 1 then do ;     ❶
    dcl hash triple     ❷
    (dataset:'Dw.AtBats(rename=(Batter_ID=Player_ID)     ❸
                        where=(Result="Triple"))') ;
    triple.defineKey ("Player_ID") ;     ❹
    triple.defineData ("Player_ID") ;     ❺
    triple.defineDone () ;
  end ;
  set Bizarro.Player_candidates ;     ❻
  where Team_SK in (193) ;
  if triple.check() = 0 ;     ❼
* if not triple.check() = 0 ;     ❽
run ;
```

Let us now see how the step above goes about its subsetting business:

❶ Ensure that the *Create* operation is done only once, on the first iteration of the DATA step implied loop, to prevent the ensuing iterations from dropping the hash table and re-creating it.

❷ Declare hash object Triple and create an instance of it.

❸ At the same time, use the DATASET argument tag to perform the indirect *Insert* operation by reading data set Dw.AtBats and load the Batter_ID key-values into the table. The data set option RENAME is used to make input variable Batter_ID consistent with the name of hash variable Player_ID defined on the next line. The data set option WHERE is used to select only the batters who have hit a triple. Note that the argument tag MULTIDATA is omitted. Hence, only the first record for each batter who has hit a triple is loaded into the table, and the rest are discarded. It fits the purpose of the task, at the same time reducing the number of hash items (and the hash memory footprint).

❹ Define the key portion of table Triple as a *simple numeric key* using a single hash variable Player_ID.

❺ For the task at hand, we do not need the data portion since we need only to discover if the key-values coming from Bizarro.Player_candidates are in the table. However, the hash object design requires that we *must* have the data portion with at least one hash variable in it. The optimal decision is to define it with a single variable to make the least impact on the hash memory footprint. As it turns out from testing on all major platforms, a single numeric variable takes up the least hash memory - in fact, even less than a $1 character variable. In this sense, using numeric variable Player_ID fits the purpose.

❻ Read a record from Bizarro.Player_candidates, bringing the key-value of Player_ID into the PDV.

❼ Perform the *Search* operation to find if this key-value is in table Triple. The CHECK method doing it is called *implicitly* because, thanks to the renaming done in ❸, the name of the searched PDV host variable Player_ID is the same as its hash table key counterpart. The CHECK method call is included in the subsetting IF statement, so when it returns a zero code (the key is found), the current PDV content is written as a record to output data set Triple.

❽ Alternatively, the subsetting IF condition can be simply reversed to output the player candidates who have not both batted and hit a triple. Note that it does not carry with it any performance implications like those that sometimes arise when the NOT IN clause is used with SQL.

Running the step above results in the following output for the player candidates from Bizarro.Player_candidates shown in the sample. We get the same output we would get if SQL were used instead:

Output 6.3 Simple Subsetting Results

Player_ID	Team_SK	First_Name	Last_Name	Position_Code	Bats	Throws
55438	193	Bobby	Miller	UT	R	L
37661	193	Bryan	Rivera	SP	S	R

The subsetting has resulted in the absence of two players from the output:

- Player_ID=52320 – because he has not batted.
- Player_ID=43400 – because, though he has batted, he has not hit a triple.

Note that the MULTIDATA:"Y" argument tag was not specified when the instance of hash table Triple was created. This is because, in the case of pure subsetting, we need to know only the unique values of Player_ID for players who have hit a triple. So, inserting multiple hash items for the same key-value into the table would be extraneous, and all the more so because doing that would only unnecessarily increase the hash memory footprint. Also, it does not matter from which record in Dw.AtBats each unique key-value kept in the table comes, and thus we can leave the choice up to the input process behind the argument tag DATASET (which, out of multiple input keys with the same value, automatically selects the first it encounters). However, as we will see soon later, keeping all hash items sharing the same key-value can be indispensable when the nature of a task calls for it.

6.2.3 Why a Hash Table and Not SQL?

Since we started with an SQL query to make the essence of the subsetting task more evident, a natural question arises: Why not just do it with SQL instead of using a hash table in the DATA step? The answer is two-fold:

- Searching a hash table explicitly *may* be faster than relying on the internal searching mechanism the SQL optimizer chooses when it decides on the path to execute the query.
- The DATA step language, being procedural, offers certain advantages over the declarative SQL in terms of programming flexibility. Combined with the flexibility of the hash object, it can be used to accomplish, in a natural manner and often in a single data pass, a number of things difficult or impossible to do via a single query (or making the query too convoluted).

From the perspective of the business users, the key advantage of the hash table approach is that multiple subsetting criteria can be easily added. The sport is well known for needing to identify how many times some rare event has happened (e.g., how many players on their first AtBat as professional, who are also pitchers, have hit a home run on the first pitch in the strike zone) in combination with other events..

6.2.4 Subsetting with a Twist: Adding a Simple Count

To illustrate the flexibility point made above, let us add a wrinkle to the task. To wit, suppose that, in addition to selecting the player candidates who have batted and hit a triple, we also want to know how many times each of these players has hit a triple. In order to do it using SQL, we would have to reformulate the query shown above from a subquery to a join and add a GROUP (BY) clause. This is because the SAS rendition of SQL cannot loop through the multiple rows per batter returned by the inner query in case the player has hit more than one triple. For example, the query could be recoded as follows:

Program 6.3 Chapter 6 Subsetting via SQL Join with Simple Count.sas

```
proc sql ;
  create table Triple_Count_SQL as
  select p.*, a.Count
  from   bizarro.Player_candidates p
     , (select Batter_ID, count(*) as Count
        from   Dw.AtBats
        where  Result = "Triple"
        group  Batter_ID) a
  where  a.Batter_ID = p.Player_ID
  and    p.Team_SK in (193)
  ;
quit ;
```

However, to achieve the same result using the hash table Triple in the DATA step explained above, we need not change existing code logic but need only to add the *Keynumerate* operation to count the hash items in every same-key item group with a matching Player_ID key (the additions to the DATA step above are shown in boldface):

Program 6.4 Chapter 6 Subsetting via Hash Table with Simple Count.sas

```
data Triple_Count_Hash ;
  if _n_ = 1 then do ;
    dcl hash triple (multidata:"Y"        ❶
                    , dataset:'Dw.AtBats(rename=(Batter_ID=Player_ID)
                          where=(Result="Triple"))'
                    ) ;
    triple.defineKey ("Player_ID") ;
  * triple.defineData ("Player_ID") ;     ❷
    triple.defineDone () ;
  end ;
  set Bizarro.Player_candidates ;
  where Team_SK in (193) ;
  if triple.check() = 0 ;        ❸
  do Count = 1 by 1 while (triple.do_over() = 0) ;      ❹
  end ;
run ;
```

Here is the summary of the changes:

❶   Adding the MULTIDATA:"Y" argument tag allows duplicate-key items and enables the *Keynumerate* operation to be performed later.

❷   This call can be omitted because, in its absence, all hash variables defined in the key portion (in this case, single variable Player_ID) are automatically defined in the data portion.

❸   Because of this subsetting IF statement, the DO loop below is executed only if an item group with Player_ID matching the current PDV value of host variable Player_ID is found in the table. Otherwise, program control moves to the top of the step, and the next record from Bizarro.Player_candidates is read.

❹   Perform the *Keynumerate* operation. The initial DO_OVER method call sets the item list for this key. It always succeeds since the preceding subsetting IF statement ensures that the key-value of host variable Player_ID is in the table. The initial and ensuing DO_OVER calls (which succeed if the player has hit more than one triple) enumerate the item group with the current key-value, thus incrementing the DO loop index variable Count at each iteration. (Note that using the index variable serves the dual purpose of both initializing and incrementing the counter. Thus, it is more compact and convenient than two separate statements to initialize the counter before the loop and then increment it inside it.) The loop terminates when the list set by the initial DO_OVER call has been exhausted because the final DO_OVER call returns a non-zero code, for there are no more items in the item group left to enumerate. After the DO loop is terminated, the implicit OUTPUT statement at the bottom of the step writes a record to output file Triple_Count with the current PDV values, including the aggregate value of Count.

Running the step above will result in adding variable Count compared to the output of simple subsetting:

Output 6.4 Result of Subsetting with Triple Hit Count Added

Player_ID	Team_SK	First_Name	Last_Name	Position_Code	Bats	Throws	Count
55438	193	Bobby	Miller	UT	R	L	2
37661	193	Bryan	Rivera	SP	S	R	1

It is worth noting that by presenting this technique, we have inadvertently waded into the territory outside of the pure *Search* operation domain. The reason is that the DO_OVER method implicitly performs the *Retrieve* operation (as part of any *Enumerate* operation) by extracting the value of data portion hash variable Player_ID into its PDV host variable counterpart. In the example above, it is a mere unavoidable side effect of calling the DO_OVER method, and it is not used for any purpose. However, it becomes the primary effect when it comes to combining data via a hash table. And to reinforce the point made above about the advantages of the procedural nature of the DATA step, metrics could be calculated for teams that have at least 1 player who has, for example, 5 triples.

Both our IT and business users appreciated this wrinkle because they both recognize that virtually every question that is asked that will require a program to be written, will have the inevitable add-on comment, "*Great, but can we also do X?*"

6.3 Combining Data

In the context of this discussion, *combining data* means adding *non-key* data (also termed *satellite* data) elements from data collection *B* to data collection *A* via a common key where the key-values in *A* and *B* match. Everything said above about the two principally different subsetting approaches applies equally to combining data. Namely, they are:

1. Sort *A* and *B* by their keys and run a sequential match algorithm either explicitly (e.g., using the MERGE statement) or implicitly (e.g., if the SQL optimizer chooses such a path).
2. Insert the key and the satellite information from *B* in a lookup table. Then read *A* one record at a time and search the table for the key-value it contains.

The reasons for choosing approach #2 over approach #1 are the same as with the subsetting. Moreover, the coding schemes used for the subsetting and combining tasks are very similar. The principal difference is that when we combine data, we need not merely to establish whether the key-values from *A* are *present* in *B*; but we also need to *extract the required satellite information* from *B* for every matching key-value. In other words, instead of performing the pure *Search* operation against the hash table, we have to perform the *Retrieve* operation instead. It can be executed in an explicit manner (e.g., by calling the FIND method) or implicitly as part of the *Keynumerate* or *Enumerate All* operations depending on the particular situation. A very common use case is adding dimension table variables to a fact table in a star schema data warehouse.

Again, for the benefit of our business users, the need to combine multiple data sources is something that virtually any program that calculates metrics will need to do. There will almost always be data or information that exists in other files that are needed for our analysis. For example, when the manager of a team considers what players to include in the line-up (i.e., as batters) he may want to evaluate how his players have done against the opposing pitcher for today's game. Given the Bizarro Ball data discussed in Chapter 5, that requires combining data from the Pitches and AtBats data using the appropriate keys to match the data.

6.3.1 Left / Right Joins

In order to see how it works, let us expand the specification for the subsetting task described above. Namely, suppose that for every player record for the Huskies team (Team_SK=193) from file Bizarro.Player_candidates, we want to not merely find from file Dw.AtBats who has batted and hit a triple but also:

- For the players who have batted and scored a triple, add variables Distance and Direction from file Dw.AtBats for all their triple runs. Thus, if a player has scored N triples, that player will have N records in the output.
- For the rest of the players, keep their records from Bizarro.Player_candidates in the output but leave variables Distance and Direction missing to indicate that these players have not *both* batted and scored a triple.

In terms of set operations, these requirements are equivalent to requesting a *left join* or *right join* of Bizarro.Player_candidates with the subset of Dw.AtBats where Result="Triple". They could be satisfied, for example, by running the following SQL query:

Program 6.5 Chapter 6 Left Join via SQL.sas

```
proc sql ;
  create table Triples_leftjoin_SQL as
  select p.*
       , b.date
       , b.Distance
       , b.Direction
  from   bizarro.Player_candidates (where = (Team_SK in (193))) P
  left join
         Dw.AtBats (where=(Result in ("Triple"))) B
  on     P.Player_ID = B.Batter_id
  ;
quit ;
```

The hash solution below attains the same goal as the SQL query above and is a subtle variation on the program used for data subsetting with a count shown earlier:

Program 6.6 Chapter 6 Left Join via Hash.sas

```
data Triples_leftjoin_Hash (drop = Count) ;
  if _n_ = 1 then do ;
    dcl hash triple (multidata:"Y"        ❶
                    , dataset:'Dw.AtBats(rename=(Batter_ID=Player_ID)
                         where=(Result="Triple"))'
                    ) ;
    triple.defineKey ("Player_ID") ;       ❷
    triple.defineData ("Distance", "Direction") ;   ❸
    triple.defineDone () ;
    if 0 then set Dw.AtBats (keep=Distance Direction) ;
  end ;
  set Bizarro.Player_candidates ;
  where Team_SK in (193) ;
  call missing (Distance, Direction) ;     ❹
```

```
do while (triple.do_over() = 0) ;     ❺
   Count = sum (Count, 1) ;
   output ;
  end ;
  if not Count then output ;     ❻
run ;
```

The callouts above mark the changes necessary to implement the required left join, as well as some other program lines that warrant explanation:

❶ Adding the MULTIDATA:"Y" argument tag allows duplicate-key items and enables the *Keynumerate* operation to be performed using the DO_OVER method downstream. In this situation, it is a must because we need to pair each player record on the side of Bizarro.Player_candidates with all of his triple-scoring records in Dw.AtBats. So, each triple-scoring record from Dw.AtBats for a given triple-scorer identified by Batter_ID must have a corresponding hash item in hash table Triple. Storing multiple hash items per key is what enables us to perform one-to-many (or, potentially, many-to-many) matching. Note that before table Triple is loaded via the DATASET argument tag, variable Batter_ID coming from file Dw.AtBats has been renamed as Player_ID for the sake of the simplicity of making the ensuing DO_OVER calls *implicit*.

❷ We are going to do the matching and retrieval by variable Player_ID; hence, it is defined here as the hash table key.

❸ We need to extract the values of satellite variables (Distance,Direction) from the hash table into the PDV for every Player_ID value from Bizarro.Player_candidates matching the key-values of Player_ID in hash table Triple. Therefore, it is necessary to define them as satellite variables in the data portion of the table. Otherwise, the *Keynumerate* operation would not be able to retrieve their hash values into the corresponding PDV host variables. The PDV host variables for hash variable Distance and Direction are created by the SET statement at the bottom of the IF-END block. Because of the IF 0 condition it reads no data from Dw.AtBats but lets the compiler place variables (Distance,Direction) in the compiler symbol table and the PDV at compile time. Thus later, at run time, it lets the DEFINEDONE method detect their names and other attributes, validating that the variables defined by the DEFINEDATA method already exist in the PDV.

❹ As we intend to perform an equivalent of a left join, satellite variables (Distance,Direction) in the output must be missing if no match between Player_ID from the current Bizarro.Player_candidates record and Player_ID in the hash table is found. Here we set (Distance,Direction) to missing values beforehand, so that in case the prior record had a match, their retained non-missing values would not persist. Also, testing either variable for a missing value after *Keynumerate* has done its job can optionally be used to judge whether it found a match or not.

❺ The DO_OVER method is called repeatedly in a DO WHILE loop. If the first call in the loop fails (i.e., returns a non-zero code), it means that no match for the current PDV value of Player_ID exists in table Triple. In such a case, program control exits the loop without executing the OUTPUT statement and incrementing auxiliary variable Count in its body due to the way the WHILE loop operates— and so, nothing is output, and the value of Count remains missing. Otherwise, the first DO_OVER call sets the item list for the matching key-value of Player_ID and *retrieves* the values of (Distance,Direction) from the first hash item in the item group with this value of Player_ID into their PDV host variables. Then this action repeats for every next item in the same-key item group until the group has no more items to enumerate. Thus, if the item group has N items with this key-value, exactly N records are output.

❻ If a record from Bizarro.Player_candidates has no match in table Triple, it has not been output yet. However, since we are doing a left join, it has to be. Variable Count indicates whether a match has been found or not. If its value is 1 or greater, it has been. Otherwise, its value remains missing, having been automatically reset at the top of the DATA step implied loop. Here we use this indicator to detect and output the records with no match in table Triples and missing values for (Distance,Direction) before the next record from Bizarro.Player_candidates is read.

For our sample players we have already used above, this program produces the following output:

Output 6.5 Left Hash Join of Bizarro.Player_candidates and Dw.AtBats

Direction	Distance	Player_ID	Team_SK	First_Name	Last_Name	Position_Code	Bats	Throws
.	.	43400	193	Andrew	Anderson	UT	L	L
17	348	55438	193	Bobby	Miller	UT	R	L
14	363	55438	193	Bobby	Miller	UT	R	L
.	.	52320	193	Jose	James	RP	R	R
17	380	37661	193	Bryan	Rivera	SP	S	R

Players 52320 and 43400 have missing values for (Distance,Direction) for the same reason they are absent from the subsetting output: Player 52320 has not batted, and player 43400, though having batted, has not scored a triple.

Note that the SQL query shown above generates the same data, except that the sequence of rows it outputs is different. This is because the hash solution keeps the relative sequence of rows in the output identical to that in the input files, while SQL has a mind of its own.

At this point our business users have expressed confusion about the difference between a left join and a right join. To our IT users, the answer to the question is both obvious and not terribly relevant. The easiest way to explain the distinction to the business users was to confirm that there really is not much of a difference. Combining the AtBats data with the Pitches data to augment the AtBats data with information from the Pitches data is a left join because we referenced the AtBats data first. If we are combining the AtBats data with the Pitches data to augment the Pitches data with information from the AtBats data, IT folks refer to that as a right join because the Pitches data is listed as the second data set (i.e., is the one on the right side). When they asked if a right join becomes a left join by just reversing the order the data sets are referenced, we replied *EXACTLY*. Adding that to IT folks, the distinction is important as it can have an impact on the efficiency of the combine operation.

6.3.2 Merging a Join with an Aggregate

Note that in the above example, variable Count played only an auxiliary role, and so it was dropped. However, if it should be desirable to *merge* its aggregate value with every output record of the player to which it pertains, it can be done in the very same DATA step by executing the DO WHILE loop twice. To achieve that, we need only to (a) keep variable Count and (b) replace the program between the callouts ❹ and ❻, inclusively, with the following construct:

Program 6.7 Chapter 6 Variation on Left Join via Hash with Count.sas

```
do while (triple.do_over() = 0) ;
  Count = sum (Count, 1) ;
end ;
call missing (Distance, Direction) ;
```

```
do while (triple.do_over() = 0) ;
  output ;
end ;
if not Count then output ;
```

Essentially, in order to merge the aggregate value of Count back with its player's record, we execute the *Keynumerate* operation against the same-key item group twice: First, to increment Count, and second, to output the needed records with Count already summed up. If in the current Bizarro.Player_candidates input record the tuple Player_ID has no match in hash table Triple, the output value of Count will remain missing. So, for our sample players, the "merged" output would look like so:

Output 6.6 Left Join Merged with Aggregate Triple Hit Count

Direction	Distance	Player_ID	Team_SK	First_Name	Last_Name	Position_Code	Bats	Throws	Count
.	.	43400	193	Andrew	Anderson	UT	L	L	.
17	348	55438	193	Bobby	Miller	UT	R	L	2
14	363	55438	193	Bobby	Miller	UT	R	L	2
.	.	52320	193	Jose	James	RP	R	R	.
17	380	37661	193	Bryan	Rivera	SP	S	R	1

An alert reader will notice that a simple count is not the only aggregate that can be merged with the join output. Indeed, nothing above precludes us from merging it with any other aggregate derived from the hash data portion values being retrieved. For example, if we wanted to also augment each output record for the triple-scorers with the values of total and average Distance, we could achieve it by recoding the *first* DO WHILE loop as follows:

Program 6.8 Chapter 6 Variation on Left Join via Hash with Multiple Aggregates.sas

```
do while (triple.do_over() = 0) ;
  Count = sum (Count, 1) ;
  Sum = sum (Sum, Distance) ;
end ;
Average = divide (Sum, Count) ;
```

As a result, the output would also contain the newly computed aggregates (note that some columns from Bizarro.Player_candidates are omitted to make the display fit the page):

Output 6.7 Left Join Merged with Multiple Aggregates

Direction	Distance	Player_ID	Team_SK	First_Name	Last_Name	Count	Sum	Average
.	.	43400	193	Andrew	Anderson	.	.	.
17	348	55438	193	Bobby	Miller	2	711	355.5
14	363	55438	193	Bobby	Miller	2	711	355.5
.	.	52320	193	Jose	James	.	.	.
17	380	37661	193	Bryan	Rivera	1	380	380.0

What makes this on-the-fly aggregation possible is the ability of the *Keynumerate* operation to visit every hash item in the item group with a given key and implicitly perform the *Retrieve* operation on each item along the way. Many other examples of using this handy functionality are included elsewhere in the book.

Given that our business users are most interested in calculating metrics which, by definition, require aggregation, they were particularly interested in this functionality.

6.3.3 Inner Joins

At this point, our business users threw up their hands because they did not understand why there was a need for another adjective to describe how to combine data. Once we explained that an inner join simply meant we kept only rows where the keys existed in both data sets, they understood the need for such a requirement and simply decided to let us move on with satisfying the requests of the IT users.

Given that we want to keep only the matching records, we need to perform an equivalent of an inner join. For our sample players above, it would mean that the records with Player_ID in (43400,52320) must not appear in the output. To make it happen, we do not need to add anything to left join code above. Indeed, the line:

```
* if not Count then output ;
```

was coded specifically to output non-matching records. So, all we need to do in order to transform the program from the left join into the inner join is to *omit* this line of code (or comment it out, as shown above). All other properties of the program are retained. Specifically, it preserves its one-to-many joining functionality (or many-to-many in case the "left" file contains records with duplicate keys), as well as the ability to merge aggregate values into the output. Hence, if in the last left join example (with the Distance aggregates added) the conditional OUTPUT statement above were omitted, the output for our sample players would appear as expected from the inner join, i.e., as follows:

Output 6.8 Inner Join with Multiple Aggregates

Direction	Distance	Player_ID	Team_SK	First_Name	Last_Name	Count	Sum	Average
17	348	55438	193	Bobby	Miller	2	711	355.5
14	363	55438	193	Bobby	Miller	2	711	355.5
17	380	37661	193	Bryan	Rivera	1	380	380.0

6.3.4 DO_OVER Versus FIND + FIND_NEXT

The DO_OVER method, used in the examples above to perform the *Keynumerate* operation, is quite handy and concise. However, it is available only starting with SAS 9.4. If a piece of code is supposed to work under SAS 9.2 and 9.3 as well as 9.4 and higher, there is a remedy already expounded on in the chapter dedicated to the hash table operations. Namely, the construct:

```
do while (triple.do_over() = 0) ;
  <code inside do-while loop>
end ;
```

can be replaced with the following programming structure (let us call it *Style 1*):

```
do _iorc_ = triple.find() BY 0 WHILE (_iorc_ EQ 0) ;
  <code inside do-WHILE loop>
  _iorc_ = triple.find_next() ;
end ;
```

The ***BY 0*** construct ensures that the loop can iterate more than once. Also note that using automatic numeric variable _IORC_ instead of RC to capture method codes is a convenient technique since: (a) the variable is freely available in the DATA step, (b) it is not used here for any other purpose (such as capturing SAS index search results), and (c) it is automatically dropped.

Alternatively, a more concise structure can be used that requires no return code variable at all (let us call it *Style 2*):

```
if triple.find() = 0 then do UNTIL (triple.find_next() NE 0) ;
  <code inside do-UNTIL loop>
end ;
```

Admittedly, using the combination of the FIND and FIND_NEXT methods is more verbose compared to the DO_OVER method. Yet, it provides the same exact functionality: The aggregation loop with the DO_OVER method shown above would look as follows if DO_OVER were replaced by the FIND + FIND_NEXT combination (using *Style 2*, for example):

```
if triple.find() = 0 then do UNTIL (triple.find_next NE 0) ;
  Count = sum (Count, 1) ;
  TotalDistance = sum (TotalDistance, Distance) ;
end ;
```

From the standpoint of the hash table operations, the difference between the DO_OVER and FIND+FIND_NEXT is only that FIND performs the direct *Retrieve* operation, while DO_OVER does it indirectly. Yet, functionally, both techniques do exactly the same. If a key-value (simple or composite) accepted by the methods exists in the table, both set the item list for the given key-value and perform *Retrieve* on the first item in the list. Otherwise, the loop is terminated without iterating once. The initial FIND is necessary because it is a prerequisite to calling FIND_NEXT - without calling FIND first, the item list for the same-key item group in question will not be set because FIND_NEXT can be called successfully only when the list is already set. Thus, the initial DO_OVER call does exactly what the priming FIND does; and the subsequent FIND_ NEXT calls do exactly what the subsequent DO_OVER calls do.

6.3.5 Unique-Key Joins

Note that all the data-combining examples above were presented with the assumption that the data to be loaded into the hash table for lookup may contain more than one item per key. Accordingly, the argument tag MULTIDATA:"Y" was used to accommodate multiple items per key-value (and thus enable the *Keynumerate* operation to retrieve the data from each same-key item). It often happens, though, that the keys in the data to be loaded is intrinsically unique; and so they will remain unique in the hash table after it is loaded. The question then arises: Do we have to change anything in the joining code schemes discussed above in order to account for the fact that the keys in the table are now unique?

The answer is "No", for the *Keynumerate* operation as coded above handles this case automatically. For a given key-value, the DO loop performing the operation executes its body exactly as many times as there are items for this key-value in the table. If the keys in the table are unique, the same-key item group has only one item, and so the body of the loop is executed only once - just as needed. Let us take another look at both variants of the DO loop which executes the operation:

```
do while (triple.do_over() = 0) ;
  Count = sum (Count, 1) ;
  output ;
end ;
```

and

```
do _iorc_ = triple.find() by 0 while (_iorc_ = 0) ;
  Count = sum (Count, 1) ;
  output ;
```

```
  _iorc_ = triple.find_next() ;
end ;
```

Let us see what happens in two possible cases: (1) the current key-values in the PDV have no match in the table and (2) they do.

1. The current PDV value of Player_ID has no match in the hash table:

   ○ In the variant with DO_OVER, the very first DO_OVER call returns a non-zero code making the WHILE condition false. Hence, program control exits the loop immediately without executing the statements in its body even once. The value of Count remains missing, and no records are output.

   ○ In the variant with the FIND+FIND_NEXT combination, exactly the same happens, except that it is the failure of the FIND method that triggers the immediate termination of the loop.

2. The current PDV value of Player_ID does have a match in the hash table:

   ○ In the DO_OVER variant, the first DO_OVER call is successful, and the body of the loop is executed, adding 1 to Count and outputting the record. Since the table has no more items for the current key, the next call to DO_OVER at the top of the loop fails and program control exits the loop. Hence, for the only item in the same-key item group, the body of the loop is executed only once.

   ○ In the variant with FIND+FIND_NEXT, the same happens, except that it is the successful FIND call that lets the loop iterate for the first time and process the only available item, and it is the non-zero code returned by the ensuing FIND_NEXT that prevents the loop from iterating further.

Thus, the *Keynumerate* joining scheme works perfectly fine regardless of whether the hash keys have duplicates or not. The only reason to deviate from it is to utilize the fact that the keys are unique to make the program simpler and terser by replacing the *Keynumerate* operation with the *Retrieve* operation. For example, in the case of the *left join*, we can omit the MULTIDATA argument tag altogether and replace the block of code:

```
call missing (Distance, Direction) ;
do while (triple.do_over() = 0) ;
  Count = sum (Count, 1) ;
  output ;
end ;
if not Count then output ;
```

with mere:

```
call missing (Distance, Direction) ;
_iorc_ = triple.find() ;
```

The assigned call to FIND is used to prevent errors appearing in the SAS log in case the value of Player_ID present in the PDV at the time of the FIND call has no match in the hash table. The OUTPUT statement is omitted because at the bottom of the DATA step it is implied. Output occurs regardless of whether FIND succeeds or fails. If it fails (i.e., there is no match), the record is output with the (Distance,Direction) pair having missing values, just as required by in a left join; otherwise, they will have the values retrieved from the hash table for the current PDV key-values.

If we are to perform the inner join rather than left join, this is further reduced to:

```
call missing (Distance, Direction) ;
if triple.find() = 0 ;
```

In this case, the subsetting IF statement outputs the record only if we have a match.

For a complete example of joining with a file where the key is unique, suppose that we want to know which player candidates in file Bizarro.Player_candidates from team Huskies (Team_SK=193) have actually pitched and in how many unique games each. In order to demonstrate this we need to use a data set created in Chapter 7. The data set Dw.Players_positions_played is a subset of bizarro.Player_candidates and includes a numeric variable Pitcher which is how many games the player has appeared as a pitcher.

Below, is a glimpse of this data set for our four sample players (only the variables in question are shown). Variable Pitcher indicates in how many unique games the player has pitched; if he has not pitched, its value is missing. The key-values of Player_ID are unique throughout the file.

Output 6.9 Sample from Data Set Dw.Players_positions_played

Player_ID	Pitcher
37661	39
43400	.
52320	180
55438	.

Let us also recall that for our sample of player candidates from team Huskies (Team_SK=193), the corresponding records look as follows:

Output 6.10 Sample from Data Set Bizarro.Player_candidates (Team Huskies)

Player_ID	Team_SK	First_Name	Last_Name	Position_Code	Bats	Throws
43400	193	Andrew	Anderson	UT	L	L
55438	193	Bobby	Miller	UT	R	L
52320	193	Jose	James	RP	R	R
37661	193	Bryan	Rivera	SP	S	R

Therefore, we can code:

Program 6.9 Chapter 6 Unique-Key Left or Inner Join via Hash.sas

```
data Pitched ;
  if _n_ = 1 then do ;
    if 0 then set dw.Players_positions_played (keep=Player_ID Pitcher) ;
    dcl hash pitch (dataset: "dw.Players_positions_played (where=(Pitcher))") ; ❶
    pitch.defineKey ("Player_ID") ;
    pitch.defineData ("Pitcher") ;
    pitch.defineDone () ;
  end ;
  set bizarro.Player_candidates ;
  where Team_SK in (193) ;
  call missing (Pitcher) ;   ❷
```

```
  _iorc_ = pitch.find() ; /*Left  join*/    ❸
* if _iorc_ = 0 ;           /*Inner join*/  ❹
* if pitch.find() = 0 ;    /*Inner join*/  ❺
run ;
```

This example warrants a few notes:

❶   The MULTIDATA:"Y" argument tag is omitted since there are no duplicate key-values in the input.

   The SET statement at the top of the IF-DO-END block is used to create the PDV host variables for hash variables Player_ID and Pitcher. Doing parameter type matching in this manner ensures that the host and the hash variables inherit their attributes from the like-named variables in file Dw.Players_positions_played. Also, doing it before the MISSING routine is called downstream guarantees that it will populate variable Pitcher with a missing value of the correct data type.

   Note that (Pitcher) is used in the WHERE clause as a *Boolean* expression. It evaluates true if the value of Pitcher is neither missing nor zero; otherwise, it evaluates false.

❷   Ensure that in the case of left join, the output values of the added satellite variables (here, just variable Pitcher) have missing values if the current PDV value of Player_ID has no match in hash table Pitch.

❸   The FIND method is called to perform the *Retrieve* operation. It is called *assigned* to prevent log errors if there is no match for the current PDV value of Player_ID in the table. If the next line is omitted (or remains commented as shown), the operation results in the left join since the output is created for every record from Bizarro.Player_candidates regardless of whether its value of Player_ID has a match in table Pitch or not.

❹   If this line of code is uncommented, the records for which there is no match are not written to the output. In other words, it results in the inner join.

❺   If the inner join is required, both lines ❸ and ❹ can be replaced with this single line. In this case, if the FIND method fails, it does not generate errors in the log because it is called as part of a conditional statement.

In the case of a left join, this is the output we get from the program for our sample:

Output 6.11 Results of Unique-Key Left Join

Player_ID	Pitcher	Team_SK	First_Name	Last_Name	Position_Code	Bats	Throws
43400	.	193	Andrew	Anderson	UT	L	L
55438	.	193	Bobby	Miller	UT	R	L
52320	180	193	Jose	James	RP	R	R
37661	39	193	Bryan	Rivera	SP	S	R

If the inner (equi-) join is opted for instead, the output will have only the rows with the non-missing values for Pitcher:

Output 6.12 Results of Unique-Key Inner (Equi-) Join

Player_ID	Pitcher	Team_SK	First_Name	Last_Name	Position_Code	Bats	Throws
52320	180	193	Jose	James	RP	R	R
37661	39	193	Bryan	Rivera	SP	S	R

6.4 Splitting Data

At times, we need to accomplish a task opposite to combining data, i.e., to *split* a data set in a number of data sets in a specific manner. For instance, suppose that we have a data set with N distinct key-values of some ID variable (or a combination of variables) and need to split it into N separate data sets whose names are identified by the corresponding key-values.

For the benefit of our business users, we decided to point out that just one of the benefits of splitting data is that it can make the calculation of certain metrics easier. We provided several examples of such use cases that we felt they could relate to: treating starting pitchers differently than relievers in terms of what pitching related metrics are of interest; treating AtBats for pinch hitters differently from starters. This point prompted the IT users to consider the possibility of different kinds of metrics to include in the project once the Proof of Concept effort is completed.

Returning to the focus of satisfying the curiosity of the IT users, as a simple example, let us look at a subset of file Bizarro.Teams where only a few teams from each league are shown:

Output 6.13 Data Set Bizarro.Teams - Sorted Sample

Team_SK	Team_Name	League_SK
317	Mountaineers	1
342	Gladiators	1
228	Storm	2
165	Saints	2
136	Pirates	2

As outlined above, the task is to split the file into two separate files named League_1 and League_2 and containing the records with League_SK=1 and League_SK=2, correspondingly. It is surely reasonable to ask *why* such a thing may be needed. Indeed, whatever has to be done against the separate files can be done against the original file using, for example, BY processing logic. The answer is that it is often demanded by business end users. For example, they may want to create a separate sheet in a workbook for each key-value present in the original file.

The traditional approach to this task seems to be very simple:

Program 6.10 Chapter 6 Splitting Data - Hard Coded.sas

```
data League_1 League_2 ;
  set Bizarro.Teams ;
  select (League_SK) ;
    when (1) output League_1 ;
```

```
    when (2) output League_2 ;
    otherwise ;
  end ;
run ;
```

The advantages of this method are: (a) apparent simplicity, and (b) no need to have the file sorted or grouped by the key variable(s), in this case by League_SK. However, there are a few reefs lurking beneath this quiet surface:

- We have to discover all the distinct ID values beforehand.
- The output data set names in the DATA and SELECT statement have to be *typed in* - in other words, *hard-coded* - with that knowledge in mind.
- Hard-coding can be avoided by pre-processing the file to find out what the distinct ID values are and use the result to construct the DATA and SELECT statements programmatically by using a macro or other means of generating code. It means, however, that we need to make two passes through the input file.
- This is true even if the input file is pre-sorted by the ID variable. Though in such a case we can use BY processing to write the records from each BY group to its own file named after the respective key-values, these files must be listed in the DATA statement - they cannot be created at run time on the fly.

With the hash object, however, it is different: It has its own I/O facilities *independent* from those of the DATA step proper; and thus it can create, name, write, and close output data sets at run time. Let us see what hash table operations we need to make it happen and how.

6.4.1 Hash Data Split - Sorted Input

When the input file to be split is sorted or grouped by the ID variable (which is the case with file Bizarro.Teams sorted by variable League_SK), the operational plan is simple:

On the first iteration of the DATA step implied loop (first DATA step execution), use the *Create* operation to create a hash table instance.

1. Process the input file one BY group at a time.
2. For each record in the current BY group, use the *Insert* operation to add its PDV values to the hash table as a new item.
3. After the last record in the current BY group has been read, use the *Output* operation to: (a) create a new output file with the current value of the ID variable (converted to a character string if it is numeric) as part of the file's name, (b) open it, (c) write the data portion content from the hash table to it, and (d) close it. Then use the *Clear* operation to purge all items from the hash table without deleting the table itself.
4. Read the next BY group, i.e., go to #1.

The most convenient programming structure to process BY groups one at a time is the DoW loop, as it naturally isolates the programming actions taken before, during, and after each BY group. Thus, we can code:

Program 6.11 Chapter 6 Splitting Sorted Data via Hash.sas

```
data _null_ ;       ❶
  if _n_ = 1 then do ;      ❷
    dcl hash h (multidata:"Y") ;     ❸
    h.defineKey ("_n_") ;     ❹
    h.defineData ("League_SK", "Team_SK", "Team_Name") ;      ❺
    h.defineDone () ;
  end ;
  do until (last.League_SK) ;     ❻
    set bizarro.Teams ;
    by League_SK ;
    h.add() ;      ❼
  end ;
  h.output (dataset: catx ("_", "work.League", League_SK)) ; ❽
  h.clear() ;      ❾
run ;
```

This program is quite instructive in spite of its brevity; so let us take a closer look:

❶ Since all output in the step will be handled by the hash object, we do not need to list any output data set names in the DATA statement, which is why it is left with the _NULL_ specification.

❷ This IF-DO-END block contains statements and method calls performing the *Create* operation. Note that this is done only at the first execution of the DATA step since we intend to keep table H created here across all iterations of the DATA step implied loop- or, to put it another way, across all BY groups we are going to process.

❸ Coding MULTIDATA:"Y" allows multiple items with the same key-value in the hash table. As we will see later, it is not mandatory. However, it comes in handy because we can have the same key-value for every item inserted into the table from a given BY group. As we already know, for the hash items in an item group sharing the same key, their logical order in the table is the same in which they are received. Thus, when the data loaded from each BY group is written from the table to its respective output file, the relative order of the records in the input will be automatically replicated in the output without the need to program for it.

❹ The program plan does not require the table to be keyed - it merely intends to use the hash table as a *queue*. However, since no hash table can exist without a key, we need to pick one. _N_ is a good choice because: (a) it is freely available, shortest possible, and automatically dropped; and (b) owing to the DoW loop, the value of _N_, after it is incremented at the top of the step in each of its executions, persists for the duration of each corresponding BY group. Thus, it persists for the duration of the first BY group _N_=1, for the second - _N_=2, and so on. So, using _N_ in this manner serves to implement the above idea of having all items related to a given BY group keyed by the same value.

Note that as an equally good alternative, automatic variable _IORC_ could be used instead of _N_, in which case the key-value for *any* BY group would be the same, namely, _IORC_=0.

❺ Include the variables to be written to the output "split" files in the data portion of the table. The like-named PDV host variables for them will be created when the compiler processes the ensuing SET statement referencing data set Bizarro.Teams whose descriptor contains them. Thus, no extra code is needed for the parameter type matching.

❻ Launch the DoW loop (comprising the lines of code between the DO and END statements, inclusively). It will iterate through every record in the current BY group and terminate after the last record in the group has been read. On this last record, the SET statement in the presence of the BY statement sets last.League_SK=0, and so program control exits the loop at its bottom because of the UNTIL condition.

❼ Call the ADD method to perform the *Insert* operation. Because of MULTIDATA:"Y", no unduplication occurs, so every input record will have its item counterpart in table H. Thus, inserting an item with a duplicate key-value cannot cause the ADD call to fail. Also, since the key-value of hash variable _N_ stays the same for the entire current BY group, the items inserted into the table end up there in the order they are received, as explained in ❸.

❽ At this point in the control flow, the data from the BY group processed in the current execution of the DATA step has been loaded into table H. Now the *Output* operation is performed by calling the OUTPUT method. The name of the file to which the data should be written is constructed on the fly by the expression supplied to the argument tag DATASET. The expression uses the CATX function to incorporate the current value of the ID variable (i.e., League_SK) in the name of the output data set. If a file with this name exists, it is deleted and re-created (i.e., over-written); if not, a new file is created. Then the method call causes it to be opened; after the data from all hash items currently in the table has been written to it, it is closed.

Note that this file is locked only during the time the *Output* operation is being performed (i.e., while records are still being written to it). After the method call has closed, the file it is no longer locked and can be used (for example, opened, browsed, edited, etc.), even though the DATA step still keeps running. This is in sharp contrast with the situation where the name of an output file is listed in the DATA statement and written into by the OUTPUT statement (explicit or implicit) because in this case, the file is not closed until the DATA step has finished running.

❾ Perform the *Clear* operation by calling the CLEAR method. Reason: The next iteration of the DATA step implied loop is going to process the next BY group. Since all our work with the current BY group has now been finished and its content written out to the corresponding output file, we no longer need the data from the current BY group in the table. Instead, we need to purge the items currently in the table without deleting the table, so that we will have the table empty and ready to receive the data from the next BY group.

Running the DATA step shown above generates the following notes in the SAS log:

```
NOTE: The data set WORK.LEAGUE_1 has 16 observations and 3 variables.
NOTE: The data set WORK.LEAGUE_2 has 16 observations and 3 variables.
NOTE: There were 32 observations read from the data set BIZARRO.TEAMS.
```

And the output content for the two files League_1 and League_2 looks as follows (only the first 2 records for League_SK=1 and the first 3 records for League_SK=2 are displayed):

Output 6.14 Result of Splitting File Bizarro.Teams for League_SK=1

League_SK	Team_SK	Team_Name
1	317	Mountaineers
1	342	Gladiators

Output 6.15 Result of Splitting File Bizarro.Teams for League_SK=2

League_SK	Team_SK	Team_Name
2	228	Storm
2	165	Saints
2	136	Pirates

Note that the relative sequence of records coming from Bizarro.Teams has been maintained intact.

Let us now consider a couple of code variations. First, we have mentioned earlier that coding MULTIDATA:"Y" is not mandatory to achieve the goal. (In fact, if you are running SAS 9.1 - which, against all odds, still may happen - the MULTIDATA argument tag is not even available.) If it is not used, it must be ensured that for each record read from a given BY group, a corresponding item is inserted into table H. But since without MULTIDATA:"Y" duplicate-key items are not allowed, we must guarantee that for each item we attempt to insert into the table, the key-value is *different*. This is easy to achieve by recoding the program as follows (the changes are shown in bold):

Program 6.12 Chapter 6 Splitting Sorted Data via Hash with Unique Key.sas
```
data _null_ ;
  if _n_ = 1 then do ;
    dcl hash h (/*multidata:"Y"*/ ordered:"A") ;     ❶
    h.defineKey ("unique_key") ;
    h.defineData ("League_SK", "Team_SK", "Team_Name") ;
    h.defineDone () ;
  end ;
  do unique_key = 1 by 1 until (last.League_SK) ;     ❷
    set bizarro.Teams ;
    by League_SK ;
    h.add() ;
  end ;
  h.output (dataset: catx ("_", "work.League", League_SK)) ;
  h.clear() ;
run ;
```

❶ The argument tag ORDERED:"A" guarantees that the items which are now uniquely keyed are inserted into table H in the order they are received from the input. In fact, coding this way programmatically *enforces* that the relative input record sequence is replicated in the output irrespective of the internal hash table order maintained by default in presence of the argument tag MULTIDATA:"Y".

❷ Incrementing unique_key up by 1 for every new record in a given BY group ensures that every item in the table is keyed by its own unique key-value.

There is another interesting variation on the same theme related to the *Clear* operation. If you are perchance still running SAS 9.0 or 9.1, the CLEAR method is not available. One alternative is to use the combination of the *Delete* (table) and *Create* operations instead to delete the entire instance of H (and its content with it) after each BY group has been processed and re-create it anew before the next BY group. Or, in the SAS language:

Program 6.13 Chapter 6 Splitting Sorted Data via Hash with DELETE Method.sas

```
data _null_ ;
* if _n_ = 1 then do ;
    /*Create operation code block*/    ❶
    dcl hash h (ordered:"A") ;
    h.defineKey ("unique_key") ;
    h.defineData ("League_SK", "Team_SK", "Team_Name") ;
    h.defineDone () ;
* end ;
  do unique_key = 1 by 1 until (last.League_SK) ;
    set bizarro.Teams ;
    by League_SK ;
    h.add() ;
  end ;
  h.output (dataset: catx ("_", "work.League", League_SK)) ;
* h.clear() ;
  /*Delete the instance of object H*/
  h.delete() ;    ❷
                  ❸
run ;
```

Here is how it works:

❶ Now that the *Create* operation code block is executed unconditionally, the operation is performed every time program control passes through the block- i.e., for each new value of _N_, thus causing the instance of object H to be created before each respective BY group is processed.

❷ After the BY group is processed and the respective "split" file is written out, the DELETE method call erases the instance of H. Note that this call is optional (see ❸ below) and can be omitted.

❸ Program control from this point loops back to ❶ - the top of the DATA step for its next execution. The instance of H just erased at the bottom of the step is re-created.

This variant will work with any SAS version starting with 9.0. Be mindful, though, that the processing cost of doing so is much higher (by an order of magnitude or so) than merely clearing the content of the persisting table. If you have but a few BY groups in the input file, you are not likely to notice any run-time difference. Yet, if they are very numerous, it may render the program annoyingly slow. So, with SAS 9.2 or higher it is much more sensible to let the hash object persist and use the *Clear* operation to purge it before every BY group.

6.4.2 Hash Data Split - Unsorted Input

If the input file is not sorted or grouped by the ID variable, using BY processing as demonstrated above is not an option. Suppose that input file Bizarro.Teams was not intrinsically sorted by League_SK and our few sample records looked as follows:

Output 6.16 Data Set Bizarro.Teams - Unsorted Sample

Team_SK	Team_Name	League_SK
165	Saints	2
342	Gladiators	1
228	Storm	2
317	Mountaineers	1
165	Saints	2

It raises the question: Is it still possible to do the data split *in a single pass* through the input file? Actually, the answer is "Yes", and it is based on the ability of the SAS hash object to store data pointing to hash object instances. Let us see what a logical plan for such a program could be:

1. Read the file.
2. Every time a *new* value of League_SK is encountered, fully define (i.e., with keys and data) and create a new instance of the hash object H for *this* value of League_SK. Store the value of League_SK, along with *some* variable *pointing* to (or *identifying*) the related instance, in *some* suitable data structure, so that later on it can be accessed to identify this instance using League_SK as a key.
3. For every record, identify the hash instance related to the current PDV value of League_SK. Then perform the *Insert* operation to add an item with the data from this record to the hash table associated with this value of League_SK.
4. After the file has been read, *go through the stored instances of* H one at a time. For each, perform the *Output* operation to output the "split" file named after the value of League_SK identifying its own instance of H.

There is only one little problem with this plan: In order to "go through the stored instances of H", we need to save the pointers identifying these separate instances *as data* in some *enumerable* data structure keyed by League_SK. A regular SAS array is an enumerable data structure. However, it is ill suited for the purpose since it can house only scalar data, while a pointer to an object instance represents non-scalar data (of type hash object). However, a hash table (which, if you will recall, is also termed an *associative* array) is a perfect container for data of this type and so, just as any hash table, can be keyed by League_SK.

We are wading into the territory of what is termed *Hash of Hashes (HoH)* treated in detail in the corresponding chapters later in the book. Therefore, here we will limit the discussion to a working program annotated only to show how it corresponds to the plan devised above:

Program 6.14 Chapter 6 Splitting Unsorted Data via Hash of Hashes.sas

```
data _null_ ;
  if _n_ = 1 then do ;
    dcl hash h ;        ❶
    dcl hash hoh() ;    ❷
```

```
      hoh.defineKey ("League_SK") ;
      hoh.defineData ("h", "League_SK") ;
      hoh.defineDone () ;
   end ;
   set bizarro.Teams end = lr ;     ❸
   if hoh.find() ne 0 then do ;     ❹
     h = _new_ hash (multidata:"Y") ;     ❺
     h.defineKey ("_iorc_") ;     ❻
     h.defineData ("League_SK", "Team_SK", "Team_Name") ;
     h.defineDone () ;
     hoh.add() ;     ❼
   end ;
   h.add() ;     ❽
   if lr ;
   dcl hiter ihoh ("hoh") ;     ❾
   do while (ihoh.next() = 0) ;
      h.output (dataset: catx ("_", "work.League", League_SK)) ;     ❿
   end ;
run ;
```

❶ Declare a hash object named H. This has the effect of creating a non-scalar (of type hash object) PDV variable H. At any point in the program, its value identifies the hash object instance associated with it.

❷ This statement and the following three method calls perform the *Create* operation to create hash table HoH. Its key is hash variable League_SK; and the data portion includes hash variables H and scalar variable League_SK. The host variable for H is created by the statement ❶; and the host variable for League_SK - by the SET statement downstream.

❸ Read the next record from the input file.

❹ The FIND method searches table HoH for the current PDV value of League_SK. If it is there, it retrieves the related value of hash variable H (i.e., the pointer to the related instance of H) into the host variable H.

❺ Otherwise, a new distinct value for PDV host variable H is created by the _NEW_ operator. The rest of the calls in this IF-DO-END block perform the *Create* operation to fully define and create a new instance of H associated with this new value.

❻ Note that all instances of hash object H are defined with key variable _IORC_ which will have the same default value _IORC_=0 for all items in every instance of H throughout the program. Thus, _IORC_ plays the role of a convenient key portion placeholder in the situation where, because of MULTIDATA:"Y", the key portion has no practical function for the instances of H used as simple FIFO (First In - First Out) queues.

❼ The ADD method is called to perform the *Insert* operation, adding the current PDV value of League_SK and the related value of H to table HoH as a new item with the new distinct key-value of League_SK. Since ADD is called only if this value is not already in the table, the method is guaranteed to succeed and thus can be called unassigned.

❽   Perform the Insert operation to add an item to the instance of H linked to the current PDV value of League_SK. The instance is identified by the current value of PDV host variable H. At the time of this ADD call, this variable has been valued in one of two ways:

(a) If the current PDV value of League_SK is new, i.e., it was not in table HoH when FIND was called at ❹, then H also has a new value assigned by the _NEW_ operator and identifying the new instance of H the operator has just created.

(b) Otherwise, it is the hash value of H just retrieved from HoH by the FIND call using the PDV value of League_SK as its key-value.

Either way, now the value of H identifies the specific instance of H linked to the current PDV value of League_SK, and thus this is the instance into which the H.ADD() call inserts a new item with the values of League_SK, Team_SK, and Team_Name from the current record.

❾   Now that all input records have been processed (which occurs when LR=1 is set), we need to scan through the instances of H uniquely related to the distinct values of League_SK and, for each, output a "split" file with the name identified by its own value of League_SK and the data content of the corresponding hash table instance. Therefore, we need to perform the *Enumerate All* operation against table HoH. Here, it is made possible by creating hash iterator iHoH explicitly linked to table HoH.

The enumeration itself is performed by the DO WHILE loop repeatedly calling the hash iterator method NEXT until it returns a non-zero code when no more items are left to enumerate. In each iteration of the loop, NEXT retrieves the interrelated hash values of H and League_SK from the data portion of the next item of table HoH into their respective PDV host variables. Thus, for each enumerated item, when the PDV host variable H is overwritten with the value of its hash counterpart, its value now points to the specific instance of H on which to operate; and it is correctly paired in the PDV with the value of League_SK to which it is related.

Note that hash variable League_SK is included in the data portion of HoH (in addition to being included in the key portion) on purpose. Otherwise, the NEXT method would not be able to retrieve its values from the table into the PDV host variable League_SK, as only the data portion hash variables are retrievable.

❿   The OUTPUT method reads the data from the hash instance identified by the PDV value of H and writes its data portion content into the "split" file identified by the PDV value of League_SK to which the instance is linked.

This program generates the same output and log notes as the prior example where BY processing is used. It is evidently more dynamic in the sense that it does not require the input file being split to be sorted or grouped by the ID variable. However, it has two shortcomings. First, it needs more memory, as eventually all the input data ends up loaded into memory before the "split" files are output. Second, being based on the "hash of hashes" concept, it is admittedly more complex. However, we feel that it can serve as a good *propaedeutic* teaser for the chapters later in the book where the "hash of hashes" techniques are explored in much more detail and put to better use. Though you will see the notions *outlined* above reiterated, perhaps at different angles, it may be a good thing: As the Latin proverb says, "*Repetitio est mater studiorum*".

6.5 Ordering and Grouping Data

The ORDER operation packaged with the hash object via the argument tag ORDERED is what makes it a handy data-ordering tool and has a number of useful applications. In addition, the intrinsic hash object structure working behind the scenes when the argument tag MULTIDATA:"Y" is specified makes it possible to *group* data without sorting it, thus avoiding the extra expense of the latter.

At this point our business users did not need an explanation of why this was an important capability. Grouping the data to calculate metrics for each batter (just one example) made perfect sense to them- as did the idea of ordering the result sets by the batter's team and/or name.

Let us say from the outset that the data sorting capability of the hash object cannot and does not replace the functionality, efficiency, and capacity of the procedures (such as SORT and SQL) specifically designed for the purpose, particularly for sorting massive data. Moreover, the hash object has no provision (at least as of SAS 9.4) for sorting in both ascending and descending order by different key components within a composite key: All that the hash object can do is sort either descending or ascending by the entire key as a whole.

However, there are many use cases where the ability of the hash object to sort or group comes in very handy. For example (the list does not pretend to be exhaustive):

- The logic of a DATA step program requires that the data in a hash table be sorted by its key. In this case, using the ORDERED argument tag makes it unnecessary to pre-sort the data before loading it into the table.
- After the hash data has been processed (e.g., after a number of table updates or computing aggregates), it is written out to a data set required to be in a sorted order. For example, the latter may be needed just to make the result more convenient to view or if the output needs to be reprocessed using the BY statement. In these cases, having the output already ordered eliminates an extra sorting step.
- It is desirable to sort a DATA step structure different from a hash table at run time but the inherent DATA step functionality to do so is either absent or limited. Sorting a set of parallel arrays might be a good example.
- BY group post-processing. The idea is to load data from every input BY group into a number of hash tables, each ordered differently, to obtain a number of group metrics by enumerating each table based on its own order. Then the tables can be cleared to process the next group identically. As a simple example, suppose that we need to compute percentiles for N different variables for each BY group. A head-on method would require re-sorting the entire input and rereading it N times, which can be prohibitively costly. Yet, using the approach outlined above, the job can be done in a single pass using N hash tables - at a mere fraction of the cost.

Since we have discussed the ins and outs of the *Order* operation at the detailed level earlier, we will limit this section to a few sparsely annotated practical examples. You will also find numerous examples of using the operation matter-of-factly elsewhere in the book, particularly in the programs generating the sample data.

6.5.1 Reordering Split Outputs

First, let us revisit the task of splitting sorted data discussed in the preceding section. There, we already had a hint of using the argument tag ORDERED to impose the replication of the input record order on the "split" output files in the absence of MULTIDATA:"Y". Now suppose that instead, we want the "split" files related to each League_SK to be ordered by Team_SK. Surely, we can do it in a later step by re-sorting them. However, there is a better way:

Program 6.15 Chapter 6 Reordering Split Outputs.sas

```
data _null_ ;
  if _n_ = 1 then do ;
    dcl hash h (multidata:"Y", ordered:"A") ;      ❶
  * h.defineKey ("_n_") ;
    h.defineKey ("Team_SK") ;      ❷
    h.defineData ("League_SK", "Team_SK", "Team_Name") ;
    h.defineDone () ;
  end ;
  do until (last.League_SK) ;
    set bizarro.Teams ;
    by League_SK ;
    h.add() ;
  end ;
  h.output (dataset: catx ("_", "work.League", League_SK)) ; ❸
  h.clear() ;
run ;
```

Now the output for our five sample teams will appear sorted by Team_SK:

Output 6.17 Result of Splitting File Bizarro.Teams for League_SK=1 Sorted by Team_SK

League_SK	Team_SK	Team_Name
1	317	Mountaineers
1	342	Gladiators

Output 6.18 Result of Splitting File Bizarro.Teams for League_SK=2 Sorted by Team_SK

League_SK	Team_SK	Team_Name
2	136	Pirates
2	165	Saints
2	228	Storm

All we had to do in order to achieve it was to ❶ add the argument tag ORDERED:"A" and ❷ replace the dummy key _N_ with Team_SK, leaving everything else intact. If we rather wanted the output to be sorted by Team_Name, we would key the hash table by this variable instead.

Let us reiterate that though the output "split" files are thus intrinsically sorted by Team_SK, the OUTPUT method does not set the SORTEDBY= data set option accordingly and does not add it to the metadata. To save processing time in case an extraneous sort step is added downstream, we can add it by changing line ❸ above as follows:

```
  h.output(dataset:catx("_", "work.League",League_SK,"(sortedby=Team_SK)")) ;
```

In this case, if an attempt were made to use the SORT procedure to order League_1 or League_2 by Team_SK, it would detect that the data was already in order and skip sorting, only printing a note to that effect in the SAS log:

```
NOTE: Input data set is already sorted, no sorting done.
```

6.5.2 Intrinsic Data Grouping

To recap our earlier discussion of the key order of the items in a hash table, they are always logically stored as *intrinsically grouped*. It means that all items with the same key-values are always stored adjacent to each other, thus forming the same-key item groups. Inside each group, the relative sequence of the items exactly replicates the sequence in which they have been inserted into the table. By default, the groups themselves are, in general, not in key order *relative to each other*. Now, if a specific order is imposed by using the properly valued argument tag ORDERED, the groups will be also in order relative to each other with respect to the key-value of each group. The matrix below visually encapsulates these differences using a sequence of 7 key-values as an example:

Figure 6.1 Ungrouped Unsorted vs Grouped Unsorted vs Grouped Sorted

Grouped	Sorted	Key-Value Sequence						
No	No	C	A	A	B	A	B	C
Yes	No	C	C	A	A	A	B	B
Yes	**Yes**	A	A	A	B	B	C	C

To see what occurs using real data, let us extract a small sample from data set Bizarro.AtBats where:

- The team is Huskies; Team_SK=193.
- The batter is a starting pitcher; Position_code="SP".
- The batter has hit a single or double; Result="Single" or Result="Double".
- The games are home games; Top_Bot="B" (the home team bats at the bottom of the inning).
- The games have been played in 2017 between March 20 and April 9.

Program 6.16 Chapter 6 Sample from AtBats.sas

```
data Sample (keep = Batter_ID Result Sequence) ;
  set bizarro.AtBats ;
  where Team_SK = 193 and Position_code = "SP" and Top_Bot = "B"
  and   date between "20mar2017"d and "09apr2017"d
  and   Result in ("Single", "Double") ;
  Sequence + 1 ;
run ;
```

We created variable *Sequence* as a unique ID for each row indicating the sequence of the observations in *Sample*. This is the order in which the observations in *Sample* will be loaded into a hash table. The content of the sample file looks as follows:

Output 6.19 Sample from Bizarro.AtBats with a Sequence Variable

Batter_ID	Result	Sequence
11530	Single	1
37661	Double	2
57696	Double	3

Batter_ID	Result	Sequence
57696	Single	4
37661	Double	5
57696	Single	6
37661	Single	7

Now let us load the file into a hash table without imposing a specific order on it and then output the content of the table to file *Grouped*:

Program 6.17 Chapter 6 Intrinsic Hash Table Grouping.sas

```
data _null_ ;
  dcl hash h (multidata:"Y") ;
  h.defineKey ("Batter_ID") ;
  h.defineData ("Batter_id", "Result", "Sequence") ;
  h.defineDone () ;
  do until (lr) ;
    set Sample end = lr ;
    h.add() ;
  end ;
  h.output (dataset: "Grouped") ;
  stop ;
run ;
```

Note that no ORDERED argument tag is specified; hence, the table is ordered by internal default. The content of file *Grouped* written by the OUTPUT method call looks as follows:

Output 6.20 Grouped Output from an Unordered Hash Table

Batter_ID	Result	Sequence
57696	Double	3
57696	Single	4
57696	Single	6
37661	Double	2
37661	Double	5
37661	Single	7
11530	Single	1

As we can see, the file is now *grouped* by Batter_ID. (Note also that the relative input sequence within each same-key group is maintained.) It means that although it is not *sorted*, if we now needed to use the BY statement to process it one BY group at a time, it would be just as good as if it were sorted, for we could code:

```
set Grouped ;
by Batter_ID NOTSORTED ;
```

without any fear that the result could be inaccurate - as it would be if the file were *neither* sorted *nor* grouped.

6.6 Summary

The goal of this chapter was to get the buy-in of the IT users. They appreciated that of many data processing problems that can be successfully addressed using the hash object, the tasks discussed in this chapter represent a small sample of rather simple scenarios. Their choice is mainly dictated by the need to illustrate the principle of applying a specific hash table operation to perform a specific programming action according to the nature of the task. In the ensuing chapters, we will see how other, more complex, tasks that address capabilities of particular interest to the business users can be tackled with the aid of the hash object. However, the hash operations approach delineated in this chapter will remain the same.

Chapter 7: Supporting Data Warehouse Star Schemas

7.1 Introduction

The Bizarro Ball business users and the data team have asked how they might be able to use the hash object to create and update a data warehouse containing the information from the transactional data about each game (e.g., the data in the AtBats, Pitches, and Runs data). Those data sets were created to specifically simulate the real-time data feeds that provide game status and updates. The reaction to that has been very positive and has led to lots of questions that Bizarro Ball fans would like to get answers to (e.g., does position in the lineup matter in terms of a batter's performance).

While just storing the existing data was originally considered by the data team, they raised concerns about the repetitive nature of the variables in the various data sets. For example, the AtBat and Pitches data sets include the player's first and last name, and that results in the data sets being larger and taking up more space. And the business users chimed in with an example of why they did not want to store the player name data that way: players occasionally ask for their names to be changed.

The net result of these discussions was the decision to create a star schema data warehouse which keeps all the needed data in a more normalized form.

A star schema is a standard data warehouse structure supported by many software products (including SAS). It consists of one or more fact tables (e.g., our AtBats or Pitches data sets) with associated dimension tables (e.g., Players, Games) that contain the additional information.

Figure 7.1: An Illustrative Star Schema

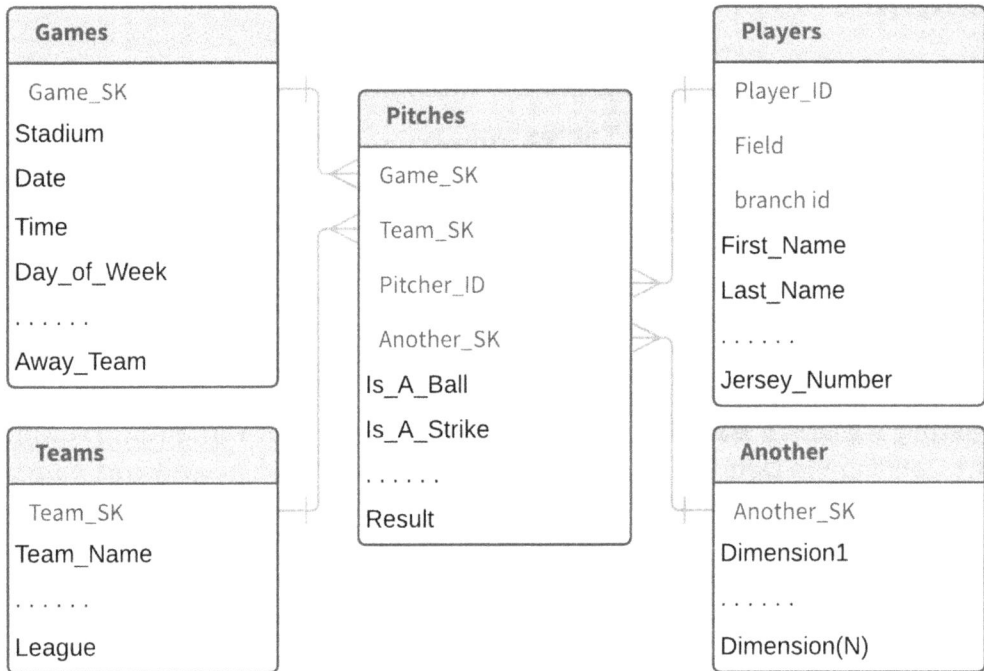

The name *star schema* is based on the table's shape: a fact table in the center surrounded by dimension tables looks like a star.

By moving the repetitive data values (e.g., player names, etc.) to dimension tables with a corresponding key we can minimize the space needed for the fact tables (which can and do have many rows) while offering the flexibility needed by our business users (e.g., player name changes).

In response to this discussion we have been asked to provide example SAS code, using hash tables, to create a star schema data warehouse, including:

- Creating and updating fact tables.
- Creating and updating slowly changing dimension tables.

We were then asked to create programs to create a data warehouse with three fact tables and all the needed dimension tables.

7.2 Creating and Updating Fact Tables

The data warehouse team tried to create the SAS programs needed to create facts tables for the transactional AtBat, Pitches, and Runs data. As you are aware, PROC APPEND is a great tool to add new data to an existing SAS table. The program to add the data for a given date was very simple:

```
proc append base = dw.AtBat data = bizzarro.AtBat;
run;

proc append base = dw.Pitches data = bizzarro.Pitches;
run;

proc append base = dw.Runs data = bizzarro.Runs;
run;
```

A problem with this approach was discovered early on in testing. Data for a given date or a given game would need to be updated. It is not uncommon for the official scorer to change a hit to an error or vice versa. Making sure the data in the fact tables reflected the updated/corrected data is essential and so after some additional research the programs were modified so that PROC SQL steps to delete data for games to be reloaded were added before each PROC APPEND. For example:

```
proc sql;
 delete from dw.AtBats
 where Game_SK in (select distinct Game_SK from Bizarro.AtBats);
quit;
```

The Game_SK field is the **S**urrogate **K**ey used to identify a single key. The above PROC SQL code deletes all the rows for any game which has been previously loaded in the dw.AtBats data and which is to be reloaded from the Bizarro.AtBats data set.

A problem with this approach was also discovered in testing. Deleting the data for the games to be reloaded prevented duplicate data, but the space for the deleted rows was not freed up by default. So whenever it was necessary to delete data for one or more games and then reload it, the fact tables grew larger. Upon consulting the business users it became clear that reloads would be a frequent occurrence.

At this point, the data team decided they needed our help to address this requirement to reload the data and do it efficiently in terms of processing time and space. We investigated two alternatives:

1. Simply replace the PROC APPEND steps the data team had used with a DATA step to concatenate the data sets. For each fact table we used an SQL step to delete the data to be reloaded and a DATA step to add the current data to the remaining rows in our facts tables. That approach solved the issue with the size of the data sets growing due to the space for the deleted rows not being reused.

2. Using hash tables to load the current contents of the fact tables into a hash object (via the indirect *Insert* operation) and then performing a direct *Insert* using the ADD method for the new data. We combined that with the *Delete* operation using the REMOVE method to delete rows for games already in the data.

Both approaches addressed the basic requirements and so the next step was performance testing. We expected the hash table approach to perform better in terms of CPU and real time since it is an in-memory data store. Our performance testing results confirmed that expectation. The SQL/DATA step approach took twice as much time as the hash object approach.

The only remaining issue to research was the memory requirements for the hash object approach. By definition, fact tables are typically narrow (i.e., fewer columns) since most of the ancillary information is stored in dimension tables. For our three fact tables we estimated the record lengths and number of rows (for a complete season) as follows:

- AtBats – a record length of slightly less than 200 bytes (and that did not factor in the space saved by defining the lengths for the numeric fields to less than 8) and roughly 300,000 rows. The net effect is that the AtBats data for all the teams for an entire season would require approximately $200*300000/(1024**3) = .06$ Gigabytes of memory.
- Pitches – a record length of approximately 100 bytes and roughly 1,000,000 rows. Doing the arithmetic, the Pitches data would require $100*1000000/(1024**3) = .09$ Gigabytes.
- Runs – a record length of approximately 64 bytes and roughly 80,000 rows, resulting in a memory requirement of $64*80000/(1024**3) = .005$ Gigabytes.

In other words, with just 1 G of memory, we could handle five years of game data for our three fact tables. As a result, the approach using hash tables was selected. And for those cases where memory might be an issue, in Chapter 11 (Hash Object Memory Management) we will show you how to partition the data and the processing in order to eliminate or minimize the issue of how much memory is needed.

In the final section of this chapter (Creating a Bizarro Ball Star Schema) we will show you the code to create and update our three fact tables along with the needed dimension tables. That star schema data warehouse will be the input source for most of the examples in the later chapters.

The program that creates our transactional data (the generatepitchandpadata.sas macro shown below) uses a similar approach. Included below are the relevant snippets from that program for the Runs data set.

Program 7.1 generatepitchandpadata.sas (partial macro listing)

```
if exist("bizarro.Runs") then data_to_load = "bizarro.Runs";      ❶
else data_to_load = "template.Runs";
declare hash facts_runs(dataset:data_to_load,ordered:"A",multidata:"Y");      ❷
facts_runs.DefineKey("Date","Game_SK");
facts_runs.DefineData("Game_SK","Date","Batter_ID","Inning"
                    ,"Top_Bot","AB_Number","Runner_ID");
facts_runs.DefineDone();
.
.
.
if lr then
do;  /* output the updated fact tables */
   facts_runs.output(dataset:"bizarro.Runs");      ❸
   facts_pitches.output(dataset:"bizarro.Pitches");
   facts_atbats.output(dataset:"bizarro.AtBats");
end; /* output the updated fact tables */

set bizarro.schedule end=lr;      ❹
where date=&date;

if game_sk ne lag(game_sk) then      ❺
do;  /* delete existing rows for this game */
   if facts_runs.check() = 0 then runs.remove();
```

```
   if facts_pitches.check() = 0 then pitches.remove();
   if facts_atbats.check() = 0 then atbats.remove();
end; /* delete existing rows for this game */
.
.
/* rest of program to generate the simulated game data */
.
facts_runs.add();      ❻
.
.
run;
```

❶ A template data set is used to define the structure of the data table. The use of such template tables also handles the case where there is no current data. An IF-THEN_ELSE statement is used to specify whether an empty table with the right structure or the current data table is used to indirectly *Load* data into the hash table. The following sections on creating the various types of dimension tables will expand upon this technique.

❷ Define a hash object for our data table. Only the statements for the Runs table are included in this snippet for illustrative purposes. The DATASET argument tag is used to perform the indirect *Insert* operation, reading one input record from the appropriate data set one row at a time. The ORDERED:"A" argument tag is used so the data is sorted based on the key variable. The MULTIDATA:"Y" argument tag is used since there are many rows for a given key. In this case, the key is just the game identifier (Game_SK).

❸ Once all the data has been generated, the OUTPUT method is used to create SAS data sets from the updated data.

❹ Read the input data set that drives the data generation process. In the final section of this chapter (Creating a Bizarro Ball Star Schema) the transactional data that contains the values for the data to be added or replaced in the hash table will be used here.

❺ Our input data is sorted by the key for the hash tables, Game_SK. The LAG function is used to see if we are reading data for a new game. If so, the CHECK method is used to perform the *Search* operation against the hash table to determine if it already contains the key. If the key is found, the REMOVE method is used to *Delete* the data from the table. These statements perform a similar operation to the PROC SQL code included above.

❻ Perform the direct *Insert* operation using the ADD method at the appropriate points in the program logic that is updating the simulated data.

7.3 Creating and Updating Slowly Changing Dimension Tables

The Bizarro Ball data warehouse has wide-ranging requirements for different kinds of fact tables, and our users (both business and IT) have asked us to provide example code for each of the various types of dimension tables which can be part of a star schema warehouse. Dimension tables provide additional information about the nature of information in one or more fact tables. For example, our PITCHES fact table contains a single surrogate key for the pitcher; additional information on the pitcher (e.g., name, left-handed vs. right handed, current team) can be maintained in a dimension table.

One of the key issues with dimension tables is the fact that the characteristics of the entity described by the columns in a fact table can change over time. For example, a player can be traded to another team. There are at least six different types of dimension tables. Such dimension tables are referred to as Slowly Changing Dimension (SCD) tables. There are a number (six to be specific) of standard variants of SCD table, including variations that include features of different types. To make sure all the bases (pun intended) are covered, our users have asked us to provide examples of each of these six types, along with possible use cases for each of the types.

The six types of SCD tables are:

1. Type 0 – Retain original value based on the first occurrence (e.g., the details about a player as of his first Bizarro Ball game).
2. Type 1 – Current data is stored and historical changes are ignored (e.g., the details about a player as of his most recent Bizarro Ball game).
3. Type 2 –Store both current and prior data by creating multiple records (e.g., when a player changes teams, his current record is updated to reflect that the data is no longer current and a new record is added with his current data).
4. Type 3 – Store current data and selected historical data. Columns are added for the selected historical data. In a Type 3 SCD, each entity (e.g., a player) has only one record (as in a Type 0 or Type 1 SCD).
5. Type 4 – Store current data in one table and complete history in a secondary table. A Type 4 SCD stores the same information as is stored in a Type 2 table, but in two different tables. A Type 4 table can be preferred when performance is a key criterion and queries that involve prior data are less important.
6. Type 6 – Is a composite of Types 1, 2, and 3 tables in that current and prior data is maintained. The key difference is that a composite key is used to distinguish the multiple rows. In addition, there is also a field that designates whether the row is current. Type 6 gets its name from the combination of Types 1, 2, and 3 (1 + 2 + 3 = 6). When Ralph Kimball coined these terms, he did not include a Type 5.

Sample code, along with a typical use case, to support each of these types of SCD tables using the hash object is discussed in the following subsections. The Player dimensional data will be used for this illustration. The surrogate key for the data is Player_ID, and we will start by keeping track of the player's name (First_Name, Last_Name), team (Team_SK), and position (Position_Code) for these examples.

There are a number of approaches to maintaining/updating dimension tables. For example, dimension tables can be updated/created:

1. In the same programs that update the related fact tables as the incoming transactional data is processed.
2. Separate programs that read the same transactional data can be used, independent of the update processing for the fact tables.
3. Completely separate input transaction tables may be used for certain updates.

Updating the Player Dimension table while processing the game lineup data or perhaps the data that reflects the results of each pitch or at bat would be examples of 1 and 2 above; an example of approach 3 might be custom transaction files for events like a player being traded to another team.

The focus of the following examples is on the specific code and methods used to update a dimension table independent of which of the above three scenarios applies.

7.3.1 Handling Type 0 Dimension Tables

Slowly Changing Dimension Type 0 (SCD Type 0) tables are not commonly used since they contain data about the entity when they were first encountered. There can be cases where an SCD Type 0 table is appropriate. The following code can be found in the sample program.

Program 7.2 Chapter 7 SCD 0.sas

```
data _null_;
 if _n_ = 1 then
 do;  /* define the hash table */
    dcl hash scd(dataset: ifc(exist("bizarro.Players_SCD0")     ❶
                             ,"bizarro.Players_SCD0"
                             ,"template.Players_SCD0"            ❷
                             )
              ,ordered:"A");
    scd.defineKey("Player_ID");
    scd.defineData("Team_SK","Player_ID","First_Name"
                ,"Last_Name","Position_Code");
    scd.defineDone();
 end; /* define the hash table */
 set bizarro.AtBats(rename=(Batter_ID=Player_ID)) end=lr;     ❸
 RC = scd.add();      ❹
 if lr;
 scd.output(dataset:"Bizarro.Players_SCD0");     ❺
 stop;
 set template.Players_SCD0;     ❻
run;
```

❶   Define a hash table object for our Players dimension table. The DATASET argument tag is used to perform the indirect *Insert* operation, reading one input record from Bizarro.Players_SCD0 at a time and loading the current dimension table into hash table SCD. The ORDERED:"A" argument tag is used since it is typically the case to have dimension tables sorted by their key.

❷   A template data set is used to define the initial structure of the data table. The use of such template tables also addresses the issue that one can't reference a data set in a SET, MERGE, MODIFY, or UPDATE statement if the same data set name is used in an OUTPUT method call. The IFC and EXIST functions are used in the expression given to the DATASET argument tag to determine if the template data set is needed. We want our code to handle the case where the dimension table does not yet exist.

❸   Read the input transactional data that contains the values for the dimension. Note that the RENAME data set option is used, so the hash object methods can be called implicitly to link the data between the hash key and data items and their PDV host variables.

❹   Perform the direct *Insert* operation if the transaction key is not in the hash table. The ADD method first searches it to see if the Player_ID key-value already exists in the table. If it does, the ADD method does nothing because the default value for the MULTIDATA argument tag is "N" which means duplicates are not allowed; if not, an item is added to the hash table using the current values of the host key and data variables in the PDV. The RC= is needed to suppress messages being written to the log whenever the data already exists in the table. Since that is a common and expected occurrence, we do not want those messages written to the SAS log.

❺   The OUTPUT method is used to perform the *Output* operation to create a SAS data table with the updated SCD Type 0 table once all the transactional data have been read. Note that we do not use the RC= here as we do not expect the OUTPUT method to encounter any errors. If an error is encountered, we want to see a message in the SAS log to that effect.

❻   The template data set is referenced in a SET statement *after* the STOP statement since its only purpose is to define the hash object host variables in the PDV. Note that since none of the variables are referenced elsewhere in the DATA step, the placement of the SET statement does not matter as far as the compiler's ability to read the data set descriptor is concerned; however, coding it after the STOP statement prevents reading the data from the data set at the execution time.

The ADD method does exactly what we need for an SCD Type 0 table – the data reflects the first occurrence of (in this case) the Player.

7.3.1.1 Performing Table Lookups Using an SCD Type 0 Table

The following sample DATA step illustrates what is commonly referred to as a table lookup into an SCD Type 0 table. The term "lookup" here refers to searching the table followed by extracting the data portion values from it into the PDV if the key is found. In other words, it refers to the direct *Retrieve* operation.

Program 7.2 Chapter 7 SCD 0.sas (continued)

```
data tableLookup;
 /* sample lookup code */
 if 0 then set bizarro.Players_SCD0;        ❶
 dcl hash scd(dataset:"bizarro.Players_SCD0");        ❷
 scd.defineKey("Player_ID");
 scd.defineData("Team_SK","Player_ID","First_Name"
               ,"Last_Name","Position_Code");
 scd.defineDone();

 /* first a key not yet in the table */
 call missing(Team_SK,First_Name,Last_Name,Position_Code);        ❸
 Player_Id = 00001;
 RC = scd.find();        ❹
 output;

 /* now a key already in the table */
 call missing(Team_SK,First_Name,Last_Name,Position_Code);        ❺
 Player_Id = 10103;
 RC = scd.find();
 output;
 stop;
run;
```

❶   Define the host variables to the PDV. Note that since we know the SCD table exists and there is no OUTPUT method referencing it, we can directly reference the SCD table in the SET statement. It is included at the top of the DATA step in order to define the variables to the PDV before they are referenced. The IF 0 condition makes sure that, while the compiler sees the data set header, the statement does not cause the data from the data set to be read.

❷ Define the hash object. Note that the ORDERED option is not used as there is no need to have the hash table ordered by the keys.

❸ Use CALL MISSING to make sure the variables we are looking up have null values. This is typically needed to handle the case where the key item is not found in the table and we want to make sure that values from a prior successful lookup are not used.

❹ Use the FIND method to do the actual lookup (i.e., perform the *Retrieve* operation). Note that *all* data portion variable values are retrieved at once. If the key is found, the PDV host variables are overwritten with these values from the item with this key. For this particular case, the key is not found.

❺ The FIND method is used as in callout 4 – except the data does exist and values will be returned.

The results can be seen in the following SAS output.

Output 7.1 Sample SCD Type 0 Lookup Results

Team SK	Player ID	First Name	Last Name	Position Code	RC
.	00001				160038
171	10103	Harry	Reed	LF	0

Note that a non-zero value for RC indicates that the key value was not found in the hash table.

7.3.2 Handling Type 1 Dimension Tables

Slowly Changing Dimension Type 1 (SCD Type 1) tables are used more often that SCD Type 0 tables. For example, given our player data, we are far more interested in what team or position he currently plays for. Knowing the team or position he played for in his first game is far less useful than knowing what his current team or position is.

The code to support an SCD Type 1 table is almost exactly the same as shown above for an SCD Type 0 table. For an SCD Type 0 table, the conditional *Insert* operation is performed if the implicit *Search* operation fails. For an SCD Type 1 table the operation is *Update All.* The only difference in the code is the use of the REPLACE method instead of the ADD method:

```
scd.replace();
```

The REPLACE method first checks if the item (the Player_ID) exists in the table. If it does not, it is added to the hash table using the current values of the host variables in the PDV; if it is found, the data portion hash variables items are updated using the current values of their PDV host variables. The REPLACE method does exactly what is needed for an SCD Type 1 table: it adds/updates the items with the current data.

The code to create the sample SCD Type 1 table is found in the sample program Chapter 7 SCD 1.sas.

7.3.2.1 Performing Table Lookups Using an SCD Type 1 Table

Since both an SCD Type 0 and SCD Type 1 table have a single row for the key values, performing a lookup into an SCD Type 1 table is exactly the same as shown above for an SCD Type 0 table: use the FIND method to perform the *Retrieve* operation and make sure to initialize the variables first to a default value (e.g., using the MISSING call routine).

Output 7.2 Sample SCD Type 1 Lookup Results

Team SK	Player ID	First Name	Last Name	Position Code	RC
.	00001				160038
228	10103	Harry	Reed	PH	0

Note that the values for the Position_Code and Team_SK fields are different from the output seen above for the SCD Type 0 table.

7.3.3 Handling Type 2 Dimension Tables

Slowly Changing Dimension Type 2 (SCD Type 2) tables are probably the most commonly used form of dimension tables, as they allow access to the current values for the dimension attributes as well as the values for any historical period.

There are several fields in our player data that we would like to keep track of. For example, if a player is traded, we want to know what team he plays for now (his current SCD Type 2 record) as well as what teams he has played for in the past (his prior records). Likewise, players can change their names as well as the positions they play.

The complicating factor in updating an SCD Type 2 table is that we have to deal with the following details:

1. Is there is a record in the dimension table for the player?
2. Find the item in the hash table with the current data.
3. Once the item with the current data is found, are the values of the data items different?
4. If there are any differences, we need to access the data items from the hash table for the current row without overwriting the new information in the PDV.

Given these extra complexities, a suggested **Best Practice** to consider might be to use separate programs that read the same transactional data as mentioned above.

The following sample DATA _NULL_ step illustrates an approach to using hash tables to manage an SCD Type 2 table.

Program 7.3 Chapter 7 SCD 2.sas

```
data _null_;
 if 0 then set template.Players_SCD2;        ❶
 if _n_ = 1 then
 do;  /* define the hash table */
    dcl hash scd(dataset:
                 ifc(exist("bizarro.Players_SCD2")     ❷
                    ,"bizarro.Players_SCD2"
                    ,"template.Players_SCD2")
                ,ordered:"A",multidata:"Y");
    scd.defineKey("Player_ID");
    scd.defineData("Player_ID","Team_SK","First_Name","Last_Name"
                  ,"Position_Code","Bats","Throws","Start_Date","End_Date");   ❸
    scd.defineDone();
 end;  /* define the hash table */
```

```
set bizarro.atbats(rename = (Batter_ID = Player_ID    ❹
                             Team_SK = _Team_SK
                             First_Name = _First_Name
                             Last_Name = _Last_Name
                             Position_Code = _Position_Code
                             Bats = _Bats
                             Throws = _Throws)
                ) end=lr;
if scd.check() ne 0 then    ❺
do;  /* need to add the player */
   scd.add(key: Player_ID    ❻
          ,data: Player_ID
          ,data: _Team_SK
          ,data: _First_Name
          ,data: _Last_Name
          ,data: _Position_Code
          ,data: _Bats
          ,data: _Throws
          ,data: Date
          ,data: &SCD_End_Date);
end; /* need to add the player */
else
do;  /* check to see if there are changes */

   RC = scd.find();     ❼
   do while(RC = 0);
      if (Start_Date le Date le End_Date) then leave;
      RC = scd.find_next();
   end;
   if catx(":", Team_SK, First_Name, Last_Name
         , Position_Code, Bats, Throws) ne
      catx(":",_Team_SK,_First_Name,_Last_Name
             ,_Position_Code,_Bats,_Throws) then  ❽
   do;  /* date out prior record and add new one */;
      if RC = 0 then scd.replaceDup(data: Player_ID
                                   ,data: Team_SK
                                   ,data: First_Name
                                   ,data: Last_Name
                                   ,data: Position_Code
                                   ,data: Bats
                                   ,data: Throws
                                   ,data: Start_Date
                                   ,data: Date-1);
      scd.add(key: Player_ID
             ,data: Player_ID
             ,data: _Team_SK
             ,data: _First_Name
             ,data: _Last_Name
             ,data: _Position_Code
             ,data: _Bats
             ,data: _Throws
             ,data: Date
             ,data: &SCD_End_Date);
   end; /* date out prior record and add new one */;
end;  /* check to see if there are changes */
```

```
  if lr;
  scd.output(dataset:"bizarro.Players_SCD2");    ❾
  stop;
run;
```

❶ Just as for the SCD Type 0 table, a template data set is used to define the variables to the PDV. It is
included at the top of the DATA step since the variables must be defined before they are referenced in
the DATA step and should to be defined before they are first referenced. This placement differs from
some of the earlier examples because, in those examples, the variable names were not explicitly used
in the DATA step (they were listed only as text strings as arguments in the definition of the hash
object/table). The non-executable (because of IF 0) SET statement provides *parameter type matching*
for the hash table SCD as well as defining the variable attributes before they are referenced in the
DATA step program. Referencing the template data set lets the compiler read its descriptor and thus
create the PDV host variables with the exactly same names as the hash variables defined for the hash
table. This way, when the DEFINEDONE method is executed, the hash variables inherit the attributes
from the template data set variables. Without the requisite PDV host variables predefined at compile
time, the method would fail and generate an error.

❷ Define the hash object for the complete SCD table.

❸ The variables Start_Date and End_Date are used to define the date range during which the values for
the dimensional variables apply.

❹ Read the transactional data. The variable names are renamed to avoid overwriting the hash data portion
variables and their PDV host counterparts as a result of the implicit method calls.

❺ The CHECK method is used to determine if the Player_ID in the transaction data set is found in the
table. Regardless of whether the key exists in the hash table, the CHECK method does not overwrite
the values of the corresponding host variables in the PDV.

❻ The ADD method is used to add the data to the hash table. It is called explicitly using the renamed
variables from the transaction data set as its arguments.

❼ We need to find out if the current PDV key-value of host variable Player_ID is in the table. If so, we
need to search through multiple items with this key to get the current data - in other words, to
Enumerate this group of items. In order to do so, the item list must be *set*. This is done by calling the
FIND method *implicitly* to make it accept the PDV key-value. The call performs 3 functions: (a) *sets
the item list* by latching on to the *logically first* item in the group, (b) extracts the data portion values
from this *first item* into the corresponding PDV host variables, and (c) moves the enumerating pointer
to the *logically second* item. Then, the FIND_NEXT method is called repeatedly in the DO WHILE
loop until either (a) the PDV value of Date is found to be between the hash values of Start_Date and
End_Date or (b) there are no more items in the group. Each FIND_NEXT call: (a) extracts the data
portion values into the PDV host variables (i.e., performs the indirect *Retrieve* operation) and (b)
moves the enumerating pointer to the next item in the list. When FIND_NEXT is called for the
logically last item in the group, it moves the enumerating pointer outside it. This action unsets the list,
causing the next FIND_NEXT call to fail with a non-zero *RC* value and thus terminating the loop.
Upon exiting, the PDV host variables now have values corresponding to the desired date value.

❽ If any of the variables are different, we need to perform the *Selective Update* operation to update the
hash table without altering the values of the PDV host variables. Hence, the method call rendering the
operation must be explicit.

First, date out the current row and overwrite only the End_Date field with a new value.

Next, add a new row using the values from the transaction data.

If any of the variables are different, we need to (a) *Update* the table by dating out the current row and overwriting only the End_Date field with the new value and (b) *Insert* a new item with the transaction data values. For (a), we cannot perform the *Update All* operation by calling the REPLACE method since it would update *all* items in the item group, whereas we need to update a specific item. Hence, we need the *Selective Update* operation - invoked here by calling the REPLACEDUP method. The method is called *explicitly* to indicate the values of which PDV variables (rather than the values of the PDV host variables used by default in implicit calls) we want to use for the update.

❾ Once all the transactions have been processed, output an updated SCD table.

The following output (5 observations) shows the results.

Output 7.3 Sample SCD Type 2 Rows

Player ID	Team SK	First Name	Last Name	Position Code	Bats	Throws	Start Date	End Date
10090	269	Harry	Long	SP	L	R	2017-04-02	2017-06-23
	115	Harry	Long	SP	L	R	2017-06-24	9999-12-31
10103	171	Harry	Reed	LF	R	R	2017-03-21	2017-07-25
	228	Harry	Reed	RF	R	R	2017-07-26	2017-08-08
	228	Harry	Reed	PH	R	R	2017-08-09	2017-08-10
	228	Harry	Reed	RF	R	R	2017-08-11	2017-08-24
	228	Harry	Reed	PH	R	R	2017-08-25	2017-08-26
	228	Harry	Reed	RF	R	R	2017-08-27	2017-08-27
	228	Harry	Reed	PH	R	R	2017-08-28	2017-08-28
	228	Harry	Reed	RF	R	R	2017-08-29	2017-09-02

This output suggests that perhaps Position_Code is not a field whose values should be tracked for such changes. Another way of tracking what position a player plays will be provided as an example of a Type 3 SCD table.

Updating multiple SCD tables with a given set of transactional data is quite common. Repeating the code in the above DATA _NULL_ step (either in a single DATA step or multiple DATA steps) can create a maintenance issue. The good news is that the code is easily parameterized. A macro could be written that uses parameters for the:

- Name of the SCD table.
- List of key variables.
- List of data variables.

This would allow for the DATA _NULL_ step to read the transactions once and have a macro call for each SCD Type 2 table that needs to be updated. Alternatively, in Chapter 9 (Hash of Hashes), a parameter-driven approach that can handle any number of SCD Type 2 tables will be illustrated.

7.3.3.1 Performing Table Lookups Using an SCD Type 2 Table

An SCD Type 2 table can have multiple rows for any given key. In order to perform a table lookup into an SCD table, both the key being searched for as well as the date is needed. In this case, it is not enough just to perform *Retrieve* on the item with a given key, but we need to be able to *Enumerate* the group of items with this key-value. In other words, it is necessary to loop through all the hash items for a given key, retrieve the values of the hash variables Start_Date and End_Date into their PDV host variables for each item, and locate the one for which the date value is between Start_Date and End_Date.

The following code illustrates this.

Program 7.3 Chapter 7 SCD 2.sas (continued)

```
data tableLookup;
 /* Sample Lookup */
 if 0 then set bizarro.Players_SCD2;
 if _n_ = 1 then
 do;
    dcl hash scd(dataset:"bizarro.Players_SCD2"
                ,multidata:"Y");          ❶
    scd.defineKey("Player_ID");
    scd.defineData("Team_SK","Player_ID","First_Name","Last_Name"
                ,"Position_Code","Bats","Throws","Start_Date","End_Date");
    scd.defineDone();
 end;
 infile datalines;
 attrib Date format = yymmdd10. informat = yymmdd10.;
 input Player_ID Date;
 RC = scd.find();          ❷
 do while(RC = 0);          ❸
    if (Start_Date le Date le End_Date) then leave;       ❹
    RC = scd.find_next();          ❺
 end;
 if RC ne 0 then call missing(Team_SK,First_Name,Last_Name       ❻
                             ,Position_Code,Bats,Throws,Start_Date,End_Date);
datalines;          ❼
10103 2017/03/23
10103 2017/07/26
99999 2017/04/15
10782 2017/03/22
10782 2017/03/21
run;
```

❶ Create the hash table and load the data. Note that the ORDERED tag is not used. Ordering by the key values has no impact on performing the lookup.

❷ Point to the logically first item in the hash table for the given key (e.g., the Player_ID).

❸ Enumerate the group of items with the given key value (e.g., the Player_ID) by looping through its items sequentially.

❹ Once an item in the hash table is found that meets the date range filter, exit the enumerating loop.

❺ If the current item is not a match on the date range, advance to the next item. Note that within the group of hash items with the same Player_ID, the items are not necessarily ordered based on either Start_Date or End_Date. Rather, their enumerating sequence follows the sequence in which they are loaded by the *Insert* operation (regardless of its variety). argument. The ORDERED argument tag controls the order of the same-key item groups.

❻ If there is no match, initialize the search results to missing since we don't want the values from the last FIND_NEXT method call to persist.

❼ The first row (Player_ID = 10091) should find a match – the initially loaded row.

The second row (Player_ID = 10091) should also find a match – an updated row.

The third row (Player_ID = 99999) does not exist in the data, so the search should fail.

The fourth row (Player_ID = 11476) does find a date range match.

The fifth row (Player_ID = 11476) exists in the hash table, but there is no matching date. So the lookup should fail.

The results of the lookup can be seen in the following output.

Output 7.4 Sample SCD Type 2 Lookup Results

Player ID	Team SK	First Name	Last Name	Position Code	Bats	Throws	Start Date	End Date	Date	RC
10103	171	Harry	Reed	LF	R	R	2017-03-21	2017-07-25	2017-03-23	0
10103	228	Harry	Reed	RF	R	R	2017-07-26	2017-08-08	2017-07-26	0
99999	.							.	2017-04-15	160038
10782	189	Jeremy	Miller	CF	R	R	2017-03-22	2017-03-23	2017-03-22	0
10782	.							.	2017-03-21	160038

7.3.4 Handling Type 3 Dimension Tables

Slowly Changing Dimension Type 3 tables are a cross between an SCD Type 0 and an SCD Type 1 table in that some of the data fields correspond to historical values and some represent current data.

Our business users have asked if there is a way to have a dimension table for Players that includes a field for both the team they currently play for as well the team they debuted for. Such a requirement is a perfect example of an SCD Type 3 table. Recall that for our sample SCD Type 0 table and SCD Type 1 table we used, respectively, the ADD and REPLACE methods *called implicitly* to make them accept the current key and data values from the PDV host variables. Supporting an SCD Type 3 table can be accomplished by using both the ADD and REPLACE methods – but with the key and data values explicitly specified to the argument tags.

The following sample program illustrates maintaining an SCD Type 3 table for which we want to store both the team that a player debuted for and his current team (as of the end of the season based on our sample data).

Program 7.4 Chapter 7 SCD 3.sas

```
data _null_;
 if 0 then set template.Players_SCD3;
 if _n_ = 1 then
 do;   /* define the hash table */
    dcl hash scd(dataset:ifc(exist("bizarro.Players_SCD3")    ❶
                       ,"bizarro.Players_SCD3"
                       ,"template.Players_SCD3"
                       )
              ,ordered:"A",multidata:"Y");
```

```
      scd.defineKey("Player_ID");
      scd.defineData("Player_ID","Debut_Team_SK","Team_SK"
                    ,"First_Name","Last_Name","Position_Code","Bats","Throws");
      scd.defineDone();
end; /* define the hash table */
set bizarro.atbats(rename=(Batter_ID = Player_ID)) end=lr;   ❷
_Team_SK = Team_SK;      ❸
if scd.find() then scd.add(Key:Player_ID      ❹
                          ,Data:Player_ID
                          ,Data:Team_SK
                          ,Data:Team_SK
                          ,Data:First_Name
                          ,Data:Last_Name
                          ,Data:Position_Code
                          ,Data:Bats
                          ,Data:Throws);
else scd.replace(Key:Player_ID      ❺
                ,Data:Player_ID
                ,Data:Debut_Team_SK
                ,Data:_Team_SK
                ,Data:First_Name
                ,Data:Last_Name
                ,Data:Position_Code
                ,Data:Bats
                ,Data:Throws);
if lr;
scd.output(dataset:"bizarro.Players_SCD3");      ❻
run;
```

❶   Define the SCD Type 3 hash table.

❷   Read AtBats data.

❸   Save a copy of the field that needs to have both the current and historical values saved.

❹   If the FIND method returns a non-zero value (which evaluates to True) the transaction represents a new key value, and the key and data values need to be added as an item to the hash table. Since this is the first time this Player_ID value is encountered, his current team and his debut team are the same value, so we load both fields with the Team_SK value for his first appearance.

❺   If the FIND method results in a 0 for its return code, the PDV values will have been updated implicitly, and the REPLACE method can be called using the value of _Team_SK, which is a copy of the original Team_SK value from the transaction data set. Note that without the RENAME statement mentioned in callout 4, the Team_SK value from the transaction file would have been overwritten.

❻   After all the transaction data has been read, the *Output* operation is performed by calling the OUTPUT method to save the data content of the updated SCD Type 3 table in data set Players_SCD3.

Output 7.5 Sample SCD 3 Rows

Player ID	Debut Team SK	Team SK	First Name	Last Name	Position Code	Bats	Throws
10090	269	115	Harry	Long	SP	L	R
10103	171	228	Harry	Reed	LF	R	R
10114	136	136	Harry	Hughes	1B	R	L
10130	176	176	Harry	Rodriguez	1B	R	R
10132	103	103	Harry	Patterson	SP	R	L

7.3.4.1 Including Facts into an SCD Table

Our business users have asked about creating another fact table that simply contains one row for each player along with how many games the player has started for each of the available positions. Such a requirement can be addressed by using an SCD table to store current information as an alternative to creating an additional fact table. Such tables are similar to SCD Type 3 tables which contain historical and current data. Instead of creating a fact table that has one row for each player, it may be desirable to store such values in an SCD 3 type table. The following sample code illustrates this:

Program 7.5 Chapter 7 SCD 3 w Facts.sas

```
data _null_;
 if 0 then set template.Players_SCD3_Facts;
 if _n_ = 1 then
 do;   /* define the hash table */
    dcl hash scd(dataset:
           ifc(exist("bizarro.Players_SCD3_Facts")        ❶
              ,"bizarro.Players_SCD3_Facts"
              ,"template.Players_SCD3_Facts"
              )
              ,ordered:"A");
    scd.defineKey("Player_ID");
    scd.defineData("Player_ID","Team_SK","First_Name","Last_Name"
                ,"First","Second","Short","Third","Left","Center","Right"
                ,"Catcher","Pitcher","Pinch_Hitter");
    scd.defineDone();
    dcl hash uniqueGames();         ❷
    uniqueGames.defineKey("Game_SK","Player_ID","Position_Code");
    uniqueGames.defineDone();
end; /* define the hash table */
set bizarro.AtBats(rename = (Batter_ID = Player_ID)) end=lr;
if scd.find() ne 0 then
    call missing(First,Second,Short,Third,Left,Center        ❸
                ,Right,Catcher,Pitcher,Pinch_Hitter);
select(Position_Code);        ❹
    when("1B")  First     + (uniqueGames.add() = 0);
    when("2B")  Second    + (uniqueGames.add() = 0);
    when("SS")  Short     + (uniqueGames.add() = 0);
    when("3B")  Third     + (uniqueGames.add() = 0);
    when("LF")  Left      + (uniqueGames.add() = 0);
    when("CF")  Center    + (uniqueGames.add() = 0);
    when("RF")  Right     + (uniqueGames.add() = 0);
    when("C" )  Catcher   + (uniqueGames.add() = 0);
```

```
        when("SP") Pitcher     + (uniqueGames.add() = 0);
        when("RP") Pitcher     + (uniqueGames.add() = 0);
        when("PH") Pinch_Hitter + (uniqueGames.add() = 0);
        otherwise;
     end;
     scd.replace();          ❺
     if lr;
     scd.output(dataset:"Bizarro.Players_SCD3_Facts");       ❻
  run;
```

❶   Define the SCD Type 3 hash table.

❷   Create a hash table that will allow us to do a unique count of how many games a player has played a given position. The details of how this works will be discussed in more detail in Chapter 8.

❸   If the FIND method returns a non-zero value (which evaluates to True) the Player_ID is not found in the hash able and so the hash data items to track the number of times a player started in the given position are initialized to missing. If the FIND method finds an item with the key-value it has accepted, the PDV host values for the summary variables are updated with current values.

❹   Update the summary variables based on the Position_Code value for the current row. If an item with the composite key (Game_SK,Player_ID,Position_Code) is not found in the uniqueGames hash table, this row represents a new game for the given player for the current position. Thus, a return code of 0 means that we need to add 1 to our unique count. The next effect of this Boolean expression is that it allows us to do a unique count. This technique to perform a count distinct calculation will be discussed further in Chapter 8.

❺   Use the REPLACE method to update the hash table data items.

❻   After all the transactions have been read, use the OUTPUT method to output an updated SCD Type 3 table.

Output 7.6 Sample SCD3 with Facts Rows

Player ID	Team SK	First Name	Last Name	First	Second	Short	Third	Left	Center	Right	Catcher	Pitcher	Pinch Hitter
10090	269	Harry	Long	.	.	.	.	.	.	.	.	38	.
10103	171	Harry	Reed	.	.	.	.	72	.	44	.	.	6
10114	136	Harry	Hughes	115	.	.	.	.	.	.	.	.	2
10130	176	Harry	Rodriguez	120	.	.	.	.	.	.	.	.	6
10132	103	Harry	Patterson	.	.	.	.	.	.	.	.	40	.

7.3.4.2 Performing Table Lookups Using an SCD Type 3 Table

Since an SCD Type 3 table has a single row for the key values, performing a lookup into an SCD Type 3 table is exactly the same as shown above for an SCD Type 0 table: use the FIND method to perform the *Retrieve* operation and make sure to initialize the variables first to a default value (e.g., using the MISSING call routine).

7.3.5 Handling Type 4 Dimension Tables

Slowly Changing Dimension Type 4 tables are a simple variant of SCD Type 2 tables. SCD Type 4 tables may be preferred over SCD Type 2 tables for performance reasons as well as simplicity of looking up the values of the dimensional variables.

Managing an SCD Type 4 table can be accomplished using code similar to what is shown above for an SCD Type 2 table. Just replace the following statement:

```
scd.output(DATASET:"Players_SCD2");
```

with:

```
scd.output(DATASET:
        "bizarro.Players_SCD4_Current(where=(End_Date= '&SCD_End_Date'")));   ❶
scd.output(DATASET:"bizarro.Players_SCD2_Historical");  ❷
```

❶ The OUTPUT method supports WHERE clauses to limit the key and data items from the hash table that are to be included in the output data set. This WHERE clause limits the output to just the current data.

❷ Use the OUTPUT method to create an output data set with both the historical and current data (i.e., an SCD Type 2 table). Using a WHERE clause and including only those rows where the End_Date is not equal to the default end date (i.e., the value of the macro variable in ❶) will create a table with only historical data.

The data set with the complete history should also be used as the value of the DATASET argument tag in order to *Insert* both the historical and current data into the hash table object.

7.3.5.1 Performing Table Lookups Using an SCD Type 4 Table

When performing a lookup into an SCD Type 4 current data table, one can take advantage of the fact that there is a single item for each key value or composite key values for the case where the key consists of more than one variable. Thus, performing a lookup into the current data SCD Type 4 table is exactly the same as shown above for an SCD Type 0 table: use the FIND method to perform the *Retrieve* operation and make sure to initialize the variables first to a default value (e.g., using the MISSING function).

When you are performing a lookup into an SCD Type 4 historical data table, the exact same approach as discussed above for an SCD Type 2 table applies.

7.3.6 Handling Type 6 Dimension Tables

A Type 6 Slowly Changing Dimension Table is conceptually similar to an SCD Type 2 table, the primary difference being:

1. A composite key is used. For the case of our player data, for example, in addition to the Player_ID field, there might be an autonumber field (alternatively the SAS UUIDGEN function could be used) that provides a secondary key value. The benefit of such an approach is that each item has a unique key.

2. In addition, another data attribute is added as a flag to indicate whether the particular item is the currently active item. In some cases, this variable can be stored both in the key portion and the data portion of the table.

There are various implementations of an SCD Type 6 table. The most common is to not have an additional explicit key variable since the Start_Date (or End_Date) field can serve the purpose of making each row have a unique key. The use of a variable that designates whether or not the current item is the active item can provide benefits as it makes it easier to identify and access the current data (much like the point of an SCD Type 4 current values table).

The following code illustrates one approach to managing an SCD Type 6 table. Because we can specify the MULTIDATA:"Y" tag when defining the hash table and use the *Selective Update* operation (via the REPLACEDUP method), we can treat the active field (Active) and a secondary key (SubKey) as data items. There is no reason or advantage in making either of these fields a key when defining the key and data portions for the hash table entry.

Program 7.6 Chapter 7 SCD 6.sas

```
data _null_;
 if 0 then set template.Players_SCD6;
 if _n_ = 1 then
 do;   /* define the hash table */
    dcl hash scd(dataset:ifc(exist("bizarro.Players_SCD6")     ❶
                             ,"bizarro.Players_SCD6"
                             ,"template.Players_SCD6"
                             )
              ,ordered:"A",multidata:"Y");
    scd.defineKey("Player_ID");     ❷
    scd.defineData("Player_ID","Active","SubKey","Team_SK"
                 ,"First_Name","Last_Name","Position_Code","Bats","Throws"
                 ,"Start_Date","End_Date");
    scd.defineDone();
 end; /* define the hash table */
 set bizarro.atbats(rename = (Batter_ID = Player_ID
                              Team_SK = _Team_SK     ❸
                              First_Name = _First_Name
                              Last_Name = _Last_Name
                              Position_Code = _Position_Code
                              Bats = _Bats
                              Throws = _Throws)
                   ) end=lr;
 if scd.check(Key:Player_ID) ne 0 then     ❹
 do;   /* player is new */
    scd.add(key: Player_ID
           ,data: Player_ID
           ,data: 1
           ,data: 1
           ,data: _Team_SK
           ,data: _First_Name
           ,data: _Last_Name
           ,data: _Position_Code
           ,data: _Bats
           ,data: _Throws
           ,data: Date
```

```
                ,data: &SCD_End_Date);
end; /* player is new */
else
do;  /* check to see if there are changes */
   RC = scd.find();      ❺
   do while(RC = 0);
      if (Start_Date le Date le End_Date) then leave;
      RC = scd.find_next();
   end;
   if RC ne 0 then
      call missing(Team_SK,First_Name,Last_Name,Position_Code,Bats,Throws);
   if catx(":", Team_SK, First_Name, Last_Name, Position_Code, Bats, Throws) ne
      catx(":",_Team_SK,_First_Name,_Last_Name,_Position_Code,_Bats,_Throws)
   then      ❻
   do;  /* date out prior record and add new one */;
      if RC = 0 then /* date out active record */    ❼
         scd.replaceDup(data: Player_ID
                        ,data: 0
                        ,data: SubKey
                        ,data: Team_SK
                        ,data: First_Name
                        ,data: Last_Name
                        ,data: Position_Code
                        ,data: Bats
                        ,data: Throws
                        ,data: Start_Date
                        ,data: Date - 1);
      /* add row with the next autonumber value */     ❽
      _SubKey = 0;
      RC = scd.find();
      do while(RC = 0);
         RC = scd.find_next();
         _SubKey = max(_SubKey,SubKey);
      end;
      scd.add(key: Player_ID    ❾
            ,data: Player_ID
            ,data: 1
            ,data: _SubKey + 1
            ,data: _Team_SK
            ,data: _First_Name
            ,data: _Last_Name
            ,data: _Position_Code
            ,data: _Bats
            ,data: _Throws
            ,data: Date
            ,data: &SCD_End_Date);
   end; /* date out prior record and add new one */;
end;  /* check to see if there are changes */
if lr;
scd.output(dataset:
         "Bizarro.Players_SCD6(index=(SCD6=(Player_ID Active SubKey)))");    ❿
run;
```

❶ Define the hash table. MULTIDATA:"Y" is used since there can be multiple inactive rows sharing the same key-value. The SubKey is not defined as a key in the hash table as the value can't be known a priori (in order to support a lookup).

❷ Neither the Active nor SubKey variable is included in the key portion; they are included in the data portion.

❸ Just as was done for the SCD Type 2 table, the dimensional variables are renamed as they are read from the transaction data.

❹ If no items with the PDV key-value of Player_ID are found in the hash table, insert an item making it Active with SubKey=1.

❺ Otherwise, if a group of items with this key-value is found, get the current active values. To do so, we need to *Enumerate* the group. First, use the FIND method to set the item list with this key, and then call the FIND_NEXT method repeatedly in the DO loop to scan the group one item at a time. If the currently active item is located, exit the loop, in which case the RC value returned by the last FIND_NEXT call remains at zero. If no currently active item is located, the loop is terminated when there are no more items in the list left to scan, and so the FIND_NEXT call fails returning RC≠0. This value indicates that though there are items with the Player_ID key-value in the table, none is currently active. CALL MISSING is used to handle such a case.

❻ Check to see if there are any changes.

❼ If an item for the current date was found, it must be dated out and the Active flag must be reset to 0 (i.e., not active). The *Selective Update* operation (activated by the REPLACEDUP method call) must be used. This is because Player_ID is the only key, which allows item groups with multiple items per key-value, and we need to update a specific item in the group.

❽ We need to find the maximum value of the SubKey value, so that we can increment its value for the new item to be added. This is done by enumerating all items in the item group with the given key-value of Player_ID.

❾ Add an item with the Active flag set to 1 (i.e., is Active) and the next autonumber value.

❿ We can use the data set option INDEX= to create an index on the output data set as part of the *Output* operation. In this case, a composite index is created on the composite key (Player_ID,SubKey,Active*)*.

7.3.6.1 Performing Table Lookups Using an SCD Type 6 Table

The table lookup into an SCD Type 6 table can be done using the exact same code as shown above for an SCD Type 2 table. However, that does not leverage the potential efficiency benefits of using the Active flag. The following code illustrates first checking to see if the Active row meets the required time/date constraint; and if it does not, the inactive rows are searched.

Program 7.6 Chapter 7 SCD 6.sas (continued)

```
data tableLookup;
 /* Sample Lookup */
 retain Player_ID;
 if 0 then set bizarro.Players_SCD6(drop=Subkey);
 if _n_ = 1 then
 do;   /* define the hash table */
```

```
      dcl hash scd(dataset:"bizarro.Players_SCD6",multidata:"Y",ordered:"D");    ❶
      scd.defineKey("Player_ID","Active");
      scd.defineData("Team_SK","Player_ID","Active","First_Name","Last_Name"
                    ,"Position_Code","Bats","Throws","Start_Date","End_Date");
      scd.defineDone();
   end; /* define the hash table */
   infile datalines;
   attrib Date format = yymmdd10. informat = yymmdd10.;
   input Player_ID Date;       ❷
   RC = scd.find(Key:Player_ID,Key:1);       ❸
   if RC = 0 and (Start_Date le Date le End_Date)
   then;       ❹
   else
   do;  /* search the inactive rows */
      RC = scd.find(Key:Player_ID,Key:0);       ❺
      do while(RC = 0);
         if (Start_Date le Date le End_Date) then leave;
         RC = scd.find_next();
      end;
   end; /* search the inactive rows */
   if RC ne 0 then call missing(Team_SK,Active,First_Name,Last_Name       ❻
                                ,Position_Code,Bats,Throws,Start_Date,End_Date);
datalines;
10103 2017/10/15
10103 2017/03/23
99999 2017/03/15
10782 2017/03/22
10782 2017/03/21
run;
```

❶ Define the hash table. Note that adding the Active field as a key can improve lookup performance, especially if the majority of lookups are for the active values.

❷ Read in the sample transactions for our lookups (this is the same data as used for the SCD Type 2 lookup example).

❸ We first use the FIND method to *Retrieve* the values for the currently active row.

❹ If there is a currently active item (RC = 0), and *Date* is in range, our search is complete.

❺ Otherwise, search for the inactive items.

❻ If the search fails because either the Player_ID key-value is not found in the hash table or there is no date range match, set the values of the dimensional variables to missing.

The results of the lookup can be seen in the following output.

Output 7.7 Sample SCD Type 6 Lookup Results

Player ID	Active	Team SK	First Name	Last Name	Position Code	Bats	Throws	Start Date	End Date	Date	RC
10103	1	228	Harry	Reed	PH	R	R	2017-10-15	9999-12-31	2017-10-15	0
10103	0	171	Harry	Reed	LF	R	R	2017-03-21	2017-07-25	2017-03-23	0
99999	.	.							.	2017-03-15	160038
10782	0	189	Jeremy	Miller	CF	R	R	2017-03-22	2017-03-23	2017-03-22	0
10782	.	.							.	2017-03-21	160038

7.4 Creating a Bizarro Ball Star Schema Data Warehouse

We will now combine some of the techniques discussed above to create a data warehouse star schema from the Bizarro Ball transactional data. Our fact tables will be normalized (i.e., repeated values like names will be removed from the AtBats data and moved to dimension tables) and will link to needed dimension tables. This sample data warehouse will be the source data for many of the examples in the rest of this book. It should be clear, given the examples so far, that the use of hash tables also allows for convenient and easy access to transactional data.

The star schema we are creating here will be limited to what is needed for the examples in the rest of the book. You will definitely encounter certain kinds of events that we are not handling in this example star schema, as the point of the book is hash tables rather than star schemas. As a SAS programmer, you should be able to add features as you see fit to this example, as well as to adapt the approach shown here to your work.

7.4.1 Defining the Data Warehouse Tables

Our star schema will contain 3 fact tables and 9 dimension tables.

The fact tables are:

- AtBats: The AtBats transactional table will be used as the source for this fact table. It will contain 1 row for each at bat. To be technical, the AtBat data really contains data for each plate appearance. We will drop fields like First_Name, Last_Name, Date, Time, and so on, as we will have keys to dimension tables to allow us to look up those values.

- Pitches: The Pitches transactional table will be used as the source for this table. It will contain 1 row for each pitch in each at bat and will contain a surrogate key to the corresponding row in the AtBats table. As was done for the AtBats table, selected fields will be dropped, as their values will be available in dimension tables.

- Runs: The Runs transactional table will be used as the source for this table. It will contain 1 row for each run scored and contain a surrogate key to the corresponding row in the AtBats table that can be used to determine which batter should be credited for the RBI (Runs Batted In) metric. As was done for the AtBats table, selected fields will be dropped as their values will be available in dimension tables.

Our dimension tables can be broadly categorized as follows:

- Dimension tables that are simply copies of tables used to create our Bizarro Ball data as discussed in Chapter 5:

  - Leagues – Contains two observations that provide the league name for each league.

  - Teams - Contains 32 observations that provide the team name for each team.

- Dimension tables that are created from the transactional data used to populate the fact tables:

  - Games: This is a slowly changing type 1 dimension table. It will be created from the Games transactional data. After consulting with the business users, an SCD Type 1 table was selected to handle the case of games that needed to be rescheduled due to postponements (e.g., rain delays). There is no need to keep information on the original date or time for such games.

  - Players: This is a Slowly Changing Type 2 dimension table as discussed above in Handling Type 2 Dimension Tables. For now, our business users have asked us to not include position played as one of the fields to track, adding that they may change their minds later on that issue. An artifact of this approach to create our Players dimension table in this Proof of Concept is that players who are never used in a game will not be found in this data set.

  - Players_Positions_Played: Our business users asked us to also create a dimension table, similar to what was discussed above in the section "Including Facts into an SCD Table." Based on their review of this table, they may decide to not track position as an SCD Type 2 fields in the Players dimension table.

7.4.2 Defining the Fact and Dimension Hash Tables via Metadata

Our star schema will require a number of hash tables. In order to minimize hard-coding, we have suggested that a metadata approach be used to define these hash tables. In addition to minimizing the number of statements to be hard-coded, such an approach offers great flexibility in adding and dropping data items in our hash tables.

We have created a SAS data set in the template library, template.Schema_Metadata, with this metadata. The following three tables show this data for our three fact tables. Note that we are showing a maximum of 6 rows only. The Runs fact table has a total of only five columns.

Output 7.8 AtBats Metadata – First 6 Rows

Member Name	Column Name	Is A Key
ATBATS	Game_SK	1
ATBATS	Batter_ID	.
ATBATS	Position_Code	.
ATBATS	Inning	.
ATBATS	Top_Bot	.
ATBATS	AB_Number	.

Output 7.9 Pitches Metadata – First 6 Rows

Member Name	Column Name	Is A Key
PITCHES	Game_SK	1
PITCHES	Pitcher_ID	.
PITCHES	Pitcher_First_Name	.
PITCHES	Pitcher_Last_Name	.
PITCHES	Pitcher_Type	.
PITCHES	Inning	.

Output 7.10 Runs Metadata – All 5 Rows

Member Name	Column Name	Is A Key
RUNS	Game_SK	1
RUNS	Inning	.
RUNS	Top_Bot	.
RUNS	AB_Number	.
RUNS	Runner_ID	.

The design of these tables (one row for each variable to include in the hash table) takes advantage of the flexibility of the DEFINEKEY and DEFINEDATA methods. First, their argument tags accept character expressions in general, not just literals. Second, they can be called multiple times, each time adding a new variable (or variables) to the definition. So, you can call either method just once with a comma-separated list of quoted variable names, or once for each variable to be included as either a key or data item, or any combination thereof.

Note that the only key is the Game_SK field. We will use the MULTIDATA:"Y" argument tag to specify that multiple items are allowed for each key value. Game_SK must be a key in order to support deleting and replacing the data for a given game.

7.4.2.1 Using a Macro to Create the Hash Object Method Calls

The following macro will be used to create the needed hash table objects in the DATA step. Given the metadata approach, the code for each hash table is the same. We just need to use an appropriate WHERE clause for each set of hash object method calls.

Program 7.7 createhash.sas Macro

```
%macro createHash      ❶
       (lib = dw
       ,hashTable = hashTable
       ,metaData = template.Schema_Metadata
       );

 if 0 then set template.&hashTable;      ❷
 dcl hash _&hashTable(dataset:"&lib..&hashtable",multidata:"Y",ordered:"A");      ❸
 lr = 0;
```

```
do while(lr=0);
   set &metadata end=lr;
   where upcase(hashTable) = "%upcase(&hashTable)";     ❹
   if is_a_key then _&hashTable..DefineKey(Column);     ❺
   _&hashTable..DefineData(Column);       ❻
end;
 _&hashTable..DefineDone();       ❼

%mend createHash;
```

❶ Define the macro with three parameters: the libref for the star schema; the name of the data set which is also used to generate the name of the hash object; and the name of the data set that contains the metadata which defines each hash object.

❷ Use a conditional SET statement (never executed) to define the needed host variables to the PDV using our template data library.

❸ Declare the hash table object using the DATASET argument tag to implicitly load the current data into the hash table; the MULTIDATA:"Y" argument tag which is needed since the only key is the Game_SK column; the ORDERED:"A" argument tag since we want the items in the table to be logically ordered by game.

❹ Loop through the schema metadata reading only the definition data for the specified hash table.

❺ If the current column is also a key, use the DEFINEKEY method to specify that the column is to be used in the key portion.

❻ Use the DEFINEDATA method to specify that the column is to be used in the data portion. Note that this macro makes the assumption that columns in the key portion should also be included in the data portion. We told both the IT users and the business users that the macro and the metadata could be updated if the flexibility of including or not including specific key variables in the data portion is a long-term requirement for any follow-on project.

❼ Use the DEFINEDONE method to finalize the creation of the hash object instance.

7.4.3 Creating the Initial Data Structures for a Star Schema

Having programs to create and/or reset the needed data structures is considered by many to be a best practice. The use of the template data sets described earlier in this chapter is one such example. This allows the programs that are run repeatedly to update our data structures without having to check if the data structures already exist. This technique has particular relevance to the use of hash tables. If there is no data, having an empty data set with the right structure allows for the use of the DATASET argument tag to insert the data using the *Implicit Insert* operation.

The following program does this for our star schema.

Program 7.8 Create Star Schema DW.sas

```
proc datasets lib = dw nolist kill;     ❶
run;
```

```
proc datasets lib = dw nolist;
 options obs = 0;
 copy in=bizarro out=dw;      ❷
 select AtBats
        Pitches
        Runs
        Games;
 copy in=template out=dw;     ❸
 select Players_Positions_Played Players;
run;
 options obs = max;      ❹
 copy in=bizarro out=dw;      ❺
 select Leagues
        Teams;
quit;
```

❶   Initialize the data library for our star schema data warehouse to make sure it contains no data sets.

❷   Use the DATASETS procedure options to create a copy of our transactional tables with no observations (the OPTIONS OBS=0;) statement for selected fact and dimensions tables. Note that since the metadata defines the columns to include in the various tables, as long as these initially empty tables include all the needed columns, any other columns will not be included once the DATA step runs to update the fact and dimension tables.

❸   Use data sets in the template library for the two Players SCD tables. The creation of these tables was discussed earlier in this chapter.

❹   Take advantage of the RUN group facility of the DATASETS procedure in order to reset the value of the OBS option.

❺   Copy the tables from the Bizarro library that were listed above as simple copies of the data tables used to create the transactional data. These tables are static and are not changed or updated by any automated process or program.

7.4.4 Updating the Fact and Dimension Tables

The following program will update our fact and dimension tables. This program is intended as a sample program to create the data tables we will use in a number of the examples in the rest of this book. It is not a robust program to create star schemas; but you can use it as a model for your efforts to create and update data warehouse tables.

Program 7.9 Update Star Schema DW.sas

```
data _null_;
 %createHash(hashTable=AtBats)      ❶
 %createHash(hashTable=Pitches)
 %createHash(hashTable=Runs)
 %createHash(hashTable=Games)
 %createHash(hashTable=Players_Positions_Played)
 %createHash(hashTable=Players)
```

```
dcl hash uniqueGames();      ❷
uniqueGames.defineKey("Game_SK","Player_ID","Position_Code");
uniqueGames.defineDone();

lr = 0;
do until(lr);
   set bizarro.AtBats(rename = (Team_SK = _Team_SK
                                First_Name = _First_Name
                                Last_Name = _Last_Name
                                Bats = _Bats
                                Throws = _Throws)
                               ) end=lr;      ❸
   if game_sk ne lag(game_sk) and _AtBats.check() = 0 then _AtBats.remove();      ❹
   _AtBats.add();      ❺
   link Games_SCD;      ❻
   Player_ID = Batter_ID;      ❼
   link Positions_Played_SCD;
   link Players_SCD;
end;
_AtBats.output(dataset:"dw.AtBats");      ❽

lr = 0;      ❾
do until(lr);
   set bizarro.Pitches end=lr;
   if game_sk ne lag(game_sk) and _Pitches.check() = 0
      then _Pitches.remove();
   _Pitches.add();
   Player_ID = Pitcher_ID;
   _Team_SK = Team_SK;
   _First_Name = Pitcher_First_Name;
   _Last_Name = Pitcher_Last_Name;
   _Bats = Pitcher_Bats;
   _Throws = Pitcher_Throws;
   link Players_SCD;
   Position_Code = Pitcher_Type;
   link Positions_Played_SCD;
end;
_Pitches.output(dataset:"dw.Pitches");

lr = 0;      ❿
do until(lr);
   set bizarro.Runs end=lr;
   if game_sk ne lag(game_sk) and _Runs.check() = 0
      then _Runs.remove();
   _Runs.add();
end;
_Runs.output(dataset:"dw.Runs");

/* output the updated dimension tables */      ⓫
_games.output(dataset:"dw.Games");
_Players_Positions_Played.output
                        (dataset:"dw.Players_Positions_Played");
```

```
_Players.output(dataset:"dw.Players");
stop;

Games_SCD:          ⓬
 Year = Year(Date);
 Month = Month(Date);
 DayOfWeek = weekday(Date);
 _games.replace();
return;

Positions_Played_SCD:      ⓭
 if _Players_Positions_Played.find() ne 0
    then call missing(First,Second,Short,Third,Left
                      ,Center,Right,Catcher,Pitcher);
 select(Position_Code);
    when("1B") First       + (uniqueGames.add() = 0);
    when("2B") Second      + (uniqueGames.add() = 0);
    when("SS") Short       + (uniqueGames.add() = 0);
    when("3B") Third       + (uniqueGames.add() = 0);
    when("LF") Left        + (uniqueGames.add() = 0);
    when("CF") Center      + (uniqueGames.add() = 0);
    when("RF") Right       + (uniqueGames.add() = 0);
    when("C" ) Catcher     + (uniqueGames.add() = 0);
    when("SP") Pitcher     + (uniqueGames.add() = 0);
    when("RP") Pitcher     + (uniqueGames.add() = 0);
    when("PH") Pinch_Hitter + (uniqueGames.add() = 0);
    otherwise;
 end;
 _Players_Positions_Played.replace();
return;

Players_SCD:          ⓮
 if _Players.check() ne 0 then
 do;  /* need to add the player */
    _Players.add(key: Player_ID
               ,data: Player_ID
               ,data: _Team_SK
               ,data: _First_Name
               ,data: _Last_Name
               ,data: _Bats
               ,data: _Throws
               ,data: Date
               ,data: &SCD_End_Date);
 end; /* need to add the player */
 else
 do;  /* check to see if there are changes */

    RC = _Players.find();
    do while(RC = 0);
       if (Start_Date le Date le End_Date) then leave;
       RC = _Players.find_next();
    end;

    if catx(":", Team_SK, First_Name, Last_Name, Bats, Throws) ne
       catx(":",_Team_SK,_First_Name,_Last_Name,_Bats,_Throws) then
    do;  /* date out prior record and add new one */
       if RC = 0 then _Players.replaceDup(data: Player_ID
                                          ,data: Team_SK
```

```
                                            ,data: First_Name
                                            ,data: Last_Name
                                            ,data: Bats
                                            ,data: Throws
                                            ,data: Start_Date
                                            ,data: Date-1);
        _Players.add(key: Player_ID
                    ,data: Player_ID
                    ,data: _Team_SK
                    ,data: _First_Name
                    ,data: _Last_Name
                    ,data: _Bats
                    ,data: _Throws
                    ,data: Date
                    ,data: &SCD_End_Date);
    end; /* date out prior record and add new one */;
  end;  /* check to see if there are changes */
 return;
run;
```

❶ Use the createHash macro to define all of our hash tables (both our fact and dimension tables).

❷ Create the hash table to allow us to perform a count distinct as discussed in the section "Including Facts into an SCD Table" earlier in this chapter. Note that we could have defined this table via metadata, but chose to create it explicitly in this program. We will revisit defining a hash table to calculate unique counts in Chapter 9, Hash of Hashes – Looping Through SAS Hash Objects.

❸ Use a DO loop to read all the transactional AtBat data. Note the rename option for the fields that will not be included in the AtBats fact table but are needed to properly create the Players Slowly Changing Type 2 table (as discussed above).

❹ If the data is for the given game (the value of the key Game_SK is found in the hash table), use the REMOVE method to remove the data. Since the input data is sorted by Date and Game_SK, the LAG function logic allows us to do that only once for each game.

❺ Use the ADD method to perform the *Insert* operation.

❻ Use a LINK statement to execute the code that updates the dimension tables that are populated from the AtBats data. The LINK statement is a useful feature that allows a SAS program to repeatedly execute the same block of code multiple times in a given DATA step without having to cut and paste the code into multiple locations. The RETURN statement at the end of the LINK block returns control to the statement immediately after the LINK statement. This use of the LINK statement allows the same code to be reused when more than one fact table can update a dimension table.

❼ We want to use the *Implicit Insert* operation for the key for our dimension tables. Since the variable name in the AtBats data is Batter_ID and the name in the dimension tables is Player_ID, we assign the Batter_ID value to Player_ID. We then use the LINK statement to execute the code to update the relevant dimension tables.

❽ Once all the AtBats transactional rows have been read, use the OUTPUT method to output the updated AtBats fact table.

⑨   Repeat the same process as used for the AtBats data for the Pitches data. Note that we update the Players dimension table (the LINK statement) only in order to handle the case of a pitcher who is replaced before his first plate appearance (e.g., a row in the AtBats data set).

⑩   Repeat the process for the Runs data. Since our programs to generate the data do not use pinch runners, this data is not used in updating either of the players dimension tables. In the follow-on project to our Proof of Concept, that could be added by simply coding in the appropriate RENAME and LINK statements.

⑪   Our DATA step reads the input data using an explicit DO loop. There is only one execution of the DATA step. At this point in the program logic, all the data has been read and so we use the OUTPUT method to output the updated dimension tables.

⑫   As discussed in "Handling Type 1 Dimension Tables," the REPLACE method performs the needed operation to either *Insert* into or *Update* the hash table. The date-related fields were not included in the bizarro.AtBats data set and so they are re-created here from the Date field. Creating date-related fields in a dimension table is a common practice and doing this here means that analysis programs do not have to recalculate these fields.

⑬   This is comparable code as discussed in "Inserting Facts into an SCD Table."

⑭   This is comparable code as discussed in "Handling Type 2 Dimension Tables."

7.5 Summary

Our business users were positive about the sample code provided in this chapter that demonstrated how SAS hash table operations could be used to support the creation of a data warehouse of the results of Bizarro Ball games. They were especially positive about the metadata-driven approach that was used in order to minimize hard-coding as they expect that they will be adding other kinds of data fields in the future.

Given that our hope and expectation is that the business users will engage us to create an application to manage their data and calculate the metrics they want to make available, the sample programs included in the following chapters will be making use of this data warehouse structure instead of the originally generated Bizarro Ball data sets in order to emphasize the applicability of using hash tables to both create a star schema data warehouse (this chapter) as well as to calculate metrics in order to answer business intelligence questions (the next chapters).

Chapter 8: Creating Data Aggregates and Metrics

8.1 Overview

Creating data aggregates and metrics is a critical component of any business intelligence effort. Our users have asked us to provide a variety of examples that can be used by their management as well as the teams, players, and especially the fans to better understand the on-the-field performance metrics that are driving the success of the game. There is a long history of standard baseball metrics that are currently in use that they have asked us to replicate. However, they have also expressed an interest in how the SAS hash object, when combined with other SAS tools, can be used to create new metrics and perform what-if type analyses.

8.2 Creating Simple Aggregates

Our first example is calculating what is commonly referred to as a player's slash line which is made up of the following 4 metrics:

- Batting Average (BA) – The rate a which a player gets a hit. It is calculated as the total number of hits divided by the total number of at bats (ABs). Note that baseball does distinguish between at bats and plate appearances (PAs). Every time a player goes *up to home plate to hit* is counted as a plate appearance. At bats exclude plate appearances where the batter reached base other than by getting a hit: for example, the batter walked, was hit by a pitch, or hit a sacrifice.

- On-Base Percentage (OBP) – The rate a which a player gets on base. The denominator in this calculation includes all plate appearances, and the numerator includes walks and hit by pitch plate appearances as well as hits (_Reached_Base in the code below).

- Slugging Percentage (SLG) – Similar to the batting average, but the numerator is the total number of bases (_Bases in the code below): a single is 1, a double is 2, a triple is 3, and a home run is 4. It is a measure of what is referred to as power hits (e.g., a double is more valuable than a single, and so on).

- On-Base Plus Slugging (OPS) – A measure that combines the power number with the frequency of getting on base and is calculated as OBP+SLG.

The following SAS program (which is also the code on the cover of the book) calculates the slash line for each player. The source data is the AtBats data (which as stated above contains one row for each plate appearance for each player).

Program 8.1 - Chapter 8 Slash Line.sas

```
data _null_;
 dcl hash slashline(ordered:"A");          ❶
 slashline.defineKey("Batter_ID");          ❷
 slashline.defineData("Batter_ID","PAs","AtBats","Hits","_Bases","_Reached_Base"
                   ,"BA","OBP","SLG","OPS");          ❸
 slashline.defineDone();
 format BA OBP SLG OPS 5.3;          ❹
 do until(lr);
    set dw.AtBats end = lr;          ❺
    call missing(PAs,AtBats,Hits,_Bases,_Reached_Base);          ❻
    rc = slashline.find();          ❼
    PAs            + 1;          ❽
    AtBats         + Is_An_AB;
    Hits           + Is_A_Hit;
    _Bases         + Bases;
    _Reached_Base  + Is_An_OnBase;
    BA = divide(Hits,AtBats);          ❾
    OBP = divide(_Reached_Base,PAs);
    SLG = divide(_Bases,AtBats);
    OPS = sum(OBP,SLG);
    slashline.replace();          ❿
 end;
 slashline.output(dataset:"Batter_Slash_Line(drop=_:)");          ⓫
run;
```

❶ Create the hash table and use the ORDERED argument tag so our output data set will be sorted by the player id number (which in this case is the Batter_ID field).

❷ Define the only key item for this table – the player id number. We will be adding other keys later in this chapter.

❸ Define the data items for this table. It includes both the final metrics that we want to include in our output along with the *sufficient statistics* that are used to do those calculations.

❹ A FORMAT statement is used so the metrics have the desired display format. Note that attributes like labels, formats, and so on are inherited by the hash variables from the corresponding PDV host variables and are carried along into the output data sets created by the OUTPUT method of the hash object.

❺ Use an explicit file-reading loop (the DoW-loop) to read all the AtBats data from the star schema data warehouse. The DW libref was defined in our autoexec.

❻ Initialize all the values we are aggregating to missing to handle the case of a Player_ID key-value not yet added to the hash table. We don't want the PDV host variables values to persist from the previous row.

❼ The FIND method performs the direct *Retrieve* operation: It searches the table for the Player_ID value currently in the PDV, and, if it is found, retrieves the values of the data portion variables from the item with this key-value. If the key-value is not found, the non-key PDV host variables remain missing; if it is found, their values are overwritten with the values of the corresponding hash variables.

❽ Aggregate the values from the AtBats data. Note that the Is_A... variables are Booleans (0/1) that can simply be aggregated to the relevant count (i.e., sum) value. Note the use of the _ prefix in the variable names for those variables we do not want included in our output data set.

❾ These metrics need to be calculated only after we have read all the data for a given player. However, by calculating them here we can avoid having to add an *enumeration* loop using the hash iterator object. As each data row is read, the calculations are updated.

❿ The REPLACE method performs the *Update All* operation by overwriting the data portion variables for the items with the Player_ID values already in the table or inserting new items for the Player_ID values not already found in the table.

⓫ Create an output data set with the results, once we have read all the data. Note the data set option **drop=_:** which excludes from the output data set any hash variable whose name begins with an underscore (_).

Running the above program will create the following output (note that only the first 5 Player_ID values are shown).

Output 8.1 Slash Line Metrics by Player

Batter ID	PAs	AtBats	Hits	BA	OBP	SLG	OPS
10090	100	92	33	0.359	0.410	0.902	1.312
10103	648	590	226	0.383	0.438	0.851	1.289
10114	667	612	251	0.410	0.459	0.958	1.416
10130	673	611	218	0.357	0.416	0.836	1.252
10132	102	95	36	0.379	0.422	0.874	1.295

Upon reviewing this output we can confirm that the calculations are correct, and our users have commented positively on the simplicity of the calculations. They have expressed the following requests:

- The name of the player needs to be included in the output results. In a traditional SAS program this would require additional PROC SORT and DATA steps or a PROC SQL step.
- They would also like to calculate the slash line for each team.
- They would like to include an additional metric – the number of games a player has played in.

We will address these requests in the following subsections.

8.2.1 Getting Variables from Other Tables

In order to address the first two concerns, we need to perform a table lookup operation – in other words, the *Search* operation, followed by the *Retrieve* operation to get the player's first and last name and his team. All three of these fields are available in our Players dimension table that was created in Chapter 7. For the purposes of this first, simple example we will assume that we need to get only the player's current name and team fields. We will present a more precise example (i.e., that deals with the possibility of players changing teams) later in this chapter.

8.2.1.1 Adding the Player Name and Team

The following program adds the player's name and the name of his team field to our output SAS data set. Note that the changes between this program and the previous one are highlighted in bold text.

Program 8.2 - Chapter 8 Slash Line with Name.sas

```
data _null_;
 dcl hash slashline(ordered:"A");
 slashline.defineKey("Last_Name","First_Name","Batter_ID");      ❶
 slashline.defineData("Batter_ID","Last_Name","First_Name","Team_SK"
                     ,"PAs","AtBats","Hits","_Bases","_Reached_Base"
                     ,"BA","OBP","SLG","OPS");
 slashline.defineDone();
 if 0 then set dw.players(rename=(Player_ID=Batter_ID));      ❷
 dcl hash players(dataset:"dw.players(rename=(Player_ID=Batter_ID))"      ❸
                  ,duplicate:"replace");
 players.defineKey("Batter_ID");
 players.defineData("Batter_ID","Team_SK","Last_Name","First_Name");
 players.defineDone();
 format BA OBP SLG OPS 5.3;
 do until(lr);
    set dw.AtBats end = lr;
    call missing(Last_Name,First_Name,Team_SK      ❹
                ,PAs,AtBats,Hits,_Bases,_Reached_Base);
    players.find();      ❺
    rc = slashline.find();
    PAs           + 1;
    AtBats        + Is_An_AB;
    Hits          + Is_A_Hit;
    _Bases        + Bases;
    _Reached_Base + Is_An_OnBase;
    BA = divide(Hits,AtBats);
    OBP = divide(_Reached_Base,PAs);
    SLG = divide(_Bases,AtBats);
    OPS = sum(OBP,SLG);
    slashline.replace();
 end;
 slashline.output(dataset:"Batter_Slash_Line(drop=_:)");
run;
```

❶ Add fields to the definition of the *slashline* hash table. Using Last_Name and First_Name as the first leading variables in the composite key along with the ORDERED:"A" argument tag will result in the output data set being sorted by the player's name.

❷ Define the player data fields to the PDV. We need to do this here so the fields are defined to the DATA step compiler before they are referenced later in this step (i.e., the CALL MISSING function).

❸ Define the hash table used to look up the needed variables and load the data using the DATASET argument tag. Since we want the current data values and we know that the dimension table is sorted in the order the rows were added for each player (by date), use the DUPLICATE argument tag, so the *last* row for each player is loaded into the hash table.

❹ Add fields to the CALL MISSING call routine function so that the values are not carried forward from previous rows.

❺ Use the FIND method call (unassigned since we know it cannot fail under the circumstances) to perform the *Retrieve* operation extracting the needed fields from the hash table.

Running the above program will create the following output (note that only the first 5 Player_ID values are shown – sorted by the player name fields).

Output 8.2 Slash Line Metrics by Player with Their Names

Batter ID	Last Name	First Name	Team SK	PAs	AtBats	Hits	BA	OBP	SLG	OPS
40122	Adams	Jack	325	743	672	263	0.391	0.450	0.878	1.328
53976	Adams	Jacoby	259	714	656	241	0.367	0.419	0.884	1.303
42120	Adams	Joe	317	682	605	225	0.372	0.443	0.881	1.324
16382	Adams	Jordan	353	128	110	37	0.336	0.430	0.918	1.348
28562	Adams	Joseph	325	102	95	36	0.379	0.422	0.863	1.285

And as we (and you, the reader) had expected, upon reviewing this, our users asked why the team level slash lines weren't there and why we did not add the team name to the table. We responded by pointing out that we were building up to that, and we will address both of those in the next section.

8.2.1.2 Adding Another Aggregation Level – Team Slash Lines

There are a number of approaches to adding levels of aggregation to our sample program. Our users have asked us to add slash lines at the Team level. The approach shown here will add Team and League level aggregation. It will do that by making the assumption that the aggregation is purely hierarchical. In other words, a player rolls up to one and only one team; and teams roll up to one and only one league. We know, however, that players can play for different teams; and that it is a regular occurrence for the real Bizarro Ball data, even if it is not common for our sample (generated data).

In this section, at the request of our users, we will describe an approach that assumes purely hierarchical roll-ups. Later, in the section "Creating Multi-Way Aggregates," we will revisit this assumption and provide alternatives that do not make that assumption and allow for non-hierarchical roll-ups.

The following program illustrates this approach based on the assumption that a player rolls up to a single team and a team rolls up to a single league. As in the previous section, the changes between this program and the previous one are highlighted in bold text.

Program 8.3 - Chapter 8 Team and Player Slash Line.sas

```
data _null_;
 dcl hash slashline(ordered:"A");
 slashline.defineKey("League","Team_Name","Last_Name"   ❶
                  ,"First_Name","Batter_ID");
 slashline.defineData("League","Team_Name","Batter_ID","Last_Name","First_Name"
                  ,"PAs","AtBats","Hits","_Bases","_Reached_Base"
                  ,"BA","OBP","SLG","OPS");
 slashline.defineDone();
 if 0 then set dw.players(rename=(Player_ID=Batter_ID))
               dw.teams      ❷
               dw.leagues;
 dcl hash players(dataset:"dw.players(rename=(Player_ID=Batter_ID))"
                  ,duplicate:"replace");
 players.defineKey("Batter_ID");
 players.defineData("Batter_ID","Team_SK","Last_Name","First_Name");
 players.defineDone();
 dcl hash teams(dataset:"dw.teams");  ❸
 teams.defineKey("Team_SK");
 teams.defineData("League_SK","Team_Name");
 teams.defineDone();
 dcl hash leagues(dataset:"dw.leagues");  ❹
 leagues.defineKey("League_SK");
 leagues.defineData("League");
 leagues.defineDone();
 format BA OBP SLG OPS 5.3;
 do until(lr);
    set dw.AtBats end = lr;
    call missing(Last_Name,First_Name,Team_SK
               ,PAs,AtBats,Hits,_Bases,_Reached_Base);
    players.find();
    teams.find();      ❺
    leagues.find();      ❻
    link slashline;      ❽
    call missing(Batter_ID,Last_Name,First_Name);      ❾
    link slashline;
    call missing(Team_Name);   ❿
    link slashline;
 end;
 slashline.output(dataset:"Batter_Slash_Line(drop=_:)");
 return;
 slashline:      ❼
    rc = slashline.find();
    PAs           + 1;
    AtBats        + Is_An_AB;
    Hits          + Is_A_Hit;
    _Bases        + Bases;
    _Reached_Base + Is_An_OnBase;
    BA = divide(Hits,AtBats);
    OBP = divide(_Reached_Base,PAs);
    SLG = divide(_Bases,AtBats);
    OPS = sum(OBP,SLG);
    slashline.replace();
 return;
run;
```

❶ Add the fields for League and Team level aggregates as both key and data portion variables in the hash table. Team_SK has been replaced by the Team_Name field.

❷ Define the League and Team data sets fields as PDV host variables.

❸ Create the hash table that will allow us to look up and extract (i.e., *Retrieve*) the team name.

❹ Create the hash table that will allow us to look up and extract (i.e., *Retrieve*) the league name.

❺ Use the FIND method (called unassigned since we know it cannot fail) to *Retrieve* the value for Team_Name, as well as League_SK (needed for the next lookup).

❻ Thanks to the prior lookup, we can now call the FIND method (unassigned since we know it cannot fail) to *Retrieve* the value for League_Name.

❼ We have moved the code to update the hash table to a LINK-RETURN section, so we can invoke it multiple times in this DATA step without replicating it. We need to run this code at the Player, Team, and League levels.

❽ Use the LINK statement to call the code that updates the Player level data in the hash object.

❾ Setting the player fields to missing allows us to create/update an item in the hash table for Team level aggregates by invoking the LINK statement.

❿ Setting the team field to missing allows us to create/update an item in the hash table for League level aggregates by invoking the LINK statement.

Running the above program will create the following output (note that only the first 5 rows are shown). Thanks to the ORDERED:"Y" argument tag and how missing values sort, note that the data is interleaved for us, so we see the League level numbers, then Teams within each League, and then the Players within each Team.

Output 8.3 Slash Line Metrics by League, Team, and Player

League	Team Name	Batter ID	Last Name	First Name	PAs	AtBats	Hits	BA	OBP	SLG	OPS
Eastern		.			143862	130485	48247	0.370	0.428	0.877	1.305
Eastern	Bears	.			9040	8222	3039	0.370	0.427	0.893	1.319
Eastern	Bears	21802	Allen	Terry	97	85	27	0.318	0.402	0.812	1.214
Eastern	Bears	31654	Brown	Austin	81	74	25	0.338	0.395	0.743	1.138
Eastern	Bears	37647	Brown	Bryan	137	123	53	0.431	0.489	0.967	1.457

Our users liked how we did this. But users, being users, did express some concerns.

First, they were confused after looking at the complete results, that all the teams and batters seemed so similar based on their calculated slash lines. They were concerned that there was something wrong in the calculations. We had to remind them that we had used random numbers to generate the data and did not add any logic to force differences between players and teams. Once we reassured them that the program would show differences after it was run on real game data, their concern was allayed.

Second, they wanted to know how this approach handled the calculations for players who had changed teams. We clarified what we meant when we said that this approach assumed hierarchical roll-ups and that in the context of this question that meant that the entire history for a player was associated with his current team. Given the standard disclaimer that *past performance is not necessarily predictive of future performance*, we further suggested that the calculation of the slash line using this approach could be viewed as a *best guess* for what to expect at team level based on the current make-ups of the teams. And we reminded them that later in this chapter we will illustrate an alternative approach that associates the player's performance with the team they played for at the time of each game or at bat.

Luckily, they liked this way of looking at the data and thought it might provide some interesting insights once the programs were run on the AtBats data from actual games.

8.2.2 Calculating Unique Counts

The third request that our users asked us to address was how could the count of the number of distinct games be added to the slash line calculation.

Counting the number of distinct values is something that is quite common; but it is not a standard metric included in the SAS data summarization tools dealing with additive descriptive statistics, such as the SUMMARY or the MEANS procedures. It can be calculated with other standard SAS tools, though. For example, using the SQL procedure, one can code something like:

```
proc sql;
 create table unique as
 count(distinct Game_SK) as Games
 from dw.AtBats
 group Player_ID;
quit;
```

Alternatively, the data could be sorted and then the DATA step FIRST.X/LAST.X logic could be used, such as:

```
proc sort data = dw.AtBats(keep=Player_ID Game_SK)
          out = AtBats nodupkey;
 by Player_ID Game_SK;
run;
data unique;
 /* note that a DoW-loop could be used here as well */
 set AtBats;
 by Player_ID;
 Games + 1;
 If last.Player_ID;
 output;
 Games = 0;
run;
```

Most SAS programmers could come up with any number of ways to count distinct values.

Another issue with counting distinct values is that the statistic's values are not additive. In other words, you cannot add up the count of distinct games calculated at the Player_ID level to get the value at the Team or League level. Code must be written to perform that calculation at each desired level.

Fortunately, the hash object, being a keyed, in-memory table, provides an efficient and easy way to implement calculating the count of distinct values. The approach is simple:

1. Define a table that contains the current list of all the values.
2. As a new data row is encountered, check to see if the key from that row already exists in the table.
3. If the key is found, do nothing.
4. Otherwise, increment the count of distinct values by 1 and add the new value to the table, so we know that we have already encountered it.

In other words:

1. Define a hash table that has the desired grouping variables as key items.
2. Issue an assigned call to the ADD method.
3. If the item with the key-value that the ADD method accepted is found, the ADD method return code is 0, and so we do nothing.
4. Otherwise, the ADD method return code is non-zero, and so we increment our count variable by 1.

The following program adds the distinct count of Games at the Player, League, and Team level. As in the previous section, the changes between this program and the previous one are highlighted in bold text. Of particular note is how few changes were needed to add the count of distinct Games to our slash line aggregate.

Program 8.4 - Chapter 8 Slash Line with Unique Count.sas

```
data _null_;
 dcl hash slashline(ordered:"A");
 slashline.defineKey("League","Team_Name","Last_Name","First_Name","Batter_ID");
 slashline.defineData("League","Team_Name","Batter_ID","Last_Name","First_Name"
                     ,"Games","PAs","AtBats","Hits","_Bases","_Reached_Base"      ❶
                     ,"BA","OBP","SLG","OPS");
 slashline.defineDone();
 if 0 then set dw.players(rename=(Player_ID=Batter_ID))
               dw.teams
               dw.leagues;
 dcl hash players(dataset:"dw.players (rename=(Player_ID=Batter_ID))"
                 ,duplicate:"replace");
 players.defineKey("Batter_ID");
 players.defineData("Batter_ID","Team_SK","Last_Name","First_Name");
 players.defineDone();
 dcl hash teams(dataset:"dw.teams");
 teams.defineKey("Team_SK");
 teams.defineData("League_SK","Team_Name");
 teams.defineDone();
 dcl hash leagues(dataset:"dw.leagues");
 leagues.defineKey("League_SK");
 leagues.defineData("League");
 leagues.defineDone();
 dcl hash u();
 u.defineKey("League","Team_Name","Batter_ID","Game_SK");      ❷
 u.defineDone();
 format BA OBP SLG OPS 5.3;
 do until(lr);
```

```
      set dw.AtBats end = lr;
      call missing(Last_Name,First_Name,Team_SK,Games      ❹
                  ,PAs,AtBats,Hits,_Bases,_Reached_Base);
      players.find();
      teams.find();
      leagues.find();
      link slashline;
      call missing(Batter_ID,Last_Name,First_Name,Games);      ❹
      link slashline;
      call missing(Team_Name,Games);      ❹
      link slashline;
   end;
   slashline.output(dataset:"Batter_Slash_Line(drop=_:)");
   return;
   slashline:
      rc = slashline.find();
      Games + (u.add() = 0);      ❸
      PAs             + 1;
      AtBats          + Is_An_AB;
      Hits            + Is_A_Hit;
      _Bases          + Bases;
      _Reached_Base + Is_An_OnBase;
      BA = divide(Hits,AtBats);
      OBP = divide(_Reached_Base,PAs);
      SLG = divide(_Bases,AtBats);
      OPS = sum(OBP,SLG);
      slashline.replace();
   return;
run;
```

❶   Add the variable Games which will be the count of unique games played to our hash table.

❷   Create a hash table that has the same keys as our slashline hash table, with the addition of the Game_SK variable. This creates a hash table whose keys represent the list of unique combinations of League, Team_Name, Batter_ID, and Game_SK. Note that since the fields Last_Name and First_Name do not contribute to the uniqueness constraint they need not be included here. They are included only as keys in the *slashline* hash table in order to have the output SAS data set sorted by the player name.

❸   If the return code from the ADD method is 0, that means that this is a new combination of League, Team_Name, Batter_ID, and Game_SK. Thus, we need to increment the value of the variable Games.

❹   Add the Games variable to the CALL MISSING routine, so it is initialized before linking to the section that includes the calculations including the incrementing of the Games fields. The first call to the link section increments the Games field by 1 if the Game_SK value is new at the Player_ID level. The second call to the link section increments the Games field by 1 if the Game_SK value is new at the Team_Name level. The third call to the link section increments the Games field by 1 if the Game_SK value is new at the League level.

Running the above program will create the following output (note that only the first 5 rows are shown). Thanks to the ORDERED:"Y" argument tag and how missing values sort, note that the data is interleaved for us, so we see the League level numbers, then teams within each league and then the players within each team.

Output 8.4 Slash Line Metrics Including Count of Unique Games

League	Team Name	Batter ID	Last Name	First Name	Games	PAs	AtBats	Hits	BA	OBP	SLG	OPS
Eastern		.			1440	143862	130485	48247	0.370	0.428	0.877	1.305
Eastern	Bears	.			180	9040	8222	3039	0.370	0.427	0.893	1.319
Eastern	Bears	21802	Allen	Terry	97	97	85	27	0.318	0.402	0.812	1.214
Eastern	Bears	31654	Brown	Austin	31	81	74	25	0.338	0.395	0.743	1.138
Eastern	Bears	37647	Brown	Bryan	53	137	123	53	0.431	0.489	0.967	1.457

Since we know that there are 16 teams in each league and each team plays each other 12 times, the values of the Games field at the Team Level (12*15=180) and at the League level (each game involves two teams (180*16/2=1440)) are exactly what we have expected.

8.2.3 Calculating Medians, Percentiles, Mode, and More

Calculating percentiles (including the median) and the mode are supported by PROC UNIVARIATE. And, of course, custom code can be written to calculate these metrics. Anticipating that there will be requests for such metrics, we proceeded to determine how to calculate these metrics using the hash object in the DATA step. Having these values available in the DATA step, so they can be integrated with other calculations, will certainly find some use cases.

Tabulating the list of unique values that was needed in order to count the number of distinct values triggered the *light bulb* with an idea for how to simply calculate the median. We can merely create a hash table that contains all the values. So let us look at the program that does just this to calculate the median of the field Distance which is a measure of the distance from home plate for each hit.

Program 8.5 - Chapter 8 Median.sas

```
data _null_;
 dcl hash medianDist(dataset:"dw.AtBats(where=(Distance > 0))"   ❶
                 ,multidata:"Y"
                 ,ordered:"A");
 medianDist.defineKey("Distance");
 medianDist.defineDone();
 dcl hiter iterM("medianDist");   ❷
 iterM.first();   ❸
 do i = 1 to .5*medianDist.num_items - 1;
    iterM.next();   ❹
 end;
 /* could add logic here to interpolate if needed */
 put "The Median is " Distance;   ❺
 stop;
 set dw.AtBats(keep=Distance);
run;
```

❶ Define a hash table and load it with all the positive values of Distance. Distance can be 0 or missing for the results of at bats that did not result in a hit. The MULTIDATA:"Y" argument tag makes sure all the values are loaded into the table; the ORDERED argument tag forces the values to be inserted in sorted order by Distance.

❷ Since we need to get to the hash item corresponding to the half-way point, we need a hash iterator to do that.

❸ The hash iterator needs to be set to point inside the table. We decided to use the FIRST method to point to the first item.

❹ Use the hash iterator to loop through the hash table to get to the midpoint. We use the NUM_ITEMS attribute to determine how many items the hash table has, multiply that value by .5, and then subtract 1 since we have already pointed to the first row. So after the i=1 loop iteration we are pointing to the second row in the hash table (and so on).

❺ For this Proof of Concept (PoC) we just write the value to the log. Also note that for the purposes of this PoC we have decided not to worry about interpolating between two values for the case when the number of rows is even. That is a nuance that our business users agreed could be dealt with later.

Running the above program will create the following output in the SAS log.

```
NOTE: There were 209970 observations read from the data set DW.ATBATS.
      WHERE Distance>0;
The Median is 204
NOTE: DATA statement used (Total process time):
      real time           0.24 seconds
      cpu time            0.12 seconds
```

A sample use case for this might be to use the median value to filter or subset the data in our current DATA step program.

8.2.3.1 Calculating Multiple Medians

Whenever we calculate metrics, we almost always need to calculate more than one metric. For example, we might want to also calculate the median value of the *Direction* field which measures the angle of the hit relative to home plate and the baseline. Alternatively, perhaps we want to calculate and compare the median distances for the different types of hits (e.g., singles, doubles, triples, home runs).

Using the approach shown above we would need a separate hash table for each required median that we wish to calculate. That could quickly escalate to the point where we have to create a lot of hash tables, which requires lots of code and perhaps too much memory.

Our first thought is to create four hash tables (e.g., single, double, triple, homerun) and four hash iterators (e.g., singleIter, doubleIter, tripleIter, homerunIter) and use arrays to loop through them. Arrays are commonly used when we need to perform the same operation on multiple variables – by simply looping through them. We expected that looping through our hash objects using an array would be a simple way to address this.

Program 8.6 Chapter 8 Multiple Medians Array Error.sas

```
data _null_;
 dcl hash single(dataset:"dw.AtBats (where=(Result='Single'))"     ❶
                ,multidata:"Y",ordered:"A");
 single.defineKey("Distance");
 single.defineDone();
 dcl hiter singleIter("single");
 .
 .
 .
 array _hashes(*) single double triple homerun;     ❷
 array _iters(*) singleIter doubleIter tripleIter homerunIter;
 stop;
run;
```

❶ Define the hash table for singles just as in the prior example. Comparable code is used (but not shown here) for doubles, triples, and home runs.

❷ Define the arrays needed to loop through the hash objects and their respective iterators.

The SAS log resulting from this approach highlights that is not an option.

```
670   array _hashes(*) single double triple homerun;
ERROR: Cannot create an array of objects.
ERROR: DATA STEP Component Object failure.
      Aborted during the COMPILATION phase.
NOTE: The SAS System stopped processing this step because of errors.
NOTE: DATA statement used (Total process time):
      real time           0.03 seconds
      cpu time            0.01 seconds

671   array _iters(*) singleIter doubleIter tripleIter homerunIter;
```

At first we were surprised at this error. But then realized that it made sense since neither a hash object, nor a hash iterator create numeric or character variables. As the error implies, they are objects (e.g., pointers to memory locations) as was discussed in Part 1.

Since we are just exploring this requirement as a PoC, for now we just present a simple example that calculates the median for each type of hit by repeating the code to create the hash objects and their iterators. Should our users decide to proceed with a follow-on effort we pointed out that we could use the macro language to minimize hard-coding. We also told them that an alternative to looping through multiple hash object would be described Chapter 9 Hash of Hashes – Looping Through SAS Hash Objects.

The following program illustrates an approach as an interim hard-coded solution that takes advantage of the LINK statement in the DATA step. The LINK statement is a useful feature that allows us to repeatedly execute the same block of code multiple times in a given DATA step; but without having to cut and paste the code into multiple locations.

Program 8.7 - Chapter 8 Multiple Medians.sas

```
data Medians;
 length Type $12;
 keep Type Distance Count;
 format Count comma9.;

 dcl hash h();        ❶
 h = _new_ hash(dataset:"dw.AtBats(where=(Result='Single'))"
                ,multidata:"Y",ordered:"A");
 h.defineKey("Distance");
 h.defineDone();
 type = "Singles";
 dcl hiter iter;      ❷
 iter = _new_ hiter("h");
 link getMedians;     ❸

 h = _new_ hash(dataset:"dw.AtBats (where=(Result='Double'))"     ❶
                ,multidata:"Y",ordered:"A");
 h.defineKey("Distance");
 h.defineDone();
 type = "Doubles";
 iter = _new_ hiter("h");       ❷
 link getMedians;     ❸

 h = _new_ hash(dataset:"dw.AtBats(where=(Result='Triple'))"     ❶
                ,multidata:"Y",ordered:"A");
 h.defineKey("Distance");
 h.defineDone();
 type = "Triples";
 iter = _new_ hiter("h");       ❷
 link getMedians;     ❸

 h = _new_ hash(dataset:"dw.AtBats(where=(Result='Home Run'))" ❶
                ,multidata:"Y",ordered:"A");
 h.defineKey("Distance");
 h.defineDone();
 type = "Home Runs";
 iter = _new_ hiter("h");       ❷
 link getMedians;     ❸

 stop;
 getMedians:
    Count = h.num_items;        ❹
    iter.first();     ❺
    do i = 1 to .5*Count - 1;
      iter.next();
    end;
```

```
    /* could add logic here to interpolate if needed */
    output;      ❻
    h.delete();      ❼
 return;

 set dw.AtBats(keep=Distance);      ❽
run;
```

❶ Create an instance of hash object H for each of the four types of hits we want to calculate the median of. By using the _NEW_ operator to create a new hash object instance, we can reuse the same name for the hash object. That facilitates reusing the same code to do the median calculation.

❷ Create the requisite corresponding instances of hash iterator object ITER. Just as for the hash object creation, use the _NEW_ operator.

❸ Use the LINK statement to branch to the code that does the actual calculation of the medians. The RETURN statement at the end of the LINK block returns control to the statement after the LINK statement.

❹ Use the NUM_ITEMS attribute to get the number of items in the hash table. By assigning that value to a SAS numeric variable, we can include in the output the number of hits of each type.

❺ Use the same DO loop construct as in the Program 8.5 to get to the midpoint of the number of items. That value for the distance from the midpoint item (subject to the interpolation comment above) is the median.

❻ Once we have determined the *median* value, output the value to a SAS data set.

❼ While not necessary for the program to run (since these hash tables are reasonably small), using the DELETE method frees up the memory (as would the CLEAR method), making it available for the rest of the DATA step. This could be important if there are lots of medians to calculate when combined with lots of distinct values.

❽ Tell the DATA step compiler to create the PDV host variables with the same names as the hash variables defined via the DEFINEKEY and DEFINEDATA methods.

Running the above program will create the following output.

Output 8.5 Multiple Medians Values

Type	Count	Distance
Singles	43,509	56
Doubles	26,462	251
Triples	8,709	350
Home Runs	17,622	436

8.2.3.2 Calculating Additional Percentiles

Recognizing that the median is simply the 50th percentile allows us to generalize the simple program above so it is not hard-coded for the 50th percentile as well as to allow for multiple percentiles. Given that we have a hash table that has all the values in sorted order, we just need to tweak the code so it can step through the desired percentiles in order and identify which *item* in the hash table represents that same percentile.

The following program does that.

Program 8.8 - Chapter 8 Percentiles.sas

```
data Percentiles;
 keep Percentile Distance;     ❶
 format Percentile percent5.;
 dcl hash ptiles(dataset:"dw.AtBats (where=(Distance gt 0))"     ❷
                ,multidata:"Y",ordered:"A");
 ptiles.defineKey("Distance");
 ptiles.defineDone();
 dcl hiter iterP("ptiles");     ❸
 array _ptiles(6) _temporary_ (.5 .05 .1 .25 .75 .95);     ❹
 call sortn(of _ptiles(*));
 num_items = ptiles.num_items;     ❺
 do i = 1 to dim(_ptiles);     ❻
    Percentile = _ptiles(i);
    do while (Counter lt Percentile*num_items);     ❼
       Counter + 1;     ❽
       iterP.next();
    end;
    /* could add logic here to read
       next value to interpolate */
    output;     ❾
 end;
 stop;
 set dw.AtBats;     ❿
run;
```

❶ Specify the variables (and format) for the variables to be included in the output data set.

❷ Define the hash table to contain all the values for the subject variable (Distance) in ascending order just as in the above example that calculated the median.

❸ Define the needed hash iterator.

❹ Specify the desired percentiles using an ARRAY. Note that the values are not in the needed sort order (we added this wrinkle as one of the key IT users asked if the values needed to be specified in sort order). Since they are needed in ascending order, use the SORTN call routine to ensure the values are in sorted order. The SORTN call could be omitted if we could safely assume that the values were listed in ascending order.

❺ Get the number of items in the list so we can use that to determine which row number corresponds to the value for the desired percentile.

❻ Loop through the percentiles in order.

❼ This loop is similar to the loop used for the median calculation. We use the NEXT method to step through the data, stopping when we have gotten to the row number corresponding to that percentile (subject to the same interpolation comment as discussed above).

❽ The Counter variable keeps track of how many rows have been read so far. That allows us to avoid restarting from the first row for each desired percentile.

❾ Once we have reached the hash item for the current percentile, the value of the PDV host variable is the value of the Distance variable for that item. So we output the values to our data set.

❿ As in the above example, a SET statement after the STOP statement is used for parameter type matching- i.e., to define the needed host variables to the PDV.

Running the above program will create the following output.

Output 8.6 Multiple Percentiles

Percentile	Distance
5%	22
10%	36
25%	76
50%	204
75%	296
95%	427

Enhancing this program to deal with multiple variables will be addressed in Chapter 9 Hash of Hashes – Looping Through SAS Hash Objects.

8.2.3.3 Calculating the Mode

The definition of the mode statistic is the value in the data that occurs most often. In order to calculate the mode, we need to know how many times each distinct value occurs. So, unlike the median which has one row for each individual value in our hash table, our hash table should contain one row for each distinct value along with a count for how many times that values was encountered. Another *structural* difference between the mode and the other metrics we've examined so far is that there can be multiple values for the mode.

The following program will determine the mode for Distance, our metric of interest.

Program 8.9 - Chapter 8 Mode.sas

```
data Modes;
 keep Distance Count;
 format Distance Count comma5.;    ❶
 dcl hash mode();    ❷
 mode.defineKey("Distance");    ❸
 mode.defineData("Distance","Count");    ❹
 mode.defineDone();
 dcl hiter iterM("mode");    ❺
 do until(lr);
```

```
      set dw.AtBats(keep=Distance) end=lr;    ❻
      where Distance gt .;
      if mode.find() ne 0 then Count = 0;    ❼
      Count + 1;    ❽
      mode.replace();    ❾
      maxCount = max(Count,maxCount);    ❿
   end;
   do i = 1 to mode.num_items;    ⓫
      iterM.next();    ⓬
      if Count = maxCount then output;    ⓭
   end;
   stop;
 run;
```

❶ Specify the variables (and format) for the variables to be included in the output data set.

❷ Define the hash table to contain all the values for the subject variable (Distance). Note that the MULTIDATA argument tag is not specified, as we want to have only one hash table item for each distinct value of Distance; likewise, the ORDER argument tag is not specified, as the internal order of the Distance values is not relevant to calculating the mode.

❸ Define the Key variable. We want only one item for each distinct value of Distance.

❹ Define the Data variables. The Count variable will be used to determine how many times each value of Distance occurs in the data. As mentioned above, the definition of the mode statistic is the value in the data that occurs most often. The desired output is the Distance value (or values) with the largest value of Count; we need to include Distance as a data variable so its value can be included in the output data set.

❺ We will need a hash iterator to perform the *Enumerate All* operation- i.e., to loop through all the items in the hash table, so that we can find all the Distance value(s) that are the mode.

❻ Use a DoW-loop to read the data, reading only those data observations where Distance has a value.

❼ If the value of Distance for the PDV host variable is not found in the hash table, initialize the Count variable to 0; if the value is found, the PDV host variable (COUNT) is updated using the hash variable value.

❽ Increment the Count variable.

❾ The REPLACE method will add the item to the hash table if it is not already found (with a value of 1 for Count); otherwise it will update the current item with the incremented value of COUNT in the hash table.

❿ As we read the data, determine what the largest value of Count is. Once we have read all the data, all the Distance items with that value for Count are mode values.

⓫ Loop explicitly through the hash table using the hash iterator.

⓬ Use the NEXT method to retrieve the Distance and Count values for the current item.

⓭ Output a row to our output SAS data set for every Distance value that has the maximum value for Count.

Running the above program will create the following output. Note that we have two values of Distance which are the mode.

Output 8.7 Modes for the Variable Distance

Distance	Count
76	829
96	829

8.2.3.4 Calculating Percentiles, Mode, Mean, and More in a Single DATA Step

As expected, our users liked the simplicity of the samples to calculate these metrics. But they commented that it would be even better if we could combine all of these calculations into a single program (which we interpreted as a single DATA step program) that could be used as template or starter program. For example:

- Removing calculations that they are not interested in for a given analysis or report.
- Adding other calculations like, for example, the interquartile range.
- Combining multiple metrics in order to create a filter. For example, looking at the distribution of the kinds of hits where the Distance is between the mean and median.
- Various What-If analyses such as the variability of these results by factors such as home vs. away team.
- And so on, and so on.

We responded by thanking them for leading us into the very next example that we had already planned to present; while reminding them that our plan is to generalize and parameterize these programs- perhaps later in this Proof of Concept (Chapter 9) or as part of an implementation project.

The following program not only combines the calculation of the mean, median and the mode into a single DATA step, it also integrates the calculation of percentiles as well as a complete frequency distribution. As such, we warned both the IT and business users that it likely would take some time to fully understand it.

Program 8.10 - Chapter 8 MeanMedianMode.sas

```
%let Var = Distance;        ❶

data ptiles;        ❷
 input Ptile;
 Metric = put(Ptile,percent6.);        ❸
 retain Value . ;
 datalines;
 .05
 .1
 .25
 .5
 .75
 .95
;

data _null_;

 format Percent Cum_Percent percent7.2;
```

```
dcl hash metrics(dataset:"ptiles"          ❹
                ,multidata:"Y"
                ,ordered:"A");
metrics.defineKey("Ptile");
metrics.defineData("Ptile","Metric","Value");
metrics.defineDone();
dcl hiter iterPtiles("metrics");          ❺

dcl hash distribution(ordered:"A");        ❻
distribution.defineKey("&Var");
distribution.defineData("&Var","Count","Percent","Cumulative","Cum_Percent");
distribution.defineDone();
dcl hiter iterDist("distribution");        ❼
do Rows = 1 by 1 until(lr);                ❽
   set dw.AtBats(keep=&Var) end=lr;
   where &Var gt .;
   if distribution.find() ne 0 then Count = 0;    ❾
   Count + 1;           ❿
   distribution.replace();          ⓫
   Total + &Var;            ⓬
   maxCount = max(Count,maxCount);
end;

iterPtiles.first();          ⓭
last = .;
do i = 1 to distribution.num_items;        ⓮
   iterDist.next();
   Percent = divide(Count,Rows);         ⓯
   _Cum + Count;          ⓰
   Cumulative = _Cum;
   Cum_Percent = divide(_Cum,Rows);
   distribution.replace();         ⓱
   if Count = maxCount then metrics.add(Key:.,Data:. ,Data:"Mode",Data:&Var);   ⓲
   if last le ptile le Cum_Percent then          ⓳
   do;  /* found the percentile */
      Value = &Var;
      if ptile ne 1 then metrics.replace();       ⓴
      if iterPtiles.next() ne 0 then ptile = 1;
   end; /* found the percentile */
   last = Cum_Percent;           ㉑
end;

metrics.add(Key:.,Data:.,Data:"Mean",Data:divide(Total,Rows));      ㉒

iterDist.first();          ㉓
metrics.add(Key:.,Data:.,Data:"Min",Data:&Var);

iterDist.last();          ㉔
metrics.add(Key:.,Data:.,Data:"Max",Data:&Var);

metrics.output(dataset:"Metrics(drop=ptile)");      ㉕
distribution.output(dataset:"Distribution");      ㉖
```

```
 stop;
 set ptiles;    ㉗
run;
```

❶ A macro variable is used to specify the variable to be analyzed. We wanted to introduce our users to leveraging other SAS facilities like the macro language.

❷ Likewise, we wanted to start illustrating the advantages of parameterizing (e.g., using data-driven approaches) our programs by reading in data values like the percentile values to be calculated.

❸ The variable Metric is created as a character variable so our DATA step can assign character values for other metrics to be calculated (e.g., "Mean").

❹ Define our hash table and load the data using the DATASET argument tag. The ORDERED option is used to ensure that the percentiles are loaded in ascending order as that is needed to support the calculation of the percentiles using the same logic as in the above programs. We set the MULTIDATA argument tag to "Y" since we are calculating other metrics and will set the value of the ptile variable to values like "Mean", "Median", etc., for those other metrics.

❺ A hash iterator is needed so we can loop through (i.e., enumerate) the percentile values in the *metrics* hash table.

❻ Create the hash table which will tabulate the frequency distribution (including the cumulative count) of the analysis variable values. The calculation of the mode requires that we create this table so it has one item per distinct value of our analysis variable. The ORDERED argument tag is needed so we can calculate the cumulative counts and percentages as well as the specified percentiles.

❼ A hash iterator is needed so we can loop through (i.e., enumerate) the values in the *distribution* hash table.

❽ This form of the DoW-loop allows us to loop through all the input data in a single execution of the DATA step and have a value for the Rows variable that is the number of observations read from the input data set.

❾ If the current value of our analysis variable is not found as a key in our Distribution hash table, we initialize the PDV host variable Count to 0; otherwise, if it is found, the value of the Count variable from our hash table updates the value of the PDV host variable.

❿ Increment the value of the PDV host Count variable by 1.

⓫ Use the REPLACE method to update (or add an item) to our hash table.

⓬ The Total and Maxcount variables are needed, respectively, so we can calculate the mean and determine the mode values.

⓭ Once we have read all the data and have the frequency count, get the first (which due to the ORDERED:"A" argument tag is the smallest one) percentile to be calculated and assign a value to the variable *last* which we will use to keep track of the previous cumulative percent.

⓮ Use a DO loop to enumerate through the values of the *distribution* hash table. Each time through the loop use the NEXT method to advance the hash pointer (on the first execution of the loop the NEXT method points to the first item).

⓯ Calculate the percentage for the current item by dividing Count by Rows (the total number of observations).

⑯ Since *Cumulative* is in the hash table, the NEXT method will overwrite the PDV host variable with a missing value (since we have not yet calculated it). So we use a temporary variable, _Cum, to accumulate the running total and then assign the values to the Cumulative PDV host variable followed by the calculation of the cumulative percentage.

⑰ Use the REPLACE method to update the variables in the distribution hash table with the values of *Cumulative* and *Cum_Percent*.

⑱ If the current value of our analysis variable is equal to the largest value, it is a mode value. Therefore, add a row to our hash table of results, *metrics*. Note that multiple mode values are possible. We use an explicit ADD method with a value of missing for *ptile*. This forces the *ptile* value to be inserted before the percentile values are already there. This has two desirable effects: the item is added before the current pointer value and so our NEXT method calls will not encounter them; the items are added to the hash table *metrics* in order within the value for the key which has a missing value.

⑲ If the current percentile that we are calculating is between the previous and current cumulative percent, the value of the analysis variable is our percentile value. Note that, as mentioned above, we could add interpolation logic. That level of precision is not needed in our PoC.

⑳ Our program needs to loop through the entire distribution, including those values beyond the largest desired percentile. These two statements do that. We update the hash table with the value for the current percentile as long as it is not the value of 1 (it is assigned by the next statement). The NEXT method advances to the next item in our *metrics* hash table to get the next percentile to be calculated. If there are no more percentiles, we assign a value of 1 (i.e., 100%) to force processing of the remaining rows in the *distribution* hash table.

㉑ Assign the current cumulative percentage to the *last* variable so it can be used in the comparison the next time through the loop.

㉒ Add the mean to the *metrics* hash table. Note that since we use an explicit ADD method to assign missing values to the argument tags, the item is added to the hash table in order.

㉓ Use the FIRST method to get the value for the first item. By definition that is the minimum value. It is added to the *metrics* hash table (in order).

㉔ Use the LAST method to get the value for the last item. By definition that is the minimum value. It is added to *metrics* hash table (in order).

㉕ Use the OUTPUT method to output our calculated metrics. Note that the *ptile* variable is dropped because we have the character representation of the value.

㉖ Use the OUTPUT method to output the complete distribution.

㉗ As in a number of the previous examples, a SET statement after the STOP statement is used to define the needed variables to the PDV.

Running the above program will create the following output data sets. Note that we have printed only the first 10 observations for the complete distribution.

Output 8.8 The Metrics Data Set

Metric	Value
Mode	76.000
Mode	96.000
Mean	199.002
Min	4.000
Max	480.000
5%	22.000

Metric	Value
10%	36.000
25%	76.000
50%	204.000
75%	296.000
95%	427.000

Output 8.9 The Distribution Data Set – First 10 Rows

Distance	Count	Percent	Cumulative	Cum Percent
4	236	0.11%	236	0.11%
5	243	0.12%	479	0.23%
6	250	0.12%	729	0.35%
7	274	0.13%	1003	0.48%
8	260	0.12%	1263	0.60%
9	286	0.14%	1549	0.74%
10	275	0.13%	1824	0.87%
11	781	0.37%	2605	1.24%
12	708	0.34%	3313	1.58%
13	802	0.38%	4115	1.96%

8.2.3.5 Determining the Distribution of Consecutive Events

Our users have now asked us to provide another example. They would like to determine how many times a team gets at least 1, 2, 3, or more hits in row. And they would also like to calculate a variation of that – how many times do we see exactly 1, 2, 3, or more hits in the row. The distinction between these is that, for example, 4 consecutive hits can be treated differently:

- If we are determining the distribution of at least N consecutive events, then 4 hits in a row would contribute a count of 1 to each of 1, 2, 3, and 4 hits in row.
- If we are determining the distribution of exactly N consecutive events, then 4 hits in a row would contribute a count of 1 to just the case of 4 hits in row.

At first glance, the first calculation seems to be somewhat easier to determine as we don't need to look ahead to see if the next at bat results in a hit. However, since the number of consecutive events is calculated, if the next at bat is not the same event, the number of consecutive events is not updated. The logic is actually calculating how many times we have exactly N consecutive events. We can then use those values to calculate N or more consecutive events.

In statistical terminology, what we are describing here as "consecutive events" is often referred to as a *run* which is defined as a series of increasing values or a series of decreasing values. It is a given that the term *run* has a well-defined meaning in baseball which is different from this definition. Thus, we will continue to refer to this as "consecutive events" even though it is indeed a *run* in statistical terminology.

Such calculations are something that could be done with a DATA step to calculate the value for N, followed by, for example, the FREQ procedure to get the distribution. However, given the example in the previous section, we know that we can create a frequency distribution in a DATA step using the hash object. Our DATA step hash table approach needs to:

- Calculate the value for the number of consecutive hits.
- Count how many times each value occurs (that gives us the count for that many or more).
- Calculate the count for how many times we see exactly N consecutive events.
- Output the results to an output SAS data set.

We decided that the SUMINC argument tag could be used to do this. The SUMINC argument tag is used on the DECLARE statement or with the _NEW_ method. The value provided for this argument tag is a character expression which is the name of a DATA step variable whose value is to be aggregated. This variable is not included as a data variable in the hash table; instead memory is allocated to accumulate (i.e., sum) the value of the designated variable behind the scenes.

There are two issues to be aware of when using SUMINC: first, only one variable can be summed; second, a METHOD call is needed to retrieve the value – it is not explicitly linked or tied to a PDV host variable. For our current example, neither of these is an issue, as can be seen in the following program which calculates the distribution of consecutive hits.

Program 8.11 - Chapter 8 Count Consecutive Events.sas

```
data Consecutive_Hits;
 keep Consecutive_Hits Exact_Count Total_Count;
 format Consecutive_Hits 8.
        Exact_Count Total_Count comma10.;
 retain Exact_Count 1;      ❶
 if _n_ = 1 then
 do;   /* define the hash table */
    dcl hash consecHits(ordered:"D",suminc:"Exact_Count");      ❷
    consecHits.defineKey("Consecutive_Hits");      ❸
    consecHits.defineDone();
 end; /* define the hash table */
 Consecutive_Hits = 0;
 do until(last.Top_Bot);      ❹
    set dw.atbats(keep=Game_SK Inning Top_Bot Is_A_Hit) end=lr;
    by Game_SK Inning Top_Bot notsorted;
    Consecutive_Hits = ifn(Is_A_Hit,Consecutive_Hits+1,0);      ❺
    if Is_A_Hit then consecHits.ref();      ❻
 end;
 if lr;      ❼
 Total_Adjust = 0;      ❽
 do Consecutive_Hits = consecHits.num_items to 1 by -1;      ❾
    consecHits.sum(sum:Exact_Count);      ❿
    Total_Count = Exact_Count + Total_Adjust;      ⓫
    output;
    Total_Adjust + Exact_Count;
 end;
run;
```

❶ Assign a value of 1 to the variable we will use as the value of the SUMINC argument tag. Since we are counting events, the value to be added to our sum (which is a count) is always 1.

❷ Define the hash table. We used ORDERED:"D" for two reasons. One, in order to calculate the count of N or more events we need the table to be ordered; and two, we want the output sorted by the descending count of consecutive events. Note also that these statements are executed only on the first execution of the DATA step. This is the first sample in this chapter for which our DATA step executes multiple times and so, unlike earlier examples, we need to add logic so the *Create* operation (defining and instantiating the hash table) is performed only once – on the first execution of the DATA step.

❸ Define the key to be the number of consecutive events. Note that since we do not have a DEFINEDATA method call, the same variable is automatically added to the data portion of the hash table. Since SUMINC is used to do the aggregation, no other variables are needed in the data portion.

❹ We use a DoW-loop coupled with the BY statement, so each execution of the DATA step reads the data for a single half-inning. We don't want our consecutive events to cross half-inning boundaries. The variable Consecutive_Hits is initialized to 0 since we want to restart our count for each half-inning.

❺ The IFN function is used to either increment our counter of consecutive events if the current observation represents a hit or to reset it to 0 if not. In response to a question from one of the IT users we confirmed that the IFN function has an optional fourth argument if the first argument has a null or missing value. Our use case does not have to handle null/missing values.

❻ When using SUMINC, the REF method (discussed in Part 1) should be used to keep the value of the PDV host variable and the hash table data item in synch. We need to do this only when we have consecutive events which caused the PDV host variable to be incremented by 1.

❼ Once we have read all the data, we are ready to post-process our hash table to create an output data of the distribution.

❽ Initialize the variable we use to increment the count of exactly N events to calculate N or more events.

❾ We can enumerate the hash table since we know the number of consecutive events will always be a sequence of integers starting with a value of 1 and incremented by 1. Thus, we can enumerate the list from the value of the NUM_ITEMS attributes to 1 (or from 1 to that value if we had specified ORDERED:"A").

❿ Since the sum calculated for the SUMINC variable is not a data item in the hash table, we need to use the SUM method to get the calculated value of our SUMINC variable. Note that since we have read all of our data, we must reuse the same variable name to contain the count. Also note that the value of the SUM argument is not a character expression – it must be the name of the PDV host variable that the SUM function will copy the retrieved value into.

⓫ Create the variable that contains the count for N or more consecutive events, output the results and accumulate the value so the next iteration will correctly adjust the count when calculating the number of N or more consecutive events.

Running this program will create the following output.

Output 8.10 Count of How Many Times for N Consecutive Hits

Consecutive Hits	Exact Count	Total Count
10	1	1
9	7	8
8	27	35
7	87	122
6	274	396
5	813	1,209
4	2,376	3,585

Consecutive Hits	Exact Count	Total Count
3	7,234	10,819
2	21,359	32,178
1	64,124	96,302

To clarify the distinction between the Exact and Total Count, we see that we have exactly 7 consecutive hits 87 times; there is 1 occurrence of 10, 7 occurrences of 9 and 27 of 8. Thus, we have $87 + 27 + 7 + 1 = 122$ occurrences of 7 or more consecutive hits.

8.2.3.5.1 Generating Distributions for More Than One Type of Event

Our users have now asked us to demonstrate how to calculate a distribution of consecutive events for multiple events, specifically consecutive at bats that resulted in a hit (the variable Is_A_Hit =1) and consecutive at bats that resulted in the batter getting on base (the variable Is_An_OnBase =1). Since the SUMINC functionality allows for only one such variable, we can't just add another variable to our hash table. And, in addition, since on each observation the values of the key (count of consecutive hits and the count of consecutive times getting on base) for these two event types are different, we need a different data structure, namely creating two hash tables instead of one.

The following program uses this approach. Note that we have highlighted in bold the important additions and changes to this program.

Program 8.12 - Chapter 8 Count Multiple Different Consecutive Events.sas

```
data Consecutive_Events;
 keep Consecutive_Events Hits_Exact_Count Hits_Total_Count
      OnBase_Exact_Count OnBase_Total_Count;
 format Consecutive_Events Hits_Exact_Count Hits_Total_Count
       OnBase_Exact_Count OnBase_Total_Count comma10.;
 retain Hits_Exact_Count OnBase_Exact_Count 1;        ❶
 if _n_ = 1 then
 do;  /* define the hash tables */
    dcl hash consecHits(ordered:"D",suminc:"Hits_Exact_Count");
    consecHits.defineKey("Consecutive_Hits");
    consecHits.defineDone();
    dcl hash consecOnBase(ordered:"D",suminc:"OnBase_Exact_Count");   ❷
    consecOnBase.defineKey("Consecutive_OnBase");
    consecOnBase.defineDone();
 end; /* define the hash tables */
 Consecutive_Hits = 0;
 Consecutive_OnBase = 0;     ❸
 do until(last.Top_Bot);
    set dw.atbats(keep=Game_SK Inning Top_Bot Is_A_Hit Is_An_OnBase) end=lr;   ❹
    by Game_SK Inning Top_Bot notsorted;
    Consecutive_Hits = ifn(Is_A_Hit,Consecutive_Hits+1,0);
    if Is_A_Hit then consecHits.ref();
    Consecutive_OnBase = ifn(Is_An_OnBase,Consecutive_OnBase+1,0);   ❺
    if Is_An_OnBase then consecOnBase.ref();
 end;
 if lr;
 Total_Adjust_Hits = 0;
 Total_Adjust_OnBase = 0;     ❻
 do Consecutive_Events = consecOnBase.num_items to 1 by -1;   ❼
    consecOnBase.sum(Key:Consecutive_Events,sum:OnBase_Exact_Count);   ❽
```

```
     rc = consecHits.sum(Key:Consecutive_Events,sum:Hits_Exact_Count);     ❾
     Hits_Total_Count = Hits_Exact_Count + Total_Adjust_Hits;
     OnBase_Total_Count = OnBase_Exact_Count + Total_Adjust_OnBase;     ❿
     output;
     Total_Adjust_Hits + Hits_Exact_Count;
     Total_Adjust_OnBase + OnBase_Exact_Count;     ⑪
 end;
run;
```

❶ We need to create a second variable for our OnBase count. Note that we have also changed the name for the variable for the number of consecutive hits. That is not a structural change and such changes will not be highlighted in the rest of this program.

❷ Create the hash table for consecutive on-base events. Note that both the key variable and the SUMINC variable are different.

❸ The variable Consecutive_OnBase is initialized to 0 since we want to restart our count for each half-inning.

❹ We need to add Is_An_OnBase to our KEEP data set option so we can use it to determine if the batter got on base.

❺ Similar statements about what is used to update the counts for the consecutive hits hash table are needed for the consecutive on-base hash table.

❻ We need to initialize the variable we use to increment the count of N events to calculate N or more events for the on-base calculation.

❼ Our looping control using the size of the *consecOnBase* hash table as we know that table, since it represents a superset of hits events, always has more rows. If the nature of our events did not satisfy this condition. we could use the max of all the NUM_ITEMS attributes.

❽ The SUM function is used to get the value for the count of the number of consecutive on-base events. Note that we use an explicit call since the name for our output variable is not the corresponding PDV host variable.

❾ The SUM method is used to get the value for the count of the number of consecutive hit events. The *rc=* is needed because there are key items in the on-base table not found in the hits hash table. An explicit call is used here as well since the name for our output variable is not the corresponding PDV host variable.

❿ Create the variable that contains the count for N or more consecutive on-base events.

⑪ After outputting the results, accumulate the value so the next iteration will correctly adjust the count when calculating the number of N or more consecutive onbase events.

Running this program will create the following output.

Output 8.11 Counts for Multiple Types of Consecutive Events

Consecutive Events	Hits Exact Count	Hits Total Count	OnBase Exact Count	OnBase Total Count
12	0	0	1	1
11	0	0	9	10
10	1	1	27	37
9	7	8	82	119
8	27	35	185	304
7	87	122	446	750

Consecutive Events	Hits Exact Count	Hits Total Count	OnBase Exact Count	OnBase Total Count
6	274	396	1,035	1,785
5	813	1,209	2,395	4,180
4	2,376	3,585	5,459	9,639
3	7,234	10,819	12,793	22,432
2	21,359	32,178	29,947	52,379
1	64,124	96,302	70,423	122,802

The results for the Hits counts are exactly the same as the prior results, as expected. As for the other examples, we will revisit this same example in Chapter 9 Hash of Hashes – Looping Through SAS Hash Objects.

8.2.3.5.2 Comparing the SUMINC Argument Tag with Directly Creating the Sum or Count

Given that using SUMINC is limited to being able to aggregate only a single variable, in addition to the fact that the current value of the hash table is not automatically linked to a corresponding PDV host variable, we decided to compare Program 8.11 (Chapter 8 Count Consecutive Events.sas) that calculates the count for a consecutive event with using SUMINC with one that specifically codes the aggregation using a hash data variable linked to a PDV host variable. The following program produces exactly the same results.

Program 8.13 - Chapter 8 Count Consecutive Events Not SUMINC.sas

```
data Consecutive_Hits;
 keep Consecutive_Hits Exact_Count Total_Count;
 format Consecutive_Hits 8. Exact_Count Total_Count comma10.;
 /* retain Exact_Count 1; */        ❶
 if _n_ = 1 then
 do;  /* define the hash table */
    dcl hash consecHits(ordered:"D");       ❷
    consecHits.defineKey("Consecutive_Hits");
    consecHits.defineData("Consecutive_Hits","Exact_Count");       ❸
    consecHits.defineDone();
 end; /* define the hash table */
 Consecutive_Hits = 0;
 do until(last.Top_Bot);
    set dw.atbats(keep=Game_SK Inning Top_Bot Is_A_Hit) end=lr;
    by Game_SK Inning Top_Bot notsorted;
    Consecutive_Hits = ifn(Is_A_Hit,Consecutive_Hits+1,0);
    if consecHits.find() ne 0 then call missing(Exact_Count);       ❹
    Exact_Count + 1;        ❺
    if Is_A_Hit then consecHits.replace();       ❻
 end;
 if lr;
 Total_Adjust = 0;
 do Consecutive_Hits = consecHits.num_items to 1 by -1;
    consecHits.find();        ❼
    Total_Count = Exact_Count + Total_Adjust;
    output;
    Total_Adjust + Exact_Count;
 end;
run;
```

❶ This line is commented out to highlight that we no longer need to use a variable for the SUMINC argument and thus do not need to initialize that PDV variable with a value of 1 in order to calculate the number of events.

❷ The DCL statement does not use the SUMINC argument tag.

❸ the DEFINEDATA method specifies two variables: Consecutive_Hits and Exact_Count. The example that used SUMINC did not include a DEFINEDATA method call which resulted in only Consecutive_Hits being added to the data portion.

❹ If there is an item for the current key value, the FIND method is used to load the current value of Exact_Count into its corresponding PDV host variable; otherwise, the Exact_Count is initialized to missing since the key value is not found in the hash table.

❺ Increment the Exact_Count variable by 1.

❻ We need to use the REPLACE method instead of the REF method.

❼ The FIND method is used instead of the SUM method to update the PDV host variable Exact_Count.

Running this program creates exactly the same output as the program that used SUMINC (Output 8.10). Given that (a) SUMINC can be used only for one variable, (b) the current value of that variable requires an explicit method call to update a PDV host variables, and (c) the program that performs the calculation is no more complex (and is perhaps easier to understand), we have suggested to our users that SUMINC probably has limited utility. As a result, we will likely not use it moving forward except, perhaps, in very limited and special cases.

8.3 Creating Multi-Way Aggregates

In the section "Adding Another Aggregation Level – Team Slash Lines," we looked at an approach that assumed purely hierarchical roll-ups. Our users have asked us for examples of how to calculate metrics at multiple different levels of aggregation (in baseball these are referred to as *splits*) similar to what we SAS programmers know can be done in PROC SUMMARY (or MEANS) with a CLASS statement, combined with the TYPES or WAYS statements as well as non-hierarchical roll-ups.

We will generalize that sample program to calculate multiple levels of aggregation, e.g., by:

- Player
- Team (since players can play for multiple teams, this is an example of a non-hierarchical roll-up)
- Month
- Day of the Week
- Player by Month

Each such *split* can be calculated using its own hash table. Any number of splits can be calculated in a single DATA step which reads the data just once. In order to illustrate this approach we will *hard-code* all of the hash tables (rather than using the macro language).

The following program is the hard-coded program that calculates these *splits*. In the next sub-section we will minimize some of the *hard-coding* through the use of parameter files. This amount of hard-coding will be further minimized in Chapter 9 Hash of Hashes – Looping Through SAS Hash Objects.

Program 8.14 - Chapter 8 Multiple Splits.sas

```
data _null_;
 /* define the lookup hash object tables */     ❶
 dcl hash players(dataset:"dw.players(rename=(Player_ID=Batter_ID))"  ❷
                   ,multidata:"Y");
 players.defineKey("Batter_ID");
 players.defineData("Batter_ID","Team_SK","Last_Name","First_Name"
                    ,"Start_Date","End_Date");
 players.defineDone();
 dcl hash teams(dataset:"dw.teams");      ❸
 teams.defineKey("Team_SK");
 teams.defineData("Team_Name");
 teams.defineDone();
 dcl hash games(dataset:"dw.games");       ❹
 games.defineKey("Game_SK");
 games.defineData("Date","Month","DayOfWeek");
 games.defineDone();
 /* define the result hash object tables */
 dcl hash h_pointer;       ❺
 dcl hash byPlayer(ordered:"A");      ❻
 byPlayer.defineKey("Last_Name","First_Name","Batter_ID");
 byPlayer.defineData("Last_Name","First_Name","Batter_ID","PAs","AtBats","Hits"
                     ,"_Bases","_Reached_Base","BA","OBP","SLG","OPS");
 byPlayer.defineDone();

 dcl hash byTeam(ordered:"A");
 byTeam.defineKey("Team_SK","Team_Name");
 byTeam.defineData("Team_Name","Team_SK","PAs","AtBats","Hits"
                   ,"_Bases","_Reached_Base","BA","OBP","SLG","OPS");
 byTeam.defineDone();

 dcl hash byMonth(ordered:"A");
 byMonth.defineKey("Month");
 byMonth.defineData("Month","PAs","AtBats","Hits"
                    ,"_Bases","_Reached_Base","BA","OBP","SLG","OPS");
 byMonth.defineDone();

 dcl hash byDayOfWeek(ordered:"A");
 byDayOfWeek.defineKey("DayOfWeek");
 byDayOfWeek.defineData("DayOfWeek","PAs","AtBats","Hits"
                        ,"_Bases","_Reached_Base","BA","OBP","SLG","OPS");
 byDayOfWeek.defineDone();

 dcl hash byPlayerMonth(ordered:"A");
 byPlayerMonth.defineKey("Last_Name","First_Name","Batter_ID","Month");
 byPlayerMonth.defineData("Last_Name","First_Name","Batter_ID","Month"
                          ,"PAs","AtBats","Hits","_Bases","_Reached_Base"
                          ,"BA","OBP","SLG","OPS");
 byPlayerMonth.defineDone();

 if 0 then set dw.players(rename=(Player_ID=Batter_ID))
                dw.teams
                dw.games;
 format PAs AtBats Hits comma6. BA OBP SLG OPS 5.3;

 lr = 0;
 do until(lr);       ❼
```

```
        set dw.AtBats end = lr;
        call missing(Team_SK,Last_Name,First_Name,Team_Name,Date,Month,DayOfWeek);  ❽
        games.find();     ❾
        players_rc = players.find();
        do while(players_rc = 0);
            if (Start_Date le Date le End_Date) then leave;
            players_rc = players.find_next();
        end;
        if players_rc ne 0
            then call missing(Team_SK,First_Name,Last_Name);
        teams.find();
        h_pointer = byPlayer;     ❿
        link slashline;
        h_pointer = byTeam;
        link slashline;
        h_pointer = byMonth;
        link slashline;
        h_pointer = byDayOfWeek;
        link slashline;
        h_pointer = byPlayerMonth;
        link slashline;
    end;
    byPlayer.output(dataset:"byPlayer(drop=_:)");       ⓫
    byTeam.output(dataset:"byTeam(drop=_:)");
    byMonth.output(dataset:"byMonth(drop=_:)");
    byDayOfWeek.output(dataset:"byDayOfWeek(drop=_:)");
    byPlayerMonth.output(dataset:"byPlayerMonth(drop=_:)");
    stop;        ⓭
    slashline:        ⓬
        call missing(PAs,AtBats,Hits,_Bases,_Reached_Base);
        rc = h_pointer.find();
        PAs           + 1;
        AtBats        + Is_An_AB;
        Hits          + Is_A_Hit;
        _Bases        + Bases;
        _Reached_Base + Is_An_OnBase;
        BA = divide(Hits,AtBats);
        OBP = divide(_Reached_Base,PAs);
        SLG = divide(_Bases,AtBats);
        OPS = sum(OBP,SLG);
        h_pointer.replace();
    return;
run;
```

❶ Define the hash tables needed to do the table lookups for the player, team, and game data.

❷ For the player lookup, we need to get the player's name and team. Note that we will be doing a Type 2 lookup since players can be traded and change teams. Since players can have multiple rows, MULTIDATA:"Y" is required.

❸ For the team lookup, we need only to look up the team name.

❹ For the game lookup, we need the date so that the player lookup will get the right values for that date. While we could use SAS date functions in our code to get the month and day of the week from the date, it is common in dimensional tables to store such fields instead of calculating them each time.

❺ We saw earlier that we can't assign the identifier value for a hash object to a scalar variable (i.e., a numeric or character variable). However, we can assign one object type hash variable to another. What it does is create another pointer (i.e., a non-scalar variable) that can be used to point to a hash table. So h_pointer is a mere placeholder allowing the program to point to a different table by using a single non-scalar variable name by simply assigning its value to be the table we want to reference. We can take advantage of this fact in this example. If we assign the value of byPlayer to h_pointer, any hash method or operator reference to h_pointer will point to the byPlayer hash object instance; then if we assign the value of byTeam to h_pointer, h_pointer will point to the byTeam hash object instance instead; and so on.

❻ Define the hash tables for each of our *splits*. Note that the only differences in these tables are the fields that are the keys for each of the *splits*. Note also that we created a hash PDV host variable so we can access any of the hash table objects using a single variable.

❼ Use the DoW loop construct to read all the data.

❽ Initialize all the fields to be looked up to missing in case they are not found in the various dimension tables. Note that due to how the data was constructed, we know that we should always find matches except in the case where the data has been corrupted. Regardless, it is a good practice to always handle non-matches explicitly.

❾ Do the Games, Players, and Teams lookups. Note that it is important to do games before the players lookup as the players lookup requires the date, given that it is a Type 2 lookup.

❿ Do the calculations for each of the *splits*. We assign the value for each hash object to our reusable PDV hash object variable (h_pointer) and then use a LINK statement to perform the same calculations for each *split* using exactly the same code.

⓫ Output a separate data set for each *split*. Note the use of **drop=_:** to exclude working variables from the output. Prefixing such variable names with an underscore (_) is simply a convention to make it simple to include variables needed for the calculations and exclude them from the output data set.

⓬ Perform the needed calculations for the *split* corresponding to the hash table whose identifier value is currently assigned to the hash object type placeholder variable *Reuse*.

⓭ Since the DoW loop was used to read all the data, stop after the first execution of the DATA step.

The first 5 observations for each of the output data sets are shown below.

Output 8.12 byPlayer Splits

Last Name	First Name	Batter ID	PAs	AtBats	Hits	BA	OBP	SLG	OPS
Adams	Jack	40122	743	672	263	0.391	0.450	0.878	1.328
Adams	Jacoby	53976	714	656	241	0.367	0.419	0.884	1.303
Adams	Joe	42120	682	605	225	0.372	0.443	0.881	1.324
Adams	Jordan	16382	128	110	37	0.336	0.430	0.918	1.348
Adams	Joseph	28562	102	95	36	0.379	0.422	0.863	1.285

Output 8.13 byTeam Splits

Team Name	Team SK	PAs	AtBats	Hits	BA	OBP	SLG	OPS
Tigers	103	8,985	8,143	3,019	0.371	0.430	0.874	1.304
Lions	115	9,036	8,184	3,041	0.372	0.431	0.884	1.315
Cardinals	130	8,842	8,042	2,903	0.361	0.419	0.862	1.280
Vikings	132	8,978	8,165	3,029	0.371	0.428	0.873	1.301
Pirates	136	9,112	8,293	3,166	0.382	0.437	0.895	1.332

Output 8.14 byMonth Splits

Month	PAs	AtBats	Hits	BA	OBP	SLG	OPS
3	15,743	14,266	5,135	0.360	0.420	0.850	1.269
4	41,651	37,894	14,108	0.372	0.429	0.879	1.308
5	43,067	39,056	14,399	0.369	0.427	0.881	1.308
6	39,885	36,247	13,400	0.370	0.427	0.882	1.309
7	42,827	38,883	14,143	0.364	0.422	0.867	1.290

Output 8.15 byDayOfWeek Splits

DayOfWeek	PAs	AtBats	Hits	BA	OBP	SLG	OPS
1	48,298	43,790	16,348	0.373	0.432	0.885	1.317
2	47,844	43,412	16,044	0.370	0.428	0.879	1.307
3	47,880	43,526	16,121	0.370	0.428	0.877	1.305
4	47,854	43,493	16,079	0.370	0.427	0.873	1.300
6	47,711	43,245	15,815	0.366	0.425	0.871	1.296

Output 8.16 byPlayerMonth Splits

Last Name	First Name	Batter ID	Month	PAs	AtBats	Hits	BA	OBP	SLG	OPS
Adams	Jack	40122	3	47	46	18	0.391	0.404	0.739	1.143
Adams	Jack	40122	4	117	101	38	0.376	0.462	0.881	1.343
Adams	Jack	40122	5	126	113	42	0.372	0.437	0.885	1.321
Adams	Jack	40122	6	116	107	35	0.327	0.379	0.729	1.108
Adams	Jack	40122	7	95	88	41	0.466	0.505	1.091	1.596

And, we pointed out to the IT users, this approach could also perhaps use the SAS macro language.

8.3.1 Using Parameter Files to Define Aggregates

The approach of hard-coding to define and calculate the *splits* can become unwieldy fairly quickly, and our business users have asked if there is a way to make this easier. They have said that there are likely cases where there are dozens of possible *splits* and they would like to be able to add new *splits* without having to add new code.

We, of course, expected that question and were prepared to demonstrate how we could use data sets (also known as parameter files) to define and calculate the same *splits* as presented in the previous section. The first step is to create a parameter file that defines the splits. A screenshot of a simple example of such a parameter file is shown below. For the purposes of this Proof of Concept, we decided to keep this parameter file simple. For example, we are assuming that any field that is part of the key is also included in the data portion. The data set Chapter8Parmfile in the TEMPLATE library includes the definitions for all the splits included in the previous example.

Output 8.17 First 12 Rows of Chapter8Parmfile

hashTable	Column	is A Key
BYDAYOFWEEK	DayOfWeek	1
BYDAYOFWEEK	PAs	0
BYDAYOFWEEK	AtBats	0
BYDAYOFWEEK	Hits	0
BYDAYOFWEEK	BA	0
BYDAYOFWEEK	OBP	0
BYDAYOFWEEK	SLG	0
BYDAYOFWEEK	OPS	0
BYDAYOFWEEK	_Bases	0
BYDAYOFWEEK	_Reached_Base	0
BYMONTH	Month	1
BYMONTH	PAs	0

The use of the macro language to generate the code to define hash table objects from metadata (i.e., a parameter file) was introduced in the Chapter 7 section, "Using a Macro to Create the Hash Object Method Calls." We will use a similar macro here. Since the next chapter will introduce what we believe is a more flexible approach that will better meet our users' needs, we decided not to try to generalize the macro logic to handle the requirements addressed in Chapter 7 as well as what we need to do here. Given the metadata approach, the code for each hash table is the same. We just need to use an appropriate WHERE clause for each set of hash object method calls.

In order to impress upon the IT users the fact that SAS has a very broad range of tools, we decided to use the macro language, a PROC SQL step, and a DATA step for this example. Let us cover each of these individually.

Program 8.15 - Chapter 8 Multiple Splits Parameterized.sas (Part 1)

```
%macro createHash      ❶
      (hashTable = hashTable
      ,parmfile = template.Chapter8ParmFile
      );

 lr = 0;
 dcl hash &hashTable(ORDERED:"A");      ❷
 do while(lr=0);
    set &parmfile end=lr;      ❸
    where upcase(hashTable) = "%upcase(&hashTable)";
    if Is_A_key then &hashTable..DefineKey(Column);      ❹
    &hashTable..DefineData(Column);      ❺
 end;
 &hashTable..DefineDone();      ❻

%mend createHash;
```

❶ Define the macro with two parameters: the name of the variable that provides the name for the hash object to be created; and the name of the data set that contains the metadata which defines each hash object.

❷ Declare the hash table. The ORDERED:"A" argument tag is used since we want the items in the table to be logically ordered by the key so the output data set is sorted by the key.

❸ Loop through the schema metadata, reading only the definition data for the specified hash table.

❹ If the current column is a key, use the DEFINEKEY method to specify that the column is to be used in the key portion.

❺ Use the DEFINEDATA method to specify that the column is to be used in the data portion. Note that this macro makes the assumption that columns in the key portion should also be included in the data portion.

❻ Use the DEFINEDONE method to finalize the creation of the hash object instance.

By using this macro we can create each hash object table with a single macro call. However, we still need a separate macro call for each hash table. We could certainly write different or additional macro code to loop through all the desired hash tables. Since our users and the developers are not fully familiar with the range of available SAS tools, we decided to take a different approach to handle this looping.

There are actually three sets of repetitive code. We need to generate code to perform the following functions for each *split* (i.e., hash table):

- Define the hash tables using the createHash macro.
- The statements to invoke the linked section to perform the calculation for each hash table.
- The method calls to create output data sets corresponding to the contents of each hash table.

The following PROC SQL statements create macro variables for each of these three repetitive blocks of code.

Program 8.15 - Chapter 8 Multiple Splits Parameterized.sas (Part 2)

```
proc sql noprint;     ❶
 select distinct cats('%createHash(hashTable='
                      ,hashTable
                      ,")"
                      )
 into:createHashCalls separated by " "
 from template.Chapter8ParmFile;     ❷
 select distinct cats("h_pointer="
                      ,hashTable
                      ,";"
                      ,"link slashline"
                      )      ❸
 into:calcHash separated by ";"
 from template.Chapter8ParmFile;
 select distinct cats(hashTable
                      ,'.output(dataset:"_'
                      ,hashTable
                      ,'(drop=_:)")'
                      )      ❹
 into:outputHash separated by ";"
 from template.Chapter8ParmFile;
quit;
```

❶ Use the SQL procedure to create the needed macro variables. The NOPRINT option is used as we don't need to see the output listing of the individual selected rows.

❷ Use the CATS function to concatenate the needed text strings, with the value of the hashTable variable from the metadata parameter file template.Chapter8ParmFile. The DISTINCT option ensures that each distinct value of the hashTable variable is included only once and the SEPARATED BY clause ensures that each generated value is included in the macro variable createHashCalls. Since macro calls do not need a semicolon we used a space as the separator character. For consistency with the next two macro variables, we chose to use a semicolon (;) here as well.

❸ A comparable CATS function call is used to create the statements to assign the hashTable object identifier to our reusable object type variable and to use the LINK statement to perform the needed calculations.

❹ A comparable CATS function call is used to create the statements that output the resulting data sets from each hash table object.

Output 8.18 Listing of Created Macro Variable Names and Values

Name	Value

CALC HASH	h_pointer=BYDAYOFWEEK;link slashline;
	h_pointer=BYMONTH;link slashline;
	h_pointer=BYPLAYER;link slashline;
	h_pointer=BYPLAYERMONTH;link slashline;
	h_pointer=BYTEAM;link slashline

CREATE HASH CALLS	%createHash(hashTable=BYDAYOFWEEK) %createHash(hashTable=BYMONTH) %createHash(hashTable=BYPLAYER) %createHash(hashTable=BYPLAYERMONTH) %createHash(hashTable=BYTEAM)
OUTPUT HASH	BYDAYOFWEEK.output(dataset:"_BYDAYOFWEEK(drop=_:)"); BYMONTH.output(dataset:"_BYMONTH(drop=_:)"); BYPLAYER.output(dataset:"_BYPLAYER(drop=_:)"); BYPLAYERMONTH.output(dataset:"_BYPLAYERMONTH(drop=_:)"); BYTEAM.output(dataset:"_BYTEAM(drop=_:)")

The DATA step that does the *splits* calculations is virtually the same as the program discussed in the previous section. The only difference is the repetitive code has been replaced by macro variable references.

Program 8.15 - Chapter 8 Multiple Splits Parameterized.sas (Part 3)

```
data _null_;
 /* define the lookup hash object tables */
 dcl hash players(dataset:"dw.players(rename=(Player_ID=Batter_ID))"
               ,multidata:"Y");
 players.defineKey("Batter_ID");
 players.defineData("Batter_ID","Team_SK","Last_Name","First_Name"
                 ,"Start_Date","End_Date");
 players.defineDone();
 dcl hash teams(dataset:"dw.teams");
 teams.defineKey("Team_SK");
 teams.defineData("Team_Name");
 teams.defineDone();
 dcl hash games(dataset:"dw.games");
 games.defineKey("Game_SK");
 games.defineData("Date","Month","DayOfWeek");
```

```
games.defineDone();
/* define the result hash object tables */
dcl hash h_pointer;

&createHashCalls      ❶

if 0 then set dw.players(rename=(Player_ID=Batter_ID))
               dw.teams
               dw.games;
format PAs AtBats Hits comma6. BA OBP SLG OPS 5.3;

lr = 0;
do until(lr);
   set dw.AtBats end = lr;
   call missing(Team_SK,Last_Name,First_Name,Team_Name,Date,Month,DayOfWeek);
   games.find();
   players_rc = players.find();
   do while(players_rc = 0);
      if (Start_Date le Date le End_Date) then leave;
      players_rc = players.find_next();
   end;
   if players_rc ne 0
      then call missing(Team_SK,First_Name,Last_Name);
   teams.find();
   &calcHash;      ❷
end;
&outputHash;      ❸
stop;
slashline:
   call missing(PAs,AtBats,Hits,_Bases,_Reached_Base);
   rc = h_pointer.find();
   PAs             + 1;
   AtBats          + Is_An_AB;
   Hits            + Is_A_Hit;
   _Bases          + Bases;
   _Reached_Base + Is_An_OnBase;
   BA = divide(Hits,AtBats);
   OBP = divide(_Reached_Base,PAs);
   SLG = divide(_Bases,AtBats);
   OPS = sum(OBP,SLG);
   h_pointer.replace();
 return;
run;
```

❶ Replace the (in this case) five macro calls with a reference to the macro variable whose value is the five macro calls.

❷ Replace the five sets of assignment and LINK statements with a reference to the macro variable whose value is the five sets of statements.

❸ Replace the five sets OUTPUT method calls with a reference to the macro variable whose value is the five method calls.

Running this program creates exactly the same output as seen above in Output 8.12-8.16. The approach of parameterizing (via metadata) the list of *splits* to be calculated has the important additional benefit that additional splits can be defined and calculated without having to change the code. All that is needed is to update the metadata table to define additional *splits*. When we present that to our users we will suggest that a web-based interface could be built (again, using SAS software) to maintain and update a flexible metadata structure to support defining such *splits*.

8.4 Summary

The reaction to these sample programs was very positive and both the business and IT users were interested in the next phase of our Proof of Concept. In Chapter 9 we will further generalize and parameterize the samples presented in this chapter. We will also leverage the sample presented in this chapter in our case studies in Chapters 12 and 13.

Part Three—Expanding the WHAT and the WHY, along with the HOW of the SAS Hash Object

Our Proof of Concept project could have wrapped up after the completion of Part Two. Virtually anyone who has worked on a project knows that there are always what one could refer to as the *Yea, **BUT*** issues and questions. Every project has numerous such issues/questions.

Part Three contains three chapters that address issues the authors have encountered in projects they have collaborated on that involve the SAS hash object.

1. Chapter 9 describes an approach to minimize hard coding via a metadata approach. Metadata (AKA parameter files) are used to revisit previous examples in order to both minimize hard-coding and to provide more flexibility in the use of the SAS hash object to address business intelligence questions.
2. Chapter 10 provides a more detailed discussion (compared to Chapter 6) of the SAS hash object as a flexible (in-memory) Dynamic Data Storage facility.
3. Chapter 11 discusses several approaches to addressing memory management issues when the size of the data exceeds the memory capacity of the available environment. Techniques are presented to minimize key length as well as to partition the data into separate buckets or groups that can be processed either sequentially or in parallel.

Part Three is targeted to both our business and IT users. For the IT users, its goal is to acknowledge that such issues are real and there are reasonably straight-forward solutions. For the business users, the intent is to provide them with information that they can use to push forward with such projects.

Chapter 9: Hash of Hashes – Looping Through SAS Hash Objects

9.1 Overview

Both our business users and their IT support staff have expressed appreciation for the examples provided so far that allow for parameterization and code generation. They commented positively on being exposed to the macro language and the capabilities of the SQL procedure.

However, they have expressed concern about having to write code for each hash table to be created, even given the overview of what could be done with the macro language and the SQL procedure. They plan on collecting more data that they would like to produce metrics on. For example, pitch speed, hit speed, run speed, and the list goes on. And they don't want to have to add new code as they introduce new data fields and metrics.

They have also expressed surprise that we have not provided any examples of looping through multiple hash objects to address this requirement. Specifically, they were surprised that it was not possible to define a DATA step array of hash objects and loop through the hash objects. The IT support staff felt that such looping would be both simpler and clearer than using the macro language or the LINK statement to reuse the same/similar code.

We explained that such looping was possible by creating a hash object with the data portion containing non-scalar variables whose values point to other hash objects (note that the key portion cannot include such non-scalar variables).We also reminded them that this effort is a Proof of Concept and was never intended to be an implementation of their complete set of requirements.

In this chapter we provide an overview of creating a hash object whose items contain pointer values to other hash objects as their data – a Hash of Hashes which we abbreviate as HoH.

9.2 Creating a Hash of Hashes (HoH) Table – Simple Example

In section 8.2.3.1 "Calculating Multiple Medians," we showed the following error message generated by the Program 8.6 Chapter 8 Multiple Medians Array Error.sas:

```
670    array _hashes(*) single double triple homerun;
ERROR: Cannot create an array of objects.
ERROR: DATA STEP Component Object failure.
       Aborted during the COMPILATION phase.
NOTE: The SAS System stopped processing this step
      because of errors.
```

As discussed in section 8.2.3.1, this error message is due to the fact that the PDV host variable that contains the pointer or identifier for a hash object is neither a numeric nor character field. As the ERROR message implies, it is an object; SAS considers numeric and character DATA step variables to be scalars, and the programming statements to deal with scalars can't properly handle objects.

As we have already seen, a hash table entry can have scalar variables for which the hash items contain numeric or character values as data. However, the *data portion* of the table can also include non-scalar variables, particularly of type hash object. For any such hash variable (named H, say), the corresponding data values represent pointers to - or, if you will, identifiers of - the instances of other hash objects. The *Retrieve* operation (either by key or via enumeration) can be used to extract these pointer values from the HoH table into PDV host variable H, also of hash object type. After a pointer value has been retrieved into H, any hash method or operator call referencing H will now be executed against the hash table instance identified by the pointer value H currently contains. So just like an array can be used to loop through multiple numeric or character variables to perform the same operations, a hash object can be used to loop through multiple other hash objects to perform the same operations on each of those hash tables. The program below creates two hash tables and uses the DATASET argument tag to load the AtBats data sets into one of them and the Pitches data set into the second one. It then loops through those hash objects to produce notes in the log to demonstrate that we accessed two different hash tables.

Program 9.1 Chapter 9 Simple HoH.sas

```
data _null_;
 length Table $41;
 dcl hash HoH();        ❶
 HoH.defineKey ("Table");
 HoH.defineData("H","Table");
 HoH.defineDone();

 dcl hash h();     ❷

 Table = "DW.AtBats";    ❸
 h = _new_ hash(dataset:Table);
 h.defineKey("Game_SK","Inning","Top_Bot","AB_Number");
 h.defineData("Result");
 h.defineDone();
 HoH.add();
```

```
Table = "DW.Pitches";        ❹
h = _new_ hash(dataset:Table);
h.defineKey("Game_SK","Inning","Top_Bot","AB_Number","Pitch_Number");
h.defineData("Result");
h.defineDone();
HoH.add();

dcl hiter i_HoH("HoH");      ❺
do while (i_HoH.next() = 0);     ❻
   Rows = h.num_items;       ❼
   put (Table Rows)(=);
end;
stop;      ❽
set dw.pitches(keep = Game_SK Inning Top_Bot AB_Number Pitch_Number Result);     ❾
run;
```

❶   Define a hash object named HoH. The data portion variables in its hash table will be: (a) the non-scalar identifier H whose values point to other hash object instances; and (b) the scalar (character) variable Table. Note that though in this example we are not going to search table HoH by key, its key portion must contain at least one scalar variable; so we used variable Table, as the PDV host variable with this name already exists.

❷   Define a hash object named H. At the same time, the statement defines a non-scalar PDV variable H of type hash object. Its purpose is two-fold. First, it is to contain a pointer whose value identifies a particular instance of hash object H (similar to the field h_pointer discussed in 8.3 "Creating Multi-Way Aggregates"). Second, it is to serve as the PDV host variable for hash variable H defined in the data portion of table HoH.

❸   Create an instance of hash object H for the AtBats data and use the DATASET argument tag to load the data into its table. Now that the instance has been created, use the *Insert* operation to add a new item to table HoH. The value of *Table* forms its key portion; and the pointer to the instance of H (the current value of PDV variable H), along with the value of scalar variable Table, form its data portion. This value of H just loaded to the HoH hash table points to the hash table containing the AtBats data. (Note the use of the _NEW_ operator to create the instance of hash object H and assign the value of the identifier for this instance to the non-scalar PDV host variable H. We don't use the DECLARE HASH statement here for instantiation, as it would require a different variable name for each hash object instance. Since we are going to insert the pointer values identifying different instances of the same hash object H into our HoH table, we want to reuse the same name.)

❹   Do the same for the AtBats data as we did for the Pitches data. Namely, first create a new instance of hash object H with the DATASET argument tag valued so as to load the data from Pitches. This value of H points to the hash table containing the Pitches data. Then use the pointer to this new instance as the value of hash variable H in another new item added to table HoH. We now have two entries in the hash table HoH with a value for H: one points to the hash table containing the AtBats data; one points to the hash table containing the Pitches data.

❺   Define a hash iterator for our Hash of Hashes (HoH) table. It enables the *Enumerate All* operation, so that we can illustrate looping through the HoH table sequentially one item at a time.

❻   Loop through our Hash of Hashes object (HOH) using the iterator, so that we can illustrate how to retrieve the pointer to each individual instance of hash object H using the same code. As we iterate through the HoH, for each item we access in the loop the PDV host values for H and Table are updated with the values from the corresponding HOH data portion hash variables.

❼   Get the number of items in the hash object instance whose identifier value, having just been retrieved from HoH, is currently stored in host variable H and assign it to variable Rows. Use the PUT statement to write the values of Rows to the log. Its point is to confirm that in each iteration through the loop, another pointer value that is stored in hash variable H is extracted into its PDV host variable H; and so each time operator H.NUM_ITEMS references hash object H, it accesses another instance of it identified by the pointer value retrieved in the current iteration.

❽ A STOP statement is used since the data is loaded into the hash tables using the DATASET argument tag. There is no need to execute the DATA step more than once.

❾ A SET statement (after the STOP statement) is used to define the needed PDV host variables. As mentioned in earlier programs, since that PDV host variables are not referenced in our DATA step program, the SET statement can appear anywhere; it does not need to be at the top of the DATA step. We needed only to reference the Pitches data set since it contains all the variables that need to be defined to the PDV. We could have just as easily referenced just the AtBats data set.

The SAS log notes and the PUT statement output are shown below.

```
NOTE: There were 287304 observations read from the data set DW.ATBATS.
NOTE: There were 875764 observations read from the data set DW.PITCHES.
Table=DW.AtBats Rows=287304
Table=DW.Pitches Rows=875764
```

We will provide a number of examples in the rest of this chapter that use the Hash of Hashes approach to parameterize and generalize a number of the examples discussed earlier in this book. These examples will illustrate that:

- Both scalar and non-scalar hash variables can coexist in the data portion of a hash table.

- The non-scalar variables of type hash object are pointers identifying hash object instances.

- The non-scalar variables of type hash iterator object can also be pointers identifying hash iterator object instances.

9.3 Calculating Percentiles, Mode, Mean, and More

Our first set of examples illustrating the Hash of Hashes approach is the reworking of section 8.2.3 "Calculating Medians, Percentiles the Mode, and More."

We emphasized to both our business and IT users that we will be building up to the full capabilities of the Hash of Hashes approach by slowly building on the examples. As we revisit each of the examples from the previous chapter we will highlight additional functionality.

9.3.1 Percentiles

The first example is calculating percentiles. It illustrates iterating through the Hash of Hashes objects as well as each hash object item it contains. We are still hard-coding our list of variables; we will illustrate eliminating such hard coding shortly.

The following program does this calculation.

Program 9.2 Chapter 9 HoH Percentiles.sas

```
data Percentiles;
 keep Variable Percentile Value;
 length Variable $32;
 format Percentile percent5. Value Best.;
 dcl hash HoH(ordered:"A");        ❶
 HoH.defineKey ("Variable");
 HoH.defineData ("H","ITER","Variable");     ❷
 HoH.defineDone();
 dcl hash h();
 dcl hiter iter;      ❸

 h = _new_ hash(dataset:"dw.AtBats(where=(Value) rename=(Distance=Value))"    ❹
                ,multidata:"Y",ordered:"A");
 h.defineKey("Value");
 h.defineDone();
 iter = _new_ hiter("H");     ❺
 Variable = "Distance";
 HoH.add();      ❻

 h = _new_ hash(dataset:"dw.AtBats(where=(Value) rename=(Direction=Value))"    ❼
                ,multidata:"Y",ordered:"A");
 h.defineKey("Value");
 h.defineDone();
 iter = _new_ hiter("H");
 Variable = "Direction";
 HoH.add();

 array _ptiles(6) _temporary_ (.05 .1 .25 .5 .75 .95);
 call sortn(of _ptiles(*));

 dcl hiter HoH_Iter("HoH");     ❽
 do while (HoH_Iter.next() = 0);      ❾
    Counter = 0;
    num_items = h.num_items;
    do i = 1 to dim(_ptiles);      ❿
       Percentile = _ptiles(i);
       do while (Counter lt Percentile*num_items);
          Counter + 1;
          iter.next();
       end;
       /* could add logic here to read next value to interpolate */
       output;
    end;
 end;
 stop;
 Value = 0;
run;
```

❶ Define our Hash of Hashes table HoH. Since we want the output data ordered by the variable name, we use the ORDERED argument tag.

❷ Our percentile calculation requires us to iterate through our hash table objects that contain the data- i.e., we need an iterator object to do that. We can define a non-scalar field to be part of the data portion whose value is the pointer for that iterator object.

❸ Just as we saw above, we can define a non-scalar PDV variable Iter of type hash iterator. It will contain a pointer whose value identifies a particular instance of hash object iterator.

❹ Use the _NEW_ method to create an instance of the hash object H and load it with the Distance values from the AtBats data set. Note that in the argument to the DATASET argument, we renamed the Distance variable since we need to process multiple different variables. Our output data set will contain a column whose value is the variable's name (Variable) and another column whose value is the variable's value (Value).

❺ Use the _NEW_ method to create an instance of the hash iterator object Iter.

❻ The ADD method is used to add our hash object containing the data for the variable Distance to our Hash of Hashes object H.

❼ Repeat steps 4 through 6 for the variable Direction. Once the ADD method is executed, our Hash of Hashes table contains two items.

❽ In order to loop (i.e., enumerate) through our Hash of Hashes object HoH, we need an iterator object.

❾ We use the NEXT method to loop through all the items in our hash object HoH. Note that the first call to the NEXT method will point to the first item in HoH.

❿ The percentile calculation uses the same logic that we used in 8.2.3.2 "Calculating Additional Percentiles," except we are now calculating percentiles for more than one variable in our input data set.

Running the above program creates the following output. Note that the percentile values for the variable *Distance* are the same as what we saw in Output 8.6.

Output 9.1 Direction Percentiles

Variable	Percentile	Value
Direction	5%	1
	10%	2
	25%	5
	50%	9
	75%	14
	95%	18

Output 9.2 Distance Percentiles

Variable	Percentile	Value
Distance	5%	22
	10%	36
	25%	76
	50%	204
	75%	296
	95%	427

9.3.2 Multiple Medians

We previously showed how to calculate medians using the hash object, including calculating medians for subgroups of the data – each type of hit. That required a separate hash object for each type of hit (similar to what many SAS users recognize as functionality that can be implemented using the BY statement) for each variable for which the median is required. As either the number of values or variables increases, the hard-coding is problematic.

This example uses parameter files and a Hash of Hashes object to provide functionality similar to what can be done with the BY statement. We need to calculate medians for both Distance and Direction for each value of the Result variable. Our first step is to create a parameter file that contains all the values of Result that have non-missing values for Distance and Direction.

While we could easily create a hash table with all the distinct values, we decided to use an SQL step to create the data we need. This approach reinforces combining the use of hash object techniques with other SAS functionality. We decided to take this approach for two reasons: to reinforce to the IT users the breadth of SAS tools; and to allow this example to focus on the Hash of Hashes approach. In response to a request from several of the IT users, we did agree to provide a variation of this program that just uses hash objects. They were curious about performance issues of this approach vs. an approach that uses only hash objects. You can access the blog entry that describes this program from the author page at http://support.sas.com/authors. Select either "Paul Dorfman" or "Don Henderson." Then look for the cover thumbnail of this book, and select "Blog Entries."

Program 9.3 Chapter 9 HoH Multiple Medians.sas (Part 1)

```
proc sql;
 create table HoH_List as
 select distinct "Distance " as Field
       ,Result
 from dw.Atbats
 where Distance is not null
 outer union corr
 select distinct "Direction" as Field
       ,Result
 from dw.Atbats
 where Distance is not null
 ;
quit;
```

The above SQL step creates the data set that specifies the 12 medians to be calculated.

Output 9.3 Medians to Be Calculated

Field	Result
Distance	Double
Distance	Error
Distance	Home Run
Distance	Out
Distance	Single
Distance	Triple
Direction	Double
Direction	Error
Direction	Home Run
Direction	Out
Direction	Single
Direction	Triple

This data set is used in the following program to create the Hash of Hashes object that contains the 12 hash objects needed to calculate the desired medians.

Program 9.3 Chapter 9 HoH Multiple Medians.sas (Part 2)

```
data Medians;
 if 0 then set dw.AtBats(keep=Result);
 length Median 8;
 keep Result Field Median;
 dcl hash HoH(ordered:"A");      ❶
 HoH.defineKey ("Result","Field");
 HoH.defineData ("Result","Field","h","iter");
 HoH.defineDone();
 dcl hash h();
 dcl hiter iter;

 do until(lr);     ❷
    set HoH_List end = lr;
    h = _new_ hash(dataset:cats("dw.AtBats"
                               || "(where=(Result='"    ❸
                               ,Result
                               ,"')"
                               ,"rename=("
                               ,field
                               ,"=Median))")
               ,multidata:"Y",ordered:"A");
    h.defineKey("Median");
    h.defineDone();
    iter = _new_ hiter("h");
    HoH.add();
 end;

 dcl hiter HoH_Iter("HoH");      ❹
 do while (HoH_Iter.next() = 0);     ❺
    Count = h.num_items;
    iter.first();
    do i = 1 to .5*Count - 1;
      iter.next();
    end;
    /* could add logic here to interpolate if needed */
    output;
 end;
 stop;
run;
```

❶ Just as in our previous examples, define our Hash of Hashes table and use the ORDERED argument tag since we want the output sorted by the values of the Result variable and the variable name.

❷ Read all the observations in the parameter file to create all the needed hash objects.

❸ Use the _NEW_ method to create an instance of the hash object H and load it with the values from the AtBats data for the subject variables using the DATASET argument. The CATS function is used to create a character expression that filters the data using a WHERE clause and renames our variable to Median. We did this so the variable name in the output data set is the same for each of the variables. For the first row in our HoH_List data set, the CATS function returns the following character string:

```
dw.AtBats(where=(Result='Double') rename=(Distance=Median))
```

❹ In order to loop (i.e., enumerate) through our Hash of Hashes object HoH, the hash iterator object HoH_Iter is created.

❺ Loop (enumerate) through the hash objects for each desired median and calculate the medians using the same logic used in section 8.2.3.1 "Calculating Multiple Medians."

Running the above program creates the following output.

Output 9.4 Calculated Medians

Result	Median	Field
Double	9	Direction
Double	251	Distance
Error	9	Direction
Error	176	Distance
Home Run	9	Direction
Home Run	436	Distance
Out	10	Direction
Out	195	Distance
Single	9	Direction
Single	56	Distance
Triple	9	Direction
Triple	350	Distance

When we reviewed this example with the IT staff, they noticed that the two loops seemed duplicative. They asked why not just calculate the medians inside the loop that read the data. We agreed that could be done and, in fact, would not need a Hash of Hashes object. We also reminded them again that this is a Proof of Concept and that we wanted to use this example to build upon the capabilities of the Hash of Hashes approach. We did agree to provide an example of such an approach (in addition to the alternative mentioned above that does not use the SQL procedure). You can access the blog entry that describes this program from the author page at http://support.sas.com/authors. Select either "Paul Dorfman" or "Don Henderson." Then look for the cover thumbnail of this book, and select "Blog Entries."

9.3.3 Percentiles, Mode, Median, and More

In 8.2.3.4 "Calculating Percentiles, Mode, Mean, and More in a Single DATA Step" we generated our percentiles and medians examples to illustrate using the hash object to, in one step, calculate percentages, and frequency distribution as well as the Mean, Median, Min, and Max.

This example demonstrates:

- More parameterization.
- The use of macro variables created using the SQL procedure.
- Using the ARRAY statement to define what Hash of Hashes objects are needed.
- Using an output data set for selected output metrics (the frequency distributions).
- Creating a hash object specifically to include the results of our calculations.
- Additional columns in the data portion of our Hash of Hashes object in order to facilitate the needed calculations.

The following DATA step program implements the Hash of Hashes approach to perform these calculations on multiple variables.

Program 9.4 Chapter 9 HoH MeanMedianMode.sas (Part 1)

```
data Ptiles;      ❶
 input Percentile;
 Metric = put(Percentile,percent8.);
 datalines;
.05
.1
.25
.5
.75
.95
;

data Variables;      ❷
 infile datalines;
 length Variable $32;
 input Variable $32.;
 datalines;
Distance
Direction
;

proc sql noprint;      ❸
 select distinct Variable
        into:Vars separated by ' '
 from Variables;
quit;
```

❶ As we did previously, a data set is used to define the desired percentiles instead of an array. Note that we have both a numeric version (for use in performing the calculations) and character version (created by the PUT function for use in labeling the results) for each desired percentile value.

❷ A data set is used to list the variables to be analyzed. We pointed out to our users that, long term, this would be parameterized and not created in each program (again reminding them that this is just a Proof of Concept).

❸ The SQL procedure is then used to create a macro variable whose value is the list of variables (this approach also works if there is only one variable) to be analyzed.

The above code will likely be replaced in the followon implementation project. We wanted to separate the discussion of what this code does from the DATA step that implements the HoH approach which follows.

Program 9.4 Chapter 9 HoH MeanMedianMode.sas (Part 2)

```
data Distributions(keep=Variable Value Count Percent Cumulative Cum_Percent);      ❶
 length Variable $32 Value 8 Metric $8;
 format Count Cumulative total comma12. Percent Cum_Percent percent7.2;
 array _Variables(*) &Vars;      ❷

 dcl hash ptiles(dataset:"ptiles",ordered:"A");      ❸
 ptiles.defineKey("Percentile");
 ptiles.defineData("Percentile","Metric");
 ptiles.defineDone();
 dcl hiter iter_ptiles("ptiles");
```

```
dcl hash results(ordered:"A",multidata:"Y");        ❹
results.defineKey("Variable","Metric");
results.defineData("Variable","Metric","Value");
results.defineDone();

dcl hash HoH(ordered:"A");        ❺
HoH.defineKey ("I");
HoH.defineData ("H","ITER","Variable","Total","Sum","maxCount");
HoH.defineDone();
dcl hash h();
dcl hiter iter;

do I = 1 to dim(_Variables);        ❻
   h = _new_ hash(ordered:"A");
   h.defineKey(vname(_Variables(i)));
   h.defineData(vname(_Variables(i)),"Count");
   h.defineDone();
   iter = _new_ hiter("H");
   Variable = vname(_Variables(i));
   HoH.add();
end;

maxCount=0;
do Rows = 1 by 1 until(lr);        ❼
   set dw.AtBats end=lr;
   do I = 1 to dim(_Variables);        ❽
      HoH.find();        ❾
      if missing(_Variables(i)) then continue;        ❿
      if h.find() ne 0 then Count = 0;
      Count + 1;
      h.replace();        ⓫
      Total + 1;
      Sum + _Variables(I);
      maxCount = max(Count,maxCount);
      HoH.replace();        ⓬
   end;
end;
do I = 1 to dim(_Variables);        ⓭
   _cum = 0;
   HoH.find();
   iter_ptiles.first();
   last = .;
   do j = 1 to h.num_items;        ⓮
      iter.next();
      Percent = divide(Count,Total);
      _Cum + Count;
      Cumulative = _Cum;
      Cum_Percent = divide(_Cum,Total);
      Value = _Variables(I); /*vvalue(_Variables(I))*/
      output;        ⓯
      if Count = maxCount
         then results.add(Key:Variable
                          ,Key:"Mode"
                          ,Data:Variable
                          ,Data:"Mode"
                          ,Data:_Variables(I) /*vvalue(_Variables(I))*/        ⓰
                          );
```

```
            if last le Percentile le Cum_Percent then
         do;  /* found the percentile */
            if percentile ne 1 then results.add();
            if iter_ptiles.next() ne 0 then percentile = 1;
         end; /* found the percentile */
         last = Cum_Percent;
      end;
      Value = divide(Sum,Total);        ⑰
      Metric = "Mean";
      results.add();
      iter.first();        ⑱
      Value = _Variables(I); /*vvalue(_Variables(I))*/
      Metric = "Min";
      results.add();
      iter.last();        ⑲
      Value = _Variables(I); /*vvalue(_Variables(I))*/
      Metric = "Max";
      results.add();
   end;
 results.output(dataset:"Metrics");        ⑳
 stop;
 set ptiles;
run;
```

❶   Create an output data set for the frequency distributions.

❷   Define an array that contains the list of variables. The number of elements in this array defines how many hash object instances to create in our Hash of Hashes table.

❸   Define a hash object containing the desired percentile values similar to what was done previously. Note that this hash object does not contain the additional variables in the data portion for the results.

❹   Define a hash object that will contain the results (other than the frequency distribution). Note that both the Variable name and the Metric name are keys. The MULTIDATA argument tag is used to allow multiple items for the same key-values. This is needed because the Mode metric can have more than one value.

❺   Define our Hash of Hashes object. Note that it contains additional columns in the data portion that are used in performing our calculations. As we iterate (enumerate) through our hash object instances, these values will be updated and kept in synch with the PDV host variables.

❻   Loop through the elements of our array and create a hash object instance for each variable in the array. Note the use of the DIM function which prevents hard-coding of the number of variables as well as the VNAME function to insert the variable name into the Hash of Hashes table.

❼   Read all the AtBats data rows in an explicit (DoW) loop.

❽   As each row is read from the AtBats data set, loop through the items in the hash object HoH. Note that we do not use a hash iterator object; instead we can loop through the items as we loop through the array.

❾   Since the hash object HoH uses the array index variable I as its key, the FIND method retrieves the data portion values for the hash item corresponding to the variable in the array. This includes the running totals for the sum of the variable (Sum) and the total number of rows (Total), as well as the maximum frequency (maxCount).

❿   Skip rows that have a missing value for the subject variable.

⓫   After the current frequency value (Count) has been incremented, update the instance of the hash object for the current variable.

⑫ After the running totals for the sum of the variable (Sum) and the total number of rows, (Total) as well as the maximum frequency (maxCount), update the data portion of the hash object HoH.

⑬ Once all the data has been read and summarized into our hash object, loop through the items in the hash object HoH to perform the needed calculations.

⑭ The instances of the hash object H contain the frequency counts. Loop through the data to calculate the percent, cumulative count, and the cumulative percent.

⑮ Output a row to the data set for the frequency distribution.

⑯ If the value of Count is equal to the maximum (maxCount), add an item to the Results hash object for the Mode. Note that this assumes that all the variables to be analyzed are numeric. If both numeric and character variables are included, the PDV host variable Value can be defined as character and the VVALUE function can be used to provide the formatted value as a character expression.

⑰ Calculate the Mean and call the ADD method to add an item to the hash object *Results*.

⑱ Since the hash object H is in acending order, the first item is the minimum. Use the FIRST iterator object method to retrieve that value and call the ADD method to add an item to the hash object *Results*.

⑲ Likewise, the last item is the maximum. Use the LAST iterator object method to retrieve that value and call the ADD method to add an item to the hash object *Results*.

⑳ Use the OUTPUT method to create a data set containing the calculated metrics to the data set Metrics.

The calculated metric values generated by the above program follow. Note that we have two Mode values for Distance. The metrics output for the Distance variable (Output 9.6) displays the same values as seen in Output 8.8 .

Output 9.5 Direction Metrics

Variable	Metric	Value
Direction	5%	1.000
	10%	2.000
	25%	5.000
	50%	9.000
	75%	14.000
	95%	18.000
	Max	18.000
	Mean	9.484
	Min	1.000
	Mode	7.000

Output 9.6 Distance Metrics

Variable	Metric	Value
Distance	5%	22.000
	10%	36.000
	25%	76.000
	50%	204.000
	75%	296.000
	95%	427.000
	Max	480.000
	Mean	199.002
	Min	4.000
	Mode	76.000
	Mode	96.000

Output 9.7 First 10 Rows for the Direction Frequency Distribution

Variable	Value	Count	Cumulative	Percent	Cum Percent
Direction	1	11,627	11,627	5.54%	5.54%
	2	11,777	23,404	5.61%	11.1%
	3	11,754	35,158	5.60%	16.7%
	4	11,780	46,938	5.61%	22.4%
	5	11,563	58,501	5.51%	27.9%
	6	11,790	70,291	5.62%	33.5%
	7	11,893	82,184	5.66%	39.1%
	8	11,418	93,602	5.44%	44.6%
	9	11,567	105,169	5.51%	50.1%
	10	11,709	116,878	5.58%	55.7%

Output 9.8 First 10 Rows for the Distance Frequency Distribution

Variable	Value	Count	Cumulative	Percent	Cum Percent
Distance	4	236	236	0.11%	0.11%
	5	243	479	0.12%	0.23%
	6	250	729	0.12%	0.35%
	7	274	1,003	0.13%	0.48%
	8	260	1,263	0.12%	0.60%
	9	286	1,549	0.14%	0.74%
	10	275	1,824	0.13%	0.87%
	11	781	2,605	0.37%	1.24%
	12	708	3,313	0.34%	1.58%
	13	802	4,115	0.38%	1.96%

9.4 Consecutive Events

This next example illustrates the use of arrays for scalar variables (i.e., numeric variables) whose values are needed to perform the needed calculations. In this case we are using a Hash of Hashes approach to count the same consecutive events as in the prior chapter. Four arrays are used to define:

- The variables which represent the event (the batter got a hit; the batter got on base).
- The calculated key for the hash object.
- The two fields (an array for each one) needed for each of the events.

The following code illustrates this approach. Note that while we could have also used parameter files for this, we have not done that in order to focus on the use of arrays. This program produces the same results as seen in section 8.2.3.5 "Determining the Distribution of Consecutive Events."

Program 9.5 Chapter 9 HoH Count Consecutive Events.sas

```
data _null_;
 format Consecutive_Hits Consecutive_OnBase 8.
        Hits_Exact_Count Hits_Total_Count
        OnBase_Exact_Count OnBase_Total_Count comma10.;
 array _Is_A(*) Is_A_Hit Is_An_OnBase;      ❶
 array _Consecutive(*) Consecutive_Hits Consecutive_OnBase;
 array _Exact(*) Hits_Exact_Count OnBase_Exact_Count;
 array _Total(*) Hits_Total_Count OnBase_Total_Count;
 if _n_ = 1 then
 do;  /* define the hash tables */
      dcl hash HoH(ordered:"A");       ❷
      HoH.defineKey ("I");
      HoH.defineData ("H","ITER","Table");
      HoH.defineDone();
      dcl hash h();
      dcl hiter iter;

      h = _new_ hash(ordered:"D");      ❸
      h.defineKey("Consecutive_Hits");
      h.defineData("Consecutive_Hits","Hits_Total_Count","Hits_Exact_Count");
      h.defineDone();
      iter = _new_ hiter("H");
      I = 1;
      Table = vname(_Consecutive(I));
      HoH.add();

      h = _new_ hash(ordered:"D");
      h.defineKey("Consecutive_OnBase");
      h.defineData("Consecutive_OnBase","OnBase_Total_Count"
                 ,"OnBase_Exact_Count");
      h.defineDone();
      iter = _new_ hiter("H");
      I = 2;
      Table = vname(_Consecutive(I));
      HoH.add();;
 end; /* define the hash table */
 do I = 1 to dim(_Consecutive);      ❹
    _Consecutive(I) = 0;
 end;
 do until(last.Top_Bot);      ❺
```

```
      set dw.atbats(keep=Game_SK Inning Top_Bot Is_A_Hit Is_An_OnBase) end=lr;
      by Game_SK Inning Top_Bot notsorted;
      do I = 1 to dim(_Consecutive);
         _Consecutive(I)=ifn(_Is_A(I),_Consecutive(I)+1,0);   ❻
      end;
      do I = 1 to HoH.num_items;      ❼
         HoH.find();
         if h.find() ne 0 then call missing(_Exact(I));
         _Exact(I) + 1;
         if _Is_A(I) then h.replace();
      end;
   end;
   if lr;    ❽
   do I = 1 to dim(_Consecutive);      ❾
      HoH.find();
      Cum = 0;
      do consec = 1 to h.num_items;
         rc = iter.next();
         Cum + _Exact(I);
         _Total(I) = Cum;
         h.replace();
      end;
      h.output(dataset:Table);      ❿
   end;
run;
```

❶   Define the four arrays. Note that the dimension of the arrays (2) corresponds to the number of event types for which consecutive occurrences are to be calculated.

❷   Define our Hash of Hashes object. Note that the key portion is the index variable for the array. The ORDERED argument tag is used to ensure that the array elements are in sync with the instances in the Hash of Hashes object. One of the columns in the data portion is the variable name. That will be used to name the output data set.

❸   Create the hash object instances for each of the types of consecutive events. This could be parameterized (as mentioned above), but we chose not to do that in this Proof of Concept.

❹   Initialize the key portion of the hash values to 0 - in other words, no consecutive events at this point in processing the input data, just as what was done in the Program 8.13 example.

❺   Use a DoW loop to make each execution of the DATA step read the data for a single half inning, so that we can count consecutive events for each team separately. Note that we did not include the team identifier in our output data set. That could be easily done by using a hash object to look up the team details.

❻   Determine the key value for each event type just as in the Program 8.13 example (except we are using an array instead of a hard-coded variable name).

❼   Again, this is the same logic as was used in the prior example, except for the use of arrays to handle multiple event types.

❽   The remaining logic is executed only after we have read all the data.

❾   Use the array dimension to enumerate our Hash of Hashes objects and calculate the value that represents the "N or more" event counts by creating a cumulative count and adding it to the current exact count.

❿   Create an output data set for each event type.

Running the above program creates the following output. These are the same values as seen in Output 8.11. The only difference is that Output 8.11 showed one table and this output is one table for each type of consecutive event.

Output 9.9 Consecutive Hit Events

Consecutive Hits	Hits Total Count	Hits Exact Count
10	1	1
9	8	7
8	35	27
7	122	87
6	396	274
5	1,209	813
4	3,585	2,376
3	10,819	7,234
2	32,178	21,359
1	96,302	64,124

Output 9.10 Consecutive OnBase Events

Consecutive OnBase	OnBase Total Count	OnBase Exact Count
12	1	1
11	10	9
10	37	27
9	119	82
8	304	185
7	750	446
6	1,785	1,035
5	4,180	2,395
4	9,639	5,459
3	22,432	12,793
2	52,379	29,947
1	122,802	70,423

9.5 Multiple Splits

Multiple splits is probably the ideal use case for the Hash of Hashes approach as there are almost a countably infinite set of splits that our users and the fans of Bizarro Ball are interested in. We included just a small sample in the previous example. There are many, many more splits that are almost a standard. For example, in addition to the ones included previously:

- Team vs. Team
- Batter vs. Team

- Pitcher vs. Team
- Batter vs. Pitcher
- Batter hits left/right/switch vs. Pitcher left/right
- Inning
- Home vs. Away
- Any combination of the above
- And so on, and so on

Given the number of different splits that may be requested, our example here leverages all the needed features that we illustrated earlier- most notably the use of parameter files to define the hash objects for both the lookup tables needed as well as for the splits calculations.

- Template.Chapter9LookupTables which are the lookup tables and which include a column that designates the name of the data set to be loaded using the DATASET argument tag. When this is implemented in the eventual application, we will likely create a parent-child table structure for this, as the variable datasetTag is not needed on every observation.
- Template.Chapter9Splits which lists the splits to be calculated. When this is implemented in the eventual application, we will likely also create a parent-child table structure for this to specify the values for the various argument tags (e.g., the ORDERED argument tag).

Output 9.11 All 20 Rows of Template.Chapter9LookupTables

hashTable	Column	Is_A_Key	datasetTag
GAMES	Game_SK	1	DW.GAMES
GAMES	Date	0	DW.GAMES
GAMES	Time	0	DW.GAMES
GAMES	Year	0	DW.GAMES
GAMES	Month	0	DW.GAMES
GAMES	DayOfWeek	0	DW.GAMES
GAMES	League	0	DW.GAMES
GAMES	Home_SK	0	DW.GAMES
GAMES	Away_SK	0	DW.GAMES
PLAYERS	Player_ID	1	DW.PLAYERS
PLAYERS	Team_SK	0	DW.PLAYERS
PLAYERS	First_Name	0	DW.PLAYERS
PLAYERS	Last_Name	0	DW.PLAYERS
PLAYERS	Bats	0	DW.PLAYERS
PLAYERS	Throws	0	DW.PLAYERS
PLAYERS	Start_Date	0	DW.PLAYERS
PLAYERS	End_Date	0	DW.PLAYERS
TEAMS	Team_SK	1	DW.TEAMS
TEAMS	Team_Name	0	DW.TEAMS
TEAMS	League_SK	0	DW.TEAMS

Output 9.12 First 20 Rows of Template.Chapter9Splits

hashTable	Column	is_A_Key
BYDAYOFWEEK	DayOfWeek	1
BYDAYOFWEEK	PAs	0
BYDAYOFWEEK	AtBats	0
BYDAYOFWEEK	Hits	0
BYDAYOFWEEK	BA	0
BYDAYOFWEEK	OBP	0
BYDAYOFWEEK	SLG	0
BYDAYOFWEEK	OPS	0
BYDAYOFWEEK	_Bases	0
BYDAYOFWEEK	_Reached_Base	0
BYMONTH	Month	1
BYMONTH	PAs	0
BYMONTH	AtBats	0
BYMONTH	Hits	0
BYMONTH	BA	0
BYMONTH	OBP	0
BYMONTH	SLG	0
BYMONTH	OPS	0
BYMONTH	_Bases	0
BYMONTH	_Reached_Base	0

The following program calculates these splits.

Program 9.6 HoH Chapter 9 Multiple Splits.sas

```
data _null_;
 dcl hash HoH(ordered:"A");        ❶
 HoH.defineKey("hashTable");
 HoH.defineData("hashTable","H","ITER","CalcAndOutput");
 HoH.defineDone();
 dcl hiter HoH_Iter("HoH");
 dcl hash h();
 dcl hiter iter;
 /* define the lookup hash object tables */
 do while(lr=0);        ❷
    set template.chapter9lookuptables
        template.chapter9splits(in = CalcAndOutput)
    end=lr;
    by hashTable;
    if first.hashTable then        ❸
    do;  /* create the hash object instance */
       if datasetTag ne ' ' then h = _new_ hash(dataset:datasetTag
                                       ,multidata:"Y");
       else h = _new_ hash(multidata:"Y");
    end; /* create the hash object instance */
    if Is_A_key then h.DefineKey(Column);        ❹
    h.DefineData(Column);        ❺
    if last.hashTable then        ❻
```

```
      do;  /* close the definition and add it to our HoH hash table */
         h.defineDone();
         HoH.add();
      end; /* close the definition and add it to our HoH hash table */
   end;
   /* create non-scalar fields for the needed lookup tables */
   HoH.find(key:"GAMES");     ❼
   dcl hash games;
   games = h;
   HoH.find(key:"PLAYERS");
   dcl hash players;
   players = h;
   HoH.find(key:"TEAMS");
   dcl hash teams;
   teams = h;

   if 0 then set dw.players
                 dw.teams
                 dw.games;
   format PAs AtBats Hits comma6. BA OBP SLG OPS 5.3;

   lr = 0;
   do until(lr);     ❽
      set dw.AtBats(rename=(batter_id=player_id)) end = lr;
      call missing(Team_SK,Last_Name,First_Name,Team_Name,Date,Month,DayOfWeek);
      games.find();
      players_rc = players.find();
      do while(players_rc = 0);
         if (Start_Date le Date le End_Date) then leave;
         players_rc = players.find_next();
      end;
      if players_rc ne 0 then call missing(Team_SK,First_Name,Last_Name);
      teams.find();
      do while (HoH_Iter.next() = 0);     ❾
         if not calcAndOutput then continue;
         call missing(PAs,AtBats,Hits,_Bases,_Reached_Base);
         rc = h.find();
         PAs            + 1;
         AtBats         + Is_An_AB;
         Hits           + Is_A_Hit;
         _Bases         + Bases;
         _Reached_Base  + Is_An_OnBase;
         BA = divide(Hits,AtBats);
         OBP = divide(_Reached_Base,PAs);
         SLG = divide(_Bases,AtBats);
         OPS = sum(OBP,SLG);
         h.replace();
      end;
   end;
   do while (HoH_Iter.next() = 0);     ❿
      if not calcAndOutput  then continue;
      h.output(dataset:hashTable||"(drop=_:)");
   end;
   stop;
run;
```

❶   As in our previous example, define our Hash of Hashes table. The key for our hash table is the name of our hash table. We have two different uses of the hash object instances added to this hash table: used as a lookup table vs. as a table to contain our split calculations. We include the column *calcAndOutput* in the data portion to distinguish between these.

❷ Read the parameter files that define the hash tables to be created. An IN variable is used to provide the value for the calcAndOutput PDV host variable that is included in the data portion of our Hash of Hashes tables. That allows us to identify those items which contain instances that are to contain calculations and are used to create output data sets vs. those that create the needed lookup tables.

❸ Use the _NEW_ operator to create an instance of the hash object H and if the instance is a lookup table, load the data into it. Note that the MULTIDATA argument tag is set to allow multiples for the lookup tables. Eventually that should be parameterized so the setting can be specific to each lookup table. We need to allow for multiple items with the same key value since the PLAYERS lookup table is the Type 2 Slowly Changing Dimension table created in section 7.4, "Creating a Bizarro Ball Star Schema Data Warehouse."

❹ If the current column is part of the key portion, define it as a key.

❺ Regardless, add the column to the data portion of the hash object. For now we are assuming that every column is added to the data portion of the hash object. Eventually this can be parameterized so that a column can be included in just the key portion, just the data portion, of both portions of the hash object.

❻ Finalize the creation of the hash table instance.

❼ In order to perform the table lookups (search-and-retrieve) needed, we need a non-scalar variable of type hash object for each lookup table. We use the FIND method to populate non-scalar variables whose values point to the instances of the lookup table hash objects for Games, Players, and Teams.

❽ Read the AtBats data in a loop and perform the table lookups (search-and-retrieve) for the needed fields from the Games, Players, and Teams lookup tables just as was done previously. The only change made here is that we have decided to use Player_ID as the key to the lookup table.

❾ Loop (enumerate) the hash object and, for every instance that is a hash that contains calculated splits, perform the calculations. Note that this looping eliminates the need for multiple assignments and LINK statements.

❿ Loop (enumerate) the hash object and for every instance that is a hash table that contains calculated splits use the OUTPUT method to create an output data set containing the results for this split.

This program calculates the same metrics as seen in Output 8.12-8.16. The output data sets are not ordered (i.e., sorted) by the key values, as the ORDERED data set tag was not used as was used for the data sets listed in Output 8.12-8.16. Because of that, the data rows shown here may be different given that only the first 5 rows are shown. We told the business users that additional metadata options could certainly be implement to specify and control the ordering of the output data sets. We also told them that those details would be determined should they decide to proceed to a full implementation after the completion of the Proof of Concept.

Output 9.13 First 5 Rows of byDayOfWeek Split Results

DayOfWeek	PAs	AtBats	Hits	BA	OBP	SLG	OPS
2	47,844	43,412	16,044	0.370	0.428	0.879	1.307
1	48,298	43,790	16,348	0.373	0.432	0.885	1.317
3	47,880	43,526	16,121	0.370	0.428	0.877	1.305
7	47,717	43,338	15,895	0.367	0.425	0.873	1.298
4	47,854	43,493	16,079	0.370	0.427	0.873	1.300

Output 9.14 First 5 Rows of byMonth Split Results

Month	PAs	AtBats	Hits	BA	OBP	SLG	OPS
5	43,067	39,056	14,399	0.369	0.427	0.881	1.308
9	41,754	37,829	14,074	0.372	0.431	0.883	1.314
3	15,743	14,266	5,135	0.360	0.420	0.850	1.269
7	42,827	38,883	14,143	0.364	0.422	0.867	1.290
4	41,651	37,894	14,108	0.372	0.429	0.879	1.308

Output 9.15 First 5 Rows of byPlayer Split Results

Last Name	First Name	Player ID	PAs	AtBats	Hits	BA	OBP	SLG	OPS
Cook	Donald	29474	650	594	227	0.382	0.435	0.926	1.361
Davis	Harold	51972	103	93	33	0.355	0.417	0.860	1.278
Russell	Ralph	56097	96	92	31	0.337	0.365	0.717	1.082
Scott	Joseph	28568	140	124	42	0.339	0.414	0.895	1.309
Alexander	Jason	48313	96	88	35	0.398	0.448	0.841	1.289

Output 9.16 First 5 Rows of byPlayerMonth Split Results

Last Name	First Name	Player ID	Month	PAs	AtBats	Hits	BA	OBP	SLG	OPS
Anderson	Patrick	25267	10	38	34	8	0.235	0.316	0.647	0.963
Coleman	Joseph	28581	3	40	38	9	0.237	0.275	0.579	0.854
Cook	Justin	15284	5	89	85	30	0.353	0.382	0.682	1.064
Gonzalez	Benjamin	14241	5	114	104	31	0.298	0.360	0.894	1.254
Gray	Harold	51986	10	58	54	18	0.333	0.379	0.833	1.213

Output 9.17 First 5 Rows of byTeam Split Results

Team Name	Team SK	PAs	AtBats	Hits	BA	OBP	SLG	OPS
Broncos	219	8,936	8,108	2,956	0.365	0.423	0.871	1.294
Vikings	132	8,978	8,165	3,029	0.371	0.428	0.873	1.301
Red Sox	319	9,021	8,198	3,043	0.371	0.429	0.884	1.312
Owls	246	9,041	8,164	3,016	0.369	0.431	0.886	1.317
Bluejays	339	9,060	8,199	3,066	0.374	0.433	0.890	1.324

9.5.1 Adding a Unique Count

The users for our Proof of Concept have confirmed that they like this approach to deal with splits. However, they have expressed a couple of concerns. The functionality to calculate unique counts (e.g., games played) is absolutely required, and they need to see an example of doing that using the Hash of Hashes approach. We agreed to do that but qualified that to say that we would not attempt to parameterize that at this point. We assured them it could be parameterized and pointed out that just modifying the existing program using a hard-coded approach would likely do a better job of illustrating how this could be done. They agreed to that limitation and so the following program illustrates adding a unique count of Games. The changes to the program are shown in bold.

Program 9.7 Chapter 9 HoH Multiple Splits with Unique Count.sas

```
data _null_;
 dcl hash HoH(ordered:"A");
 HoH.defineKey("hashTable");
 HoH.defineData("hashTable","H","ITER","calcAndOutput","U");      ❶
 HoH.defineDone();
 dcl hiter HoH_Iter("HoH");
 dcl hash h();
 dcl hash u();      ❷
 dcl hiter iter;
 /* define the lookup hash object tables */
 do while(lr=0);
    set template.chapter9lookuptables
        template.chapter9splits(in = calcAndOutput)
    end=lr;
    by hashTable;
    if first.hashTable then
    do;  /* create the hash object instance */
       if datasetTag ne ' ' then      ❸
       do;  /* create the lookup table hash object */
          h = _new_ hash(dataset:datasetTag,multidata:"Y");
          u = _new_ hash(); /* not used */
       end; /* create the lookup table hash object */
       else
       do;  /* create the two hash objects for the calculations */
          h = _new_ hash();
          u = _new_ hash();
       end; /* create the two hash objects for the calculations */
    end; /* create the hash object instance */
    if Is_A_Key then
    do;  /* define the keys for the two hash objects for the calculations */
       h.DefineKey(Column);
       u.DefineKey(Column);      ❹
       if calcAndOutput then u.DefineKey("Game_SK");
    end; /* define the keys for the two hash objects for the calculations */
    h.DefineData(Column);
    if last.hashTable then
    do;  /* close the definition and add it to our HoH hash table */
       if calcAndOutput then h.DefineData("N_Games");      ❺
       h.defineDone();
       u.defineDone();
       HoH.add();
    end; /* close the definition and add it to our HoH hash table */
 end;
 /* create non-scalar fields for the needed lookup tables */
 HoH.find(key:"GAMES");
 dcl hash games;
 games = h;
 HoH.find(key:"PLAYERS");
 dcl hash players;
 players = h;
 HoH.find(key:"TEAMS");
 dcl hash teams;
 teams = h;
```

```
if 0 then set dw.players
               dw.teams
               dw.games;
format PAs AtBats Hits comma6. BA OBP SLG OPS 5.3;

lr = 0;
do until(lr);
   set dw.AtBats(rename=(batter_id=player_id)) end = lr;
   call missing(Team_SK,Last_Name,First_Name,Team_Name,Date,Month,DayOfWeek);
   games.find();
   players_rc = players.find();
   do while(players_rc = 0);
      if (Start_Date le Date le End_Date) then leave;
      players_rc = players.find_next();
   end;
   if players_rc ne 0 then call missing(Team_SK,First_Name,Last_Name);
   teams.find();
   do while (HoH_Iter.next() = 0);
      if not calcAndOutput then continue;
      call missing(N_Games,PAs,AtBats,Hits,_Bases,_Reached_Base);     ❻
      rc = h.find();
      N_Games + (u.add() = 0);     ❼
      PAs            + 1;
      AtBats         + Is_An_AB;
      Hits           + Is_A_Hit;
      _Bases         + Bases;
      _Reached_Base + Is_An_OnBase;
      BA = divide(Hits,AtBats);
      OBP = divide(_Reached_Base,PAs);
      SLG = divide(_Bases,AtBats);
      OPS = sum(OBP,SLG);
      h.replace();
   end;
end;
do while (HoH_Iter.next() = 0);
   if not calcAndOutput  then continue;
   h.output(dataset:hashTable||"(drop=_:)");
end;
stop;
run;
```

❶ Add a non-scalar field U (i.e., a hash pointer) to the data portion of our Hash of Hashes object HoH. The hash object instance identified by the value of U will be used to support the calculation of a distinct (or unique) count.

❷ Define a non-scalar PDV variable U of type hash.

❸ Create an instance of the hash object U in addition to the instances of the hash object H. Note that we don't need a hash object instance of U for the lookup tables. However, as it is not clear that an existing non-scalar field can be set to null, we create a new one (which has virtually no impact on performance). This prevents the lookup table hash table items from having a value for U that points to an instance that was generated previously in the loop for a splits hash table.

❹ The hash object instance U is defined to have the same keys as the splits hash object as well as the variable (Game_SK) which is the unique key for each game.

❺ Add the field N_Games to the data portion of our splits hash objects.

❻ Add N_Games to the list of calculated fields that need to be set to null before the FIND method call.

❼ Add 1 to the current value of N_Games if the current key value for the split (plus Game_SK) is not found in the hash object instance pointed to by the value of U. If the key is not found, that means this is a new game and the unique count of games should be incremented. This is exactly the same logic as used in earlier examples.

This program produces the same results for all of the fields in the previous example along with the additional field N_Games. The following output shows two of the output tables created.

Output 9.18 First 5 Rows of byDayOfWeek

DayOfWeek	PAs	AtBats	Hits	BA	OBP	SLG	OPS	N Games
2	47,844	43,412	16,044	0.370	0.428	0.879	1.307	480
1	48,298	43,790	16,348	0.373	0.432	0.885	1.317	480
3	47,880	43,526	16,121	0.370	0.428	0.877	1.305	480
7	47,717	43,338	15,895	0.367	0.425	0.873	1.298	480
4	47,854	43,493	16,079	0.370	0.427	0.873	1.300	480

The value of *N_Games* is the same for all the rows due to the nature of the schedule of the games.

Output 9.19 First 5 Rows of byMonth

Month	PAs	AtBats	Hits	BA	OBP	SLG	OPS	N Games
5	43,067	39,056	14,399	0.369	0.427	0.881	1.308	432
9	41,754	37,829	14,074	0.372	0.431	0.883	1.314	416
3	15,743	14,266	5,135	0.360	0.420	0.850	1.269	160
7	42,827	38,883	14,143	0.364	0.422	0.867	1.290	432
4	41,651	37,894	14,108	0.372	0.429	0.879	1.308	416

The above results are the same as what is shown in Output 9.13 and 9.14 with the exception of the additional column N_Games. Note that the results are not sorted by the key here as well. Also note that the number of games is different for each month since each month has a different number of games due to the fact that months don't start on a week boundary and there are no games played on Thursday.

9.5.2 Multiple Split Calculations

The second concern our users expressed is how to deal with splits that either required different key values (e.g., pitching splits instead of or in addition to batting splits), or different calculations. For different calculations we pointed out that one approach (again, via parameterization) was separate programs or perhaps the use of the macro language. We did agree to augment our original multiple splits program to illustrate performing both batting and pitching splits. Shown below is a program that calculates metrics for how each batter does against all the *pitchers* he has faced while at the same time calculating metrics for how each pitcher does against all the *batters* he has faced.

Our first step is to augment our parameter file that defines the splits. Our users agreed to let us hard-code that update into our sample programs instead of building another parameter file once we pointed out that the changes would be clearer if we took that approach.

Program 9.8 Chapter 9 HoH Multiple Splits Batter and Pitcher.sas (Part 1)

```
data chapter9splits;
 set template.chapter9splits
     template.chapter9splits(in=asPitcher);      ❶
 by hashTable;
 if asPitcher
    then hashTable = catx("_",hashTable,"PITCHER");     ❷
    else hashTable = catx("_",hashTable,"BATTER");
 output;
 if last.hashTable;
 Column = "IP";      ❸
 output;
 Column = "ERA";
 output;
 Column = "WHIP";
 output;
 Column = "_Runs";
 output;
 Column = "_Outs";
 output;
 Column = "_Walks";
 output;
run;

proc sort data = chapter9splits out = chapter9splits equals;
 by hashTable;
run;

proc sql;
 create table pitchers as      ❹
 select distinct game_sk, top_bot, ab_number, pitcher_id
 from dw.pitches;
quit;
```

❶ Interleave the data set with itself and create a copy in the WORK library that has each split definition repeated – once for batter splits and once for pitcher splits.

❷ The hashTable field is updated to include either PITCHER or BATTER in the name. We will use that (hard-coded for now) to perform the needed calculations for each type of split.

❸ The pitching calculations included all the same calculations as done for batters, plus some additional ones:

  ○ IP – Innings Pitched.

  ○ ERA – Earned Run Average. Note that our sample data does not distinguish between earned and unearned runs.

  ○ WHIP – Walks plus Hits per Inning Pitched.

  ○ Runs, _Outs and _Walks. These are fields we need to aggregate in order to do the needed additional calculations. Note that their names begin with underscores (_) so they will be excluded from the output data set.

❹ In order to create split calculations we need to know who the pitcher was for each row in the AtBats data set. That field is not available in the AtBats data set. But we can look it up in the Pitches data set using the appropriate set of keys (Game_SK, Top_Bot, AB_Number). The following SQL procedure step creates the data needed for the lookup table.

The above code will likely be replaced in the followon implementation project. We wanted to separate the discussion of what this code does from the DATA step that implements the HoH approach which follows.

Output 9.20 Modified Chapter9Splits Data Set – First 26 Rows

hashTable	Column	is A Key
BYDAYOFWEEK_BATTER	DayOfWeek	1
BYDAYOFWEEK_BATTER	PAs	0
BYDAYOFWEEK_BATTER	AtBats	0
BYDAYOFWEEK_BATTER	Hits	0
BYDAYOFWEEK_BATTER	BA	0
BYDAYOFWEEK_BATTER	OBP	0
BYDAYOFWEEK_BATTER	SLG	0
BYDAYOFWEEK_BATTER	OPS	0
BYDAYOFWEEK_BATTER	_Bases	0
BYDAYOFWEEK_BATTER	_Reached_Base	0
BYDAYOFWEEK_PITCHER	DayOfWeek	1
BYDAYOFWEEK_PITCHER	PAs	0
BYDAYOFWEEK_PITCHER	AtBats	0
BYDAYOFWEEK_PITCHER	Hits	0
BYDAYOFWEEK_PITCHER	BA	0
BYDAYOFWEEK_PITCHER	OBP	0
BYDAYOFWEEK_PITCHER	SLG	0
BYDAYOFWEEK_PITCHER	OPS	0
BYDAYOFWEEK_PITCHER	_Bases	0
BYDAYOFWEEK_PITCHER	_Reached_Base	0
BYDAYOFWEEK_PITCHER	IP	0
BYDAYOFWEEK_PITCHER	ERA	0
BYDAYOFWEEK_PITCHER	WHIP	0
BYDAYOFWEEK_PITCHER	_Runs	0
BYDAYOFWEEK_PITCHER	_Outs	0
BYDAYOFWEEK_PITCHER	_Walks	0

Note that the first 10 rows in the above table are the same except for the _BATTER suffix as seen in Output 9.12. The next 10 are also the same except for the _PITCHER suffix. The next 6 rows are the additional Pitcher split metrics to be calculated.

Program 9.8 Chapter 9 HoH Multiple Splits Batter and Pitcher.sas (Part 2)

```
data _null_;
 dcl hash HoH(ordered:"A");
 HoH.defineKey ("hashTable");
 HoH.defineData ("hashTable","H","ITER","CalcAndOutput");
 HoH.defineDone();
 dcl hiter HoH_Iter("HoH");
 dcl hash h();
 dcl hiter iter;
 /* define the lookup hash object tables */
 do while(lr=0);
    set template.chapter9lookuptables
        chapter9splits(in=CalcAndOutput)      ❶
    end=lr;
    by hashTable;
    if first.hashTable then
    do;  /* create the hash object instance */
       if datasetTag ne ' ' then h = _new_ hash(dataset:datasetTag
                                            ,multidata:"Y");
       else h = _new_ hash(multidata:"Y");
    end; /* create the hash object instance */
    if Is_A_key then h.DefineKey(Column);
    h.DefineData(Column);
    if last.hashTable then
    do;  /* close the definition and add it to our HoH hash table */
       h.defineDone();
       HoH.add();
    end; /* close the definition and add it to our HoH hash table */
 end;
 /* create non-scalar fields for the needed lookup tables */
 HoH.find(key:"GAMES");
 dcl hash games;
 games = h;
 HoH.find(key:"PLAYERS");
 dcl hash players;
 players = h;
 HoH.find(key:"TEAMS");
 dcl hash teams;
 teams = h;
 dcl hash pitchers(dataset:"Pitchers");      ❷
 pitchers.defineKey("game_sk","top_bot","ab_number");
 pitchers.defineData("pitcher_id");
 pitchers.defineDone();

 if 0 then set dw.players
               dw.teams
               dw.games
               dw.pitches;      ❸
 format PAs AtBats Hits comma6.
        BA OBP SLG OPS 5.3
        IP comma6. ERA WHIP 6.3;      ❹

 lr = 0;
 do until(lr);
    set dw.AtBats end = lr;      ❺
    call missing(Team_SK,Last_Name,First_Name,Team_Name,Date,Month,DayOfWeek);
    games.find();
```

```
        pitchers.find();      ❻
        do while (HoH_Iter.next() = 0);
           if not calcAndOutput then continue;
           call missing(PAs,AtBats,Hits,_Bases,_Reached_Base,_Outs,_Runs);      ❼
           if upcase(scan(hashTable,-1,"_")) = "BATTER" then player_id = batter_id; ❽
           else player_id = pitcher_id;
           players_rc = players.find();
           do while(players_rc = 0);
              if (Start_Date le Date le End_Date) then leave;
              players_rc = players.find_next();
           end;
           if players_rc ne 0 then call missing(Team_SK,First_Name,Last_Name);
           teams.find();
           rc = h.find();
           PAs            + 1;
           AtBats         + Is_An_AB;
           Hits           + Is_A_Hit;
           _Bases         + Bases;
           _Reached_Base  + Is_An_OnBase;
           _Outs          + Is_An_Out;      ❾
           _Runs          + Runs;
           _Walks         + (Result = "Walk");
           BA = divide(Hits,AtBats);
           OBP = divide(_Reached_Base,PAs);
           SLG = divide(_Bases,AtBats);
           OPS = sum(OBP,SLG);
           if _Outs then      ❿
           do;  /* calculate pitcher metrics suppressing missing value note */
              IP = _Outs/3;
              ERA = divide(_Runs*9,IP);
              WHIP = divide(sum(_Walks,Hits),IP);
           end; /* calculate pitcher metrics suppressing missing value note */
           h.replace();
        end;
     end;
  do while (HoH_Iter.next() = 0);
     if not calcAndOutput  then continue;
     h.output(dataset:hashTable||"(drop=_:)");
  end;
  stop;
run;
```

❶ Use our modified parameter file instead of the original one in the template library.

❷ Create the hash table needed to look up the pitcher details for each row in the AtBats data set.

❸ Add the Pitchers data to define the variables to the PDV.

❹ Add formats for our new calculations.

❺ We are not renaming Batter_ID to Player_ID as we need to use both Batter_ID and Pitcher_ID to look up the player information.

❻ Call the FIND method to find and retrieve the Pitcher_ID value for the current row in the AtBats data set.

❼ Call the MISSING call routine to set the additional fields used in the calculations to null.

❽ The value used for Player_ID needs to be the value of Batter_ID for batter split calculations and Pitcher_ID for pitcher split calculations. Because we are changing this value as we loop through the hash object instances, the player and team detail lookups are now done inside the loop.

⑨   Add the counts/sums needed for the additional pitching splits calculations.

⑩   Calculate the values for the additional pitching metrics.

This program creates 5 output data sets of batting splits and 5 of pitching splits. The first 5 rows of the team splits for batters and pitches are shown below.

Output 9.21 First 5 Rows of Batter Splits

Last Name	First Name	Player ID	PAs	AtBats	Hits	BA	OBP	SLG	OPS
Cook	Donald	29474	650	594	227	0.382	0.435	0.926	1.361
Davis	Harold	51972	103	93	33	0.355	0.417	0.860	1.278
Russell	Ralph	56097	96	92	31	0.337	0.365	0.717	1.082
Scott	Joseph	28568	140	124	42	0.339	0.414	0.895	1.309
Alexander	Jason	48313	96	88	35	0.398	0.448	0.841	1.289

Comparing this output with the byPlayer splits seen in Output 9.15 above confirms that the results are the same. This is as we expected since the point of this example was to leave the batter metrics unchanged while calculating additional metrics for each pitcher.

Output 9.22 First 5 Rows of Pitcher Splits with Additional Metrics

Last Name	First Name	Player ID	PAs	AtBats	Hits	BA	OBP	SLG	OPS	IP	ERA	WHIP
Davis	Harold	51972	1,079	981	361	0.368	0.425	0.885	1.310	195	14.077	2.821
Scott	Joseph	28568	1,488	1,346	496	0.368	0.429	0.871	1.300	270	13.100	2.441
Alexander	Jason	48313	1,029	931	344	0.369	0.430	0.903	1.333	185	13.378	2.568
Reed	Terry	21799	999	922	352	0.382	0.429	0.910	1.339	180	14.350	2.433
Wilson	Christian	40826	973	895	313	0.350	0.402	0.811	1.213	180	11.850	2.450

This is a new output result table, as it summarized the data based on who the pitcher was instead of the batter. It also includes the additional metrics (IP, ERA, and WHIP) that our business users were interested in calculating for the pitchers. The business users were also particularly pleased that the standard batting metrics could be so easily calculated for how pitchers performed against all the batters that they have faced.

9.6 Summary

Both the business users and their IT staff were very positive about the flexibility offered by the Hash of Hashes approach. They recognized that it added some complexity. But the ability to add split calculations as well as define different metrics for different splits using a metadata (i.e., parameter file) was something they really liked. In response to questions from both the business and IT users, we confirmed that many of their additional requirements could likely be addressed by an approach that used a metadata-driven Hash of Hashes approach.

Chapter 10: The Hash Object as a Dynamic Data Structure

10.1 Introduction

There exist a great variety of practical data-processing scenarios where programming actions must be taken depending on whether or not a current value in the program *has already been* encountered during its execution. In order to do it, some kind of data structure must be in place where each distinct current value can be stored for later inquiry and then looked up downstream as quickly as possible.

SAS provides a number of such structures, such as the arrays, string-searching functions, the LAG function, etc. In the simplest case, even a single temporary variable may fill the role. However, in terms of structures that are *both* convenient and fast, no other structure rivals the hash table because of its ability to store, look up, retrieve, and update data in place in *O(1)* time. Moreover, since the hash table can grow and shrink dynamically at run time, we need not know *a priori* how many items will have to be added to it.

In this book, we have already seen many cases of exploiting this capability without stressing it. In this chapter, examples will be given with the emphasis on this feature of paramount importance, as well as the hash operations needed to perform related tasks.

For the benefit of our business users, we explained that the rationale here is simply that the hash object allows us to answer a number of questions more quickly. For example, how many times have both teams scored more than some number of runs in the first inning; how many times have all nine batters gotten a hit. And the question is not only how often has it happened: The information about each occurrence also needs to be presented.

For the benefit of our IT users, this chapter expands upon the data tasks discussion in Chapter 6.

10.2 Stable Unduplication

To *unduplicate a file* means to select, from all records with the same key-value, just one and discard the rest. *Stable* means that in the output file to which the selected records are written, the original relative sequence of the input records is preserved. Which record with a given key-value should be selected depends on the nature of the inquiry. Here are just *some* possible variations:

1. The records have been entered into the file chronologically, and it may be of interest to know which record with the given key-value occurs first or last.

2. A file (e.g., extracted from a data warehouse) may contain multiple records with the same key-value, each marked by the date of its insertion, and for each key-value, we need to keep the record with the latest date since it has the most recent information.

3. Whole records in a file may have been duplicated for one or another reason, and it may be necessary to discard the duplicates in order to cleanse the data. In this case, it does not matter which same-key record to pick.

In scenarios #1 and #2, our business use case is to find all of the times some event has happened subject to some constraints.

10.2.1 Basic Stable Unduplication

Let us look, as an example, at our sample data set Dw.AtBats. Its records represent events entered into the file chronologically – i.e., the event reflected by any record occurred later than the event reflected by any preceding record.

Suppose that for the players with Batter_ID 32390, 51986, and 60088 (all from team Huskies) we want to do the following:

1. Find all the games in which these players hit a triple in the first inning in their home ballpark.

2. Find the chronologically first games in which these events occurred.

3. List these unique games in the same order as they are listed in Dw.AtBats.

All occurrences for these players satisfying condition #1 come from the records in Dw.AtBats filtered out according to the clause:

```
where Batter_ID in (32390,51986,60088)
and    Result   = "Triple"
and    Top_Bot = "B"
and    Inning   = 1
```

These records are presented in the following output:

Output 10.1 Chapter 10 Subset from Dw.AtBats with Duplicate Batter_ID

Game_SK	Batter_ID	Inning	Top_Bot	Result
18A17B24451F48BA22ED2C655E27D1B4	**60088**	1	B	Triple
39AFB84A371E99FA2334553FCE63AF87	**32390**	1	B	Triple
6CAF3CC4614963A4110B761F2F83F709	32390	1	B	Triple
7915F98D5FE34D72AE465A03A0340A06	32390	1	B	Triple
8412BB852289BCA1E15D494A84FFD762	**51986**	1	B	Triple
8AB80A36A1C9A8AC5631AA0F64919B4E	51986	1	B	Triple
9AECDBE069FD5839C59605A179A057CB	60088	1	B	Triple
AA7941D2CDC87A8856517F1094B3C129	51986	1	B	Triple
F390113CBF0964BE71BF019050382182	60088	1	B	Triple

Technically speaking, to get the information we want, we need to select the records with the sequentially first occurrence of each distinct value of Batter_ID (shown above in bold) and discard the rest. In other words, we need to *unduplicate* the file Dw.AtBats by Batter_ID after filtering it using the WHERE clause above.

Traditionally, the SORT procedure with the NODUPKEY option is used for the purpose. Optionally, it also writes the discarded duplicate records to a separate file:

Program 10.1 Chapter 10 Unduplication via PROC SORT.sas

```
proc sort nodupkey
  data   = dw.AtBats   (keep = Game_SK Batter_ID Result Top_Bot Inning)
  out    = nodup_sort (keep = Game_SK Batter_ID)
  dupout = dupes_sort (keep = Game_SK Batter_ID)
  ;
  where Batter_ID in (32390,51986,60088)
  and   Result   = "Triple"
  and   Top_Bot = "B"
  and   Inning   = 1
  ;
  by Batter_ID ;
run ;
```

The Nodup_sort output from this program is shown below:

Output 10.2 Chapter 10 Unduplication Results: PROC SORT

Game_SK	Batter_ID
39AFB84A371E99FA2334553FCE63AF87	32390
8412BB852289BCA1E15D494A84FFD762	51986
18A17B24451F48BA22ED2C655E27D1B4	60088

This technique certainly works. However, it also has its drawbacks:

- The unduplication effect we need is a *side effect* of sorting. Sorting really large files (unlike our sample file) can be costly resource-wise and much more protracted than merely reading a file.
- Because the sort is done by Batter_ID, it *permutes* the input records according to its key-values. Hence, the records in *Nodup_sort* are out of the original chronological order with which they come from file Dw.AtBats. Speaking more formally, the process is *not stable* since, from the standpoint of our task, the original relative sequence of the input records is not kept.
- Restoring the original order in the output would require us to keep the input observation numbers in a separate variable and re-sort the output by it.

Solving the same problem with the aid of the hash object requires a single pass through the file and no sorting. The scheme of using it is simple:

1. Create a hash object instance and key its table by Batter_ID.
2. Read the next record from the file. If the input buffer is empty (i.e., all records have already been read), stop.
3. Search for the value of Batter_ID coming with the current record in the hash table.
4. If it is not in the table, this is the first record with this value. Therefore, write the record to file *Nodup_hash*, and then insert an item with this key-value into the table.
5. Otherwise – i.e., if it is already in the table – write the record to file *Dupes_hash*.
6. Go to #2.

Let us now translate this plan of attack into the SAS language:

Program 10.2 Chapter 10 Stable Unduplication via Hash.sas

```
data nodup_hash dupes_hash ;
  if _n_ = 1 then do ;        ❶
    dcl hash h () ;
    h.defineKey ("Batter_ID") ;
    h.defineDone () ;
  end ;
  set dw.AtBats (keep = Game_SK Batter_ID Result Top_Bot Inning) ;    ❷
  where Batter_ID in (32390,51986,60088)
  and    Result   = "Triple"
  and    Top_Bot = "B"
  and    Inning   = 1
  ;
  if h.check() ne 0 then do ;     ❸
    output nodup_hash ;     ❹
    h.add() ;
  end ;
  else output dupes_hash ;     ❺
  keep Game_SK Batter_ID ;
run ;
```

❶ At the first iteration of the DATA step implied loop, perform the *Create* operation to create hash table H. For the purposes of the program, only the key Batter_ID is needed, but since the data portion is mandatory, we define the same variable there by omitting a call to the DEFINEDATA method.

❷ Read the next record from a subset of file Bizarro.AtBats filtered according to our specifications.

❸ Perform the *Search* operation by calling the CHECK method implicitly and thus forcing it to accept the current PDV value of Batter_ID as the search-for key-value.

❹ If the value is not in the table, write the record to *Nodup_hash* output file. Then insert a new item with this value into the table, so that when any subsequent record is read, we will be able to find that it is already there.

❺ Otherwise, if it is already in the table, the record is deemed duplicate and written to output file *Dupes_hash*.

Running the program generates the following notes in the SAS log:

```
NOTE: There were 9 observations read from the data set DW.ATBATS.
      WHERE Batter_ID in (32390, 51986, 60088) and (Result='Triple') and
      (Top_Bot='B') and (Inning=1);
NOTE: The data set WORK.NODUP_HASH has 3 observations and 2 variables.
NOTE: The data set WORK.DUPES_HASH has 6 observations and 2 variables.
```

Its *Nodup_hash* output looks as follows:

Output 10.3 Chapter 10 Unduplicated Results: Hash Object

Game_SK	Batter_ID
18A17B24451F48BA22ED2C655E27D1B4	60088
39AFB84A371E99FA2334553FCE63AF87	32390
8412BB852289BCA1E15D494A84FFD762	51986

As we see, the original relative order of the input records is now replicated. Data-wise, the hash method produces the output with the same content as PROC SORT. However, it does so:

- Without the extra cost of sorting (or re-sorting if required).
- In a single pass through the input file.
- Preserving the original record sequence – i.e., the process is *stable*.

Note that this scheme would work without any coding changes if we needed to eliminate all *completely duplicate* records but one. It would pick up the first such record in the file and discard the rest; and since all of them are exactly the same, the first duplicate record would be as good as any other. However, it would also require defining all input file variables as a composite key in the hash table and hence can be quite costly in terms of memory usage. Therefore, the description of eliminating complete duplicate records is relegated to Chapter 11, which deals specifically with memory-saving techniques.

10.2.2 Selective Unduplication

In the case of basic unduplication, we keep the records with the first occurrence of each duplicate key encountered in a file. Hence, its inherent order dictates which event is deemed "first". A somewhat more involved situation arises when the file is unordered and, out of all the records with the same key, we need to select a record to keep based on the order controlled by the values of a different variable.

As an example, in the data set Dw.Games variables Home_SK and Away_SK identify the home team and away team, respectively. First, suppose that we are interested in games the home teams 203, 246, and 281 played in May on Saturdays. The records from Dw.Games representing these games satisfy the clause:

```
where Home_SK in (203,246,281) and Month=5 and DayOfWeek=7
```

These records look as follows:

Output 10.4 Chapter 10 Dw.Games Input for Selective Unduplication

Game_SK	Date	Home_SK	Away_SK
616DE36D54BDC9DCD76594ACB02297DF	2017-05-06	203	259
6D060A0F519B896C1FD788B2681E9CED	**2017-05-13**	**203**	**147**
9F048C38E6B5EB9E168D7F1A3A648ECD	2017-05-13	281	348
9FBDCA5638623D41D162B9A42FDF2684	2017-05-06	281	193
A384ECE8841782DC263E4F8BADC56DA0	**2017-05-20**	**246**	**344**
C216B945822DCC19D51DB072B6DAF6B5	2017-05-20	281	203
DE05ED8C95C2DE52943FE790ED9AD24B	**2017-05-27**	**281**	**246**
FB4F4DB75C9D3FFB31723F1B7F7C274D	2017-05-13	246	339

Second, suppose that for each home team in the subset, we want to know *which away team it played last* – i.e., which team it played a game against in its home ballpark. In other words, we need to *unduplicate* the records for each key-value of Home_SK by selecting its record with the latest value of Date. For the home teams listed in the snapshot above, their most recent games, shown in bold, are those we want to keep.

Note that the file *is not sorted* by any of the variables. Hence, the inquiry cannot be answered simply by picking some record for each key-value of Home_SK, based on the record sequence existing in the file.

The well-beaten path most often used to solve this kind of problem is a two-step approach:

1. Sort the file by (Home_SK,Date).
2. Select the last record in each BY group identified by Home_SK.

This approach is exemplified by the program below:

Program 10.3 Chapter 10 Selective Unduplication via PROC SORT + DATA Step.sas

```
proc sort
  data = dw.Games    (keep=Date Home_SK Away_SK Month DayOfWeek)
  out  = games_sorted (keep=Date Home_SK Away_SK)
  ;
  by Home_SK Date ;
  where Home_SK in (203,246,281) and Month=5 and DayOfWeek=7 ;
run ;

data Last_games_sort ;
  set games_sorted ;
  by Home_SK ;
  if last.Home_SK ;
run ;
```

The program generates the following output:

Output 10.5 Chapter 10 Selective Unduplication via PROC SORT + DATA Step

Date	Home_SK	Away_SK
2017-05-13	203	147
2017-05-20	246	344
2017-05-27	281	246

This method suffers from the same major flaws as using the SORT procedure for basic unduplication – i.e., the need to sort as well as dense I/O traffic. Again, using the hash object instead addresses these problem areas. Let us lay out a plan first:

1. Create a hash table keyed by Home_SK and having Date and Away_SK in the data portion.
2. Read the next record.
3. Perform the *Retrieve* operation using the value of Home_SK in the record as a key.
4. If this key-value is not in the table, i.e., this is the first time this value of Home_SK is seen, insert a new item with the values of Home_SK, Date, and Away_SK in the current record into the table.
5. Otherwise, if it is in the table, compare the value of Date from the record with the hash value of Date stored in the table for this Home_SK. If *Date* from the record is more recent, use the *Update* operation to replace the hash values of Date and Away_SK in the table with those from the record.
6. If there are more records to read, go to #1.
7. Otherwise, at this point the hash table contains the data being sought – i.e., all distinct values of Home_SK paired with the latest date on which it was encountered and the respective value of Away_SK. So, all we need to do now is use the *Output* operation, write the content of the table to an output file, and stop the program.

It can be now expressed in the SAS language:

Program 10.4 Chapter 10 Selective Unduplication via Hash.sas

```
data _null_ ;
  dcl hash h (ordered:"A") ;        ❶
  h.defineKey   ("Home_SK") ;
  h.defineData ("Home_SK", "Date", "Away_SK") ;     ❷
  h.defineDone () ;
  do until (lr) ;
    set dw.Games (keep=Date Home_SK Away_SK Month DayOfWeek) end=lr ;
    where Home_SK in (203,246,281) and Month=5 and DayOfWeek=7 ;
    _Date    = Date ;       ❸
    _Away_SK = Away_SK ;
    if h.find() ne 0 then h.add() ;      ❹
    else if _Date > Date then
    h.replace(key:Home_SK, data:Home_SK, data:_Date, data:_Away_SK) ;      ❺
  end ;
  h.output (dataset: "Last_games_hash") ;     ❻
  stop ;
run ;
```

Since the program plan translates into its code almost verbatim, only certain technical details remain to be annotated:

❶ Coding the argument tag ORDERED:"Y" is optional and is needed only if we want the home teams in the output in their ascending order by Home_SK.

❷ Home_SK is added to the data portion of table H, so that it could appear in the output.

❸ Save the values of Date and Away_SK coming from the current record in auxiliary variables. This is necessary because if the FIND method called next succeeds, the values of these PDV host variables will be overwritten with the values of their hash counterparts. (This is an adverse side effect of the hash object design: Unlike an array, it does not allow looking at the value of a hash variable without extracting it into its PDV host variable first.)

❹ If the FIND method has succeeded for the current PDV value of Home_SK, the host value of Date is the same as its hash value. Now it can be compared with the value that came with the record and that was saved in variable _Date.

❺ If the value of _Date is more recent than that of Date, the hash item for the current Home_SK must be updated. An explicit REPLACE call is used to overwrite the hash variables with the saved values of _Date and _Away_SK.

❻ At this point, table H contains all distinct values of Home_SK coupled with the *latest* values of Date and the respective values of Away_SK. So, the OUTPUT method is called to dump its content to the output file.

Running the program generates exactly the same output (Output 10.4) in the same record order as the (PROC SORT + DATA step) approach if the ORDERED:"A" argument tag is used as shown above. If it were either omitted altogether or valued as "N", the output record sequence would be determined by the internal "undefined" hash table order.

An alert reader may have noticed that this program is not without flaws itself in terms of how it scales. For example, imagine that instead of the single satellite variable Away_SK we have in this simple example, there are a few dozen. It is easy to perceive that it would complicate things because:

- All these satellite variables would have to be included in the data portion of the table.

- Instead of the single reassignment _Away_SK = Away_SK in ❸, we would need as many as the number of the satellite variables.

- Though both points above could be addressed using proper code automation via the dictionary tables, the numerous variables in the data portion would increase the hash entry size accordingly. This could have negative implications from the standpoint of hash memory footprint – especially if we had a very large file with very few duplicates, which in real data processing life happens often.

The good news is that all these concerns can be addressed by using a hash index. It will be discussed in Chapter 11 as part of the discussion devoted to hash memory management.

10.3 Testing Data for Grouping

As we know, the data written from a hash table to a file always comes out *grouped* by the table key, whether it is simple or composite. Hence, it can be processed correctly by this key without sorting the file by using the NOTSORTED option with the BY statement.

Suppose, however, that we have a data set and would like to process it in this manner because we are *suspecting* (for example, from the way it looks in the SAS viewer) that the file may be intrinsically *grouped* by the variables we intend to use in the BY statement. If this were the case, we could save both computer resources and time by not sorting the file before doing the BY processing, particularly if the file is long and wide. So, the question arises: How can we determine, reading only the keys in question and in a single pass through the file, whether it is indeed *grouped* or not?

10.3.1 Grouped vs Non-Grouped

To arrive at an answer, let us observe that if a file is grouped by variable Key, it can contain one and only one sequence of same-key adjacent records for each distinct value of Key. Thus, if the values of Key in the file were:

```
2    2    1    3    3    3    3
```

then it would be *grouped* since the records with a given key-value are bunched together into one and only one sequence. But if the same keys were positioned in the file this way:

```
2    2    3    3    1    3    3
```

then the file would be *non-grouped* since now there would be two separate sequences with *Key*=3.

Therefore, we can check to see whether the data is grouped if, for each separate sequence of adjacent keys with the same key-value, we could determine whether this key-value has already been seen before in another sequence. To do so, each distinct key-value encountered in the process needs to be stored somewhere. The ideal container for such storage (and subsequent search) is a hash table. And in response to a question from one of our IT users, we agreed that, yes, most problems can be addressed using a hash table approach.

10.3.2 Using a Hash Table to Check for Grouping

Hence, we can devise a programmatic plan to determine if a file is grouped as follows:

1. Create a hash table keyed by the variables whose grouping we intend to check.
2. Read each consecutive sequence of same-key adjacent records one at a time.
3. At the end of each sequence, check to see if its key-value is in the table.
4. If it is not in the table, store it there and proceed to the next sequence.
5. Otherwise, if it is already in the table, we have found two separate sequences with the same key-value. Therefore, the file is not grouped, so we can flag it as such and stop reading it.
6. If the program proceeds past the last record without having been stopped, the file is grouped.

Following this plan, let us write a program to check to see if data set Dw.Runs is grouped by the composite key (Game_SK, Inning,Top_Bot) since, from a glimpse of it, it seems that it *might* be.

Program 10.5 Chapter 10 Testing Data for Grouping via Hash.sas

```
data _null_ ;
  if _n_ = 1 then do ;
    dcl hash h () ;
    h.definekey ("Game_SK", "Inning", "Top_Bot") ;      ❶
    h.definedone () ;
  end ;
  if LR then call symput ("Grouped", "1") ;      ❻
  do until (last.Top_Bot) ;      ❷
    set dw.Runs end = LR ;
    by Game_SK Inning Top_Bot notsorted ;      ❸
  end ;
  if h.check() ne 0 then h.add() ;      ❹
  else do ;
    call symput ("Grouped", "0") ;      ❺
    stop ;
  end ;
run ;
%put &=Grouped ;      ❼
```

Running the program results in the following line printed in the log:

```
GROUPED=1
```

Hence, the file is indeed grouped by the composite key (Game_SK,Inning,Top_Bot). Needless to say, by extension it is also grouped by (Game_SK,Inning) and (Game_SK) alone. Since the program follows the pseudocode almost verbatim, the annotated explanations below just fill in the gaps and underscore certain details:

❶   Key the hash table with the variables whose grouping we are tracking in the hierarchy we need. Note that the data portion is not specified as it is not needed. So the data portion variables are not defined, and therefore it contains the same variables as defined in the key portion.

❷   The DoW loop here is an ideal vehicle to read one sequence of same-key records per execution of the DATA step. Note that the key indicated in the UNTIL condition is the *last* component of the composite key whose grouping we are exploring.

❸   The NOTSORTED option on the BY statement ensures that the last record in the sequence read in the loop is the record whose key is *simply different* than the key in the next record.

❹   The end of the record sequence just read by the DoW loop has been reached. So, it is time to check if the key of this sequence is already in table H – in other words, whether any preceding sequence has had the same key. If yes, the file is not grouped, and so we act accordingly. If no, we store the key in the table as one more item.

❺   If we have found that the file is not grouped, macro variable *Grouped* is set to 0.

❻   If program control has reached this condition and it evaluates true, it means that after the last record was read the STOP statement has not been executed. So, the file is grouped, and macro variable

Grouped is set to 1 accordingly. (Note that in this case, the DATA step is stopped not by the STOP statement but by the attempt to read from the empty buffer after all the records have been read.)

❼ In this simple example, the value of macro variable *Grouped* is simply displayed in the SAS log. The program can be run in a test mode to decide, depending on the resulting value of *Grouped*, whether to sort the file or use its inherent grouped order for BY processing with the NOTSORTED option. Alternatively, the program can be used as a step in the process to make the decision and construct code needed downstream programmatically.

This program can be easily parameterized and encapsulated as a macro. We leave it as an exercise for inquisitive readers. We did tell the business and IT users that such parameterization and encapsulation could be done in the expected follow-on project.

10.4 Hash Tables as Other Data Structures

The SAS hash object, representing in and by itself a specific data structure (i.e., the hash table), can also be used to implement other data structures. Doing so makes sense in two situations:

- The data structure the programmer needs to accomplish a task is not yet available in SAS as an encapsulated software module.
- It is available – for example, in the form of a function or procedure. However, in terms of the specific task at hand, re-implementing it using the hash object offers better functionality and/or performance.

Historically, using existing SAS data structures to implement data structures absent from the software is not uncommon. In fact, the first practically usable hash tables in SAS were implemented using a different data structure – namely, the SAS array. Conversely, now that the hash object is available, its dynamic nature, the ability to shrink and grow at run time, and search performance all offer numerous advantages in terms of using it to implement other data structures. For example:

- The original name for the hash object is an *associative array*. And indeed, the hash object can be viewed as an implementation of an *array* differing from the standard SAS array in two aspects: (a) its elements can be added to (or removed from it) dynamically at run time, and (b) it can be "subscripted" not only by an integer index but by an arbitrary composite key consisting of a number of variables with different data types.
- The hash object constructed with the argument tag HASHEXP:0 uses a single *AVL binary tree*. It has the property of being balanced in such a way that it can be always searched in $O(log(N))$ time, regardless of the distribution of its input keys. Thus, though SAS does not offer the AVL tree as a separate structure explicitly, the hash object effectively makes it readily available for use in the DATA step. Before the hash object became available, the AVL tree, if needed, would have to be coded using arrays (an exercise most SAS users would find too unpalatable to attempt).

In principle, the rich feature set of the SAS hash object provides for the implementation of just about any dynamic data structure. In this section, we will limit the discussion to the hash table operations required to implement two simple structures not yet offered as dynamic (i.e., shrink-and-grow) canned routines.

10.4.1 Stacks and Queues

Before the advent of the SAS hash object, DATA step data memory-resident structures like stacks and queues had existed in the SAS software in a certain way. In particular, a queue is implemented in the form of the LAG and DIF functions, and a stack (as well as a queue) can be organized using arrays coupled with proper custom coding. There are two main problems with these implementations:

- The number of items they store is preset at compile time. For example, the number of items available in the LAG<*N*> queue is statically pre-defined by its "number of lags" *N*; and the number of items in an array is pre-defined by its dimensions. More elements and more memory required for them cannot be allocated at run time. Likewise, their existing elements cannot be removed and memory occupied by them – released at run time. Thus, the number of lags or array dimensions must be allocated as either "big enough" or determined programmatically by an extra pass through data.

- To organize a stack or a queue with items comprising more than one variable, *parallel* arrays or p*arallel* LAGs are needed. Though this is possible, it may be unwieldy, as it requires handling the variables in such parallel structures separately.

The hash object is devoid of these shortcomings, as its items can be added and removed (and the memory they occupy – acquired and released) at run time, so that any data structure based on the hash object can grow and shrink. Let us first take a very brief look at the essence of the stack and queue and their terminology.

A *stack* is a *last-in-first-out*, or *LIFO*, data structure. When we add an item to a stack, we perform an action called *push*. If we push items 1, 2, and 3 onto an initially empty stack *S* one at a time (from right to left below), the successive states of the stack (shown in brackets) look as follows:

S = [**1**] then [**1 2**] then [**1 2 3**]

When items are taken off the stack, an action called *pop* is performed. When we pop the stack, we surface the item we have *pushed last* and remove it from the stack. Hence, popping stack *S*=[**1 2 3**], for example, 2 times, we will first surface and remove item 3, and then item 2. The successive states of *S* after each pop will look like:

S = [**1 2**] then [**1**]

If we pop the stack one more time, item 1 will be surfaced, and the stack will become empty. Hence, a stack is useful when, after pushing items onto it one at a time, we need to access them in the order *opposite to insertion*. An apt example could be reading a file *backward*.

By contrast, a queue is a *first-in-first-out*, or *FIFO,* data structure. When we insert an item into a queue, we perform an action called *enqueue*. If we enqueue items 1, 2, and 3 into an initially empty queue *Q* one at a time (from left to right below), the successive states of the queue will look like so:

Q = [**1**] then [**2 1**] then [**3 2 1**]

When items are taken from the queue, an action called *dequeue* is performed. By dequeueing, we surface the item queued *first* and remove it from the queue. Thus, if we dequeue queue *Q* above 2 times, we will surface and remove item 1, and then item 2; and the successive states of Q after each dequeue will look as follows:

Q = [**3 2**] then [**3**]

If we dequeue once more, item 3 will be surfaced, and the queue will be emptied. A good example of the dequeuing order could be reading a file *forward*.

Implementing a stack or a queue based on a hash table in *formal* accordance to the rules above is conceptually simple:

- Have a hash table as a container for a stack or a queue and pair it with a hash iterator.
- To either *push* an item onto the stack or *queue* the queue with an item, simply insert it into the table. Do it as many times as necessary before an item needs to be either *popped* off the stack or *dequeued* from the queue.
- To *pop* an item off the stack, access the logically **last** item, retrieve its content, and remove it from the table.
- To *dequeue* an item from the queue, access the logically **first** item, retrieve its content, and remove it from the table.

Note that thanks to the fact that the hash table facilitates access to its items from both ends, the same hash table can be used both as a stack and a queue. The difference between the two materializes only when an item is surfaced: To use the table as a stack, we pop the item added to the table *most recently*; and to use it as a queue, we dequeue the item added *least recently*. Let us now see how it can be implemented programmatically, first for a stack and then for a queue.

10.4.2 Implementing a Stack

The program presented below does the following:

- Creates a hash table as a container for a stack.
- Pushes 3 items on the stack.
- Pops 1 item off the stack.
- Pushes 2 new items.
- Pops 2 items.
- After each action above, writes a record to an output demo file showing the PDV data value pushed or popped and the items on the stack.

Note that the *italicized* lines of code below are unrelated to the stack actions and serve only to render the demo output.

Program 10.6 Chapter 10 Implementing a Hash Stack.sas

```
data Demo_Stack (keep = Action PDV_Data Items) ;
  dcl hash h (ordered:"A") ;      ❶
  h.defineKey  ("Key") ;
  h.defineData ("Key", "Data") ;
  h.definedone () ;
```

```
dcl hiter ih ("h") ;        ❷
Data = "A" ; link Push ;     ❸
Data = "B" ; link Push ;
Data = "C" ; link Push ;
              link Pop ;
Data = "D" ; link Push ;
Data = "E" ; link Push ;
              link Pop ;
              link Pop ;
stop ;
Push: Key = h.num_items + 1 ;    ❹
      h.add() ;
      Action = "Push" ;
      link List ;
return ;
Pop:  ih.last() ;        ❺
      rc = ih.next() ;
      h.remove() ;
      Action = "Pop" ;
      link List ;
return ;
List: PDV_Data  = Data ;     ❻
      Items = put ("", $64.) ;
      do while (ih.next() = 0) ;
        Items = catx (" ", Items, cats ("[", Key, ",", Data, "]"));
      end ;
      output ;
  return ;
run ;
```

Let us dwell on a few points in the program that might need explanation:

❶ Hash table H created to hold the stack/queue (a) is ordered, (b) allows no same-key items, and (c) has its key also added to the data portion. The sole reason for this triple arrangement is the "classic" requirement that after an item is popped, it must be removed. This is because to remove only the first or last item – and not any other item – its key-value must be unique and we need to know it. But to retrieve it using the hash iterator alone without extraneous bookkeeping (which is the intent), the key must be also defined in the data portion. (As we will see later, if the removal requirement is dropped, none of the three measures is needed.)

❷ Hash iterator IH coupled with table H is declared in order to access the first or last item regardless of their keys.

❸ This and the next 7 lines of code call the corresponding LINK routines to: (1) push 3 items on the stack, (2) pop 1 item, (3) push 2 new items, (4) pop 2 items.

❹ The *Push* routine first creates a new unique key by adding 1 to the value of the hash attribute *H*.NUM_ITEMS. Then a new item is inserted by calling the ADD method accepting the new value of the key and the current PDV value of host variable Data.

❺ The *Pop* routine performs the *Enumerate All* operation by using the iterator to access the last item. (Since the item with the highest key is always the one pushed onto the stack last, the routine thus conforms to the LIFO principle.) The hash value of Data from the last item retrieved into the PDV host variable Data by the LAST method call is the value the *pop* action surfaces (and it will also be used in the REMOVE call later). The call to the iterator method NEXT moves the iterator pointer out of the table, thus releasing its lock on the last item, so that it can be removed. The call is *assigned* to avoid log errors because it is guaranteed to fail: if the iterator dwells on the last item, there is no next item to enumerate. Finally, the *Delete* operation is performed by calling the REMOVE method which accepts the PDV value of Key the LAST method extracted earlier.

❻   The LINK routine *List* called after each stack action assembles the [Key , Data] value pairs from the currently available hash (stack) items into a long single variable Items, and outputs the corresponding demo record. Its function is purely auxiliary, and it is unrelated to the stack actions.

Running the program generates the output below. Note that in the Items column, the rightmost and leftmost items represent the top and bottom of the stack, respectively.

Output 10.6 Chapter 10 Using a Hash Table as a Stack

Action	PDV_Data	Items
Push	A	[1,A]
Push	B	[1,A] [2,B]
Push	C	[1,A] [2,B] [3,C]
Pop	C	[1,A] [2,B]
Push	D	[1,A] [2,B] [3,D]
Push	E	[1,A] [2,B] [3,D] [4,E]
Pop	E	[1,A] [2,B] [3,D]
Pop	D	[1,A] [2,B]

The output confirms that in the program above, the behavior of hash table H conforms with the LIFO principle and thus operates as a stack, i.e.:

- After every *Push* action, the current value of PDV host variable Data ends up at the top of the stack as the value of hash variable Data in the newly inserted hash item.
- Conversely, after every *Pop* action, the host variable Data receives the value of hash variable Data from the item at the top of the stack, after which the item is removed.

10.4.3 Implementing a Queue

Now let us consider a very similar program for organizing and operating a *queue*:

Program 10.7 Chapter 10 Implementing a Hash Queue.sas

```
data Demo_Queue (keep = Action PDV_Data Items) ;
  dcl hash h (ordered:"A") ;
  h.defineKey  ("Key") ;
  h.defineData ("Key", "Data") ;
  h.definedone () ;
  dcl hiter ih ("h") ;
  Data = "A" ; link Queue ;
  Data = "B" ; link Queue ;
  Data = "C" ; link Queue ;
              link DeQueue ;
  Data = "D" ; link Queue ;
  Data = "E" ; link Queue ;
              link DeQueue ;
              link DeQueue ;
```

```
stop ;
Queue: Key + 1 ;     ❶
        h.add() ;
        Action = "Queue   " ;
        link List ;
return ;

DeQueue: ih.first() ;    ❷
        rc = ih.prev() ;
        h.remove() ;
        Action = "DeQueue" ;
        link List ;
return ;
List: PDV_Data  = Data ;
      Items = put ("", $64.) ;
      do while (ih.next() = 0) ;
        Items = catx ("  ", Items, cats ("[", Key, ",", Data, "]")) ;
      end ;
      output ;
  return ;
run ;
```

Its output looks as follows:

Output 10.7 Chapter 10 Using a Hash Table as a Queue

Action	PDV_Data	Items
Queue	A	[1,A]
Queue	B	[1,A] [2,B]
Queue	C	[1,A] [2,B] [3,C]
DeQueue	A	[2,B] [3,C]
Queue	D	[2,B] [3,C] [4,D]
Queue	E	[2,B] [3,C] [4,D] [5,E]
DeQueue	B	[3,C] [4,D] [5,E]
DeQueue	C	[4,D] [5,E]

The program and output for the queue look quite similar to those for the stack. However, there are a few significant distinctions (in the program, they are shown in bold):

❶ The *queue* action is different from *push* in that a new item is queued into the *back* of the queue. Thus, we want each newly queued item at the *bottom* of the table. This is done by incrementing the key and inserting the item being queued using the ADD method.

❷ The *dequeue* action is similar to the *pop* action of the stack. However, instead of retrieving the Data value from the last item (i.e., the *back* of the queue), it retrieves it from the first item (i.e., the *front*) and then removes it. To do so, the iterator lock on the first item needs to be released first. This is done by calling the PREV method which moves the iterator pointer out of the table, thus releasing the first item.

10.4.4 Using a Hash Stack to Find Consecutive Events

As a practical example, suppose that we want to find, from the content of file Dw.AtBats, the latest 6 games in the season where 4 or more back-to-back home runs were scored, along with the actual back-to-back home run count. This is a typical LIFO requirement, meaning that we need a stack.

The inquiry stated above can be answered by making use of the fact that the records in Dw.AtBats are entered *chronologically*. Therefore, if a series of 4 or more records marked as Result="Home Run" is *both preceded and followed* by a *different value* of Result, we have identified a back-to-back event we are looking for. For example, let us take a look at this group of consecutive records from the file. (The column From_obs is not a field in the file; it just indicates which observation in Dw.AtBats the records shown below come from.)

Output 10.8 Chapter 10 Dw.AtBats Record Sample with Back-to-Back Events

Game_SK	Result	From_obs
31707E6AB51EEE6FA328F960E8344F27	Strikeout	54246
31707E6AB51EEE6FA328F960E8344F27	Home Run	54247
31707E6AB51EEE6FA328F960E8344F27	Home Run	54248
31707E6AB51EEE6FA328F960E8344F27	Home Run	54249
31707E6AB51EEE6FA328F960E8344F27	Home Run	54250
31707E6AB51EEE6FA328F960E8344F27	Out	54251
31707E6AB51EEE6FA328F960E8344F27	Strikeout	54252
31707E6AB51EEE6FA328F960E8344F27	Strikeout	54253
31707E6AB51EEE6FA328F960E8344F27	Out	54254

Above, the records 54227-54250 and 54252-54253 represent consecutive back-to-back events. In particular, the records where Result="Home Run" represents a back-to-back, 4-home-runs event. Every time we encounter such an event, we can push the corresponding item on the stack. After the entire file has been processed, we can simply pop 6 items off the stack and output their contents to a file.

Let us now program it in SAS. Note that this program incorporates coding for the stack without calling LINK routines and deviates from the stack demo program above in some details to better fit the specific task and make use of specific data. However, the general scheme is kept intact.

Program 10.8 Chapter 10 Using a Hash Stack to Find Consecutive Events.sas

```
data StackOut (keep = Game_SK Result Count) ;
  if _n_ = 1 then do ;
    dcl hash stack (ordered:"A") ;        ❶
    stack.defineKey   ("_N_") ;           ❷
    stack.defineData ("_N_", "Game_SK", "Result", "Count") ;
    stack.defineDone () ;
    dcl hiter istack ("stack") ;          ❸
  end ;
  do until (last.Result) ;                ❹
    set dw.AtBats (keep = Game_SK Result) end = LR ;
    by Game_SK Result notsorted ;
    Count = sum (Count, 1) ;
  end ;
```

```
    if Result = "Home Run" and Count => 4 then stack.add() ;    ❺
    if LR ;
    do pop = 1 to 6 while (istack.last() = 0) ;    ❻
      output ;
      rc = istack.next() ;
      stack.remove() ;
    end ;
run ;
```

❶ Create a hash table *Stack* to be used as a stack – ordered ascending by its key.

❷ *Stack* is keyed by the automatic variable _N_. The idea is to key the table by a variable that naturally ascends along with each new event of 4 back-to-back home runs; and _N_ fits the purpose by its very nature of *invariably* ascending with each execution of the DATA step. Then _N_ is added to the data portion to facilitate the removal of every popped item via the REMOVE method.

Note that using _N_ as a key (rather than Key in the demo program) takes advantage of the intrinsic DATA step mechanics and rids us of the extra necessity to use the NUM_ITEMS attribute in the *push* action.

❸ Pair table *Stack* with a hash iterator instance needed to facilitate the stack *pop* operation later on.

❹ This DoW loop uses the NOTSORTED option to *isolate* each series of consecutive records with the same value of Result within Game_SK into a *separate* DATA step execution. This way, after the DoW loop has finished iterating, the PDV value of Result corresponds to its value in the series the loop isolates; and the value of Count indicates the number of records in the series – or, in other words, the number of times the DoW loop has iterated.

❺ The DoW loop has finished iterating over the current same-key series. If it is a series of home runs and its length is 4 or greater, push a new item on the stack *S*. Note that, unlike Key in the demo program, _N_ does not have to be incremented before calling the ADD method as a key, as it gets a proper ascending unique value automatically.

❻ At this point, all input records have been processed, so we need to pop the required number of items off the hash stack. Including the LAST iterator method call in the WHILE condition ensures that if the stack is empty – meaning that no 4 or more consecutive home runs have been scored – program control exits the DO loop without executing its body even once, and so no output record is written out. Otherwise, the call retrieves the values of the data portion hash variables into their PDV host counterparts, the next record is output; and the rest of the requisite *pop* action activities are performed. If the number of 4 or more back-to-back home run chains $N < 6$, the loop stops after its Nth iteration and having output N records according to the number of items N on the stack. (The value of 6 is chosen arbitrarily just as an example; any other value can be used.)

Running the program, we get the following output:

Output 10.9 Chapter 10 Latest Games with 4 or More Back-to-Back Home Runs

Game_SK	Result	Count
D5E3DDCE8A641634E46B14BE833523FC	Home Run	4
D49471B387AD4F4B1A3D55419E4BA95C	Home Run	5
B7CDC97DD61DA15D37A3AD3637205B71	Home Run	4
ACC134E4504E61E228C4E2CBD2D50782	Home Run	4

Game_SK	Result	Count
9A5B41520936B6EF7CB64548760EDBC9	Home Run	4
3824CFD0489B921DA304AEBFCC6182B5	Home Run	4

Since the stack is popped in the LIFO manner, the latest games where 4 or more consecutive home runs were scored come in the output first, as should be expected.

Note that we have provided alternative examples of consecutive events in section 8.2.3.5 "Determining the Distribution of Consecutive Events" and in section 9.4 "Consecutive Events." Those examples created a distribution of the number of consecutive events. Using a hash stack is an alternative approach and Program 10.8 could be modified to create a distribution in addition to creating a list of occurrences of multiple events. This example also allowed us to point out to our users that one of the great things about SAS is that there are multiple different ways to solve virtually any given problem.

10.5 Array Sorting

Sorting an array may appear as a largely academic exercise. Yet, judging from the number of inquiries about the ways to do it on various SAS forums, it must have a number of practical applications. However, custom coding an efficient array sorting routine is not an exercise for faint-hearted. Before the advent of the canned array sorting routines, some SAS programmers had successfully endeavored to code their own custom routines based on well-known algorithms, such as the quick sort or radix sort. Currently, the efficient call routines SORTN and SORTC are offered to do the job. However, they are not without shortcomings; specifically:

- They are data type specific (unlike the SORT procedure, for example).
- They are sorted only in ascending order.
- They offer no options to handle duplicate keys.
- They are not suited for sorting *parallel* arrays in sync.

10.5.1 Using a Hash Table to Sort Arrays

The *Order* hash object operation can be easily used to address these problem areas. The basic scheme of using the hash object to sort an array is quite simple:

1. Load the key and satellite elements from the array(s) into a hash table ordered ascending or descending.
2. Unload the hash table back into the array(s).

In a nutshell, that is it. Now let us look at some implementation details. Suppose we have 3 arrays valued as follows:

```
array kN [9]     ( 7   7   7   7   5   5   5   3   3 ) ;
array kC [9] $1 ('F' 'E' 'F' 'E' 'D' 'C' 'D' 'A' 'B') ;
array dN [9]     ( 8   6   9   7   4   3   5   1   2 ) ;
```

The elements of arrays kN and kC represent a *composite key*, by which they have to be sorted into ascending or descending order. As a result, the elements of array dN should be *permuted* in sync with the new index positions of the elements of the key arrays. Provisions should also be made to eliminate all duplicate-key elements, except for the first-in or last-in, if necessary. For instance, if the arrays are sorted ascending by (kN,kC) and duplicate keys are allowed, their resulting array values would look like this:

```
kN: 3 3 5 5 5 7 7 7 7
kC: A B C D D E E F F
dN: 1 2 3 4 5 6 7 8 9
```

Or, if we need to sort in descending order and only the first element with a duplicate key must be kept, they should look as follows:

```
kN: 7 7 5 5 3 3 . . .
kC: F E D C B A
dN: 8 6 4 3 2 1 . . .
```

Making use of the hash object, we can thus write the following simple array-sorting program:

Program 10.9 Chapter 11 Array Sorting via Hash

```
%let order = A ; * Sort: A/D (Ascending/Descending ;    ❶
%let dupes = N ; * Dups: Y/N (Yes/No) ;
%let which = L ; * Dupe to select: F/L (First/Last) ;

data _null_ ;
  array kN [9]    ( 7   7   7   7   5   5   5   3   3 ) ;
  array kC [9] $1 ('F' 'E' 'F' 'E' 'D' 'C' 'D' 'A' 'B') ;
  array dN [9]    ( 8   6   9   7   4   3   5   1   2 ) ;
  if _n_ = 1 then do ;
     dcl hash h (multidata:"Y", ordered:"&order") ;    ❷

     h.defineKey  ("_kN", "_kC") ;    ❸
     h.defineData ("_kN", "_kC", "_dN") ;
     h.defineDone () ;

     dcl hiter hi ("h") ;    ❹
  end ;
  do _j = 1 to dim (kN) ;
     _kN = kN[_j] ; _kC = kC[_j] ; _dN = dN[_j] ;    ❺

     if       "&dupes" = "Y" then h.add() ;    ❻

     else if "&which" = "F" then h.ref() ;    ❼

     else                      h.replace() ;    ❽
  end ;
  call missing (of kN[*], of kC[*], of dN[*]) ;    ❾

  do _j = 1 by 1 while (hi.next() = 0) ;    ❿
     kN[_j] = _kN ; kC[_j] = _kC ; dN[_j] = _dN ;
  end ;
  h.clear() ;
  put "kN: " kN[*] / "kC: " kC[*] / "dN: " dN[*] ;
run ;
```

If run with the parameterization as shown above, the DATA step will print the following in the log:

```
kN: 3 3 5 5 7 7 . . .
kC: A B C D E F
dN: 1 2 3 5 7 9 . . .
```

Let us see how hash object operations fit in this code to accomplish the task and how its logic flows:

❶ The %LET statements provide parameterization useful to encapsulate the routine in a macro if need be.

❷ Hash table H is created with MULTIDATA always valued as "Y" (i.e., this value is not parameterized on purpose). Duplicate keys are handled later via choosing a specific table-loading method. The _N_=1 condition is just a precaution in case the sorting routine is to be also performed for _N_>1, such as for new observations or new BY groups. Compare this with the CLEAR method call in ❿ below.

❸ The key and data portions of H defined with temporary variables whose PDV host variables are created downstream by assignment. (See item ❺.)

❹ We need a hash iterator linked to table H to enumerate its sorted data and put it back into the arrays. So, a hash iterator named HI is declared here.

❺ Forward-scan the arrays in a DO loop. Assign each element to the respective temporary PDV variable. The assignment statements both value them and make them take on the data type (numeric or character) corresponding to that of the respective array. This way, the hash variables get their host variables in the PDV at compile time and inherit the proper data types (according to the arrays they relate to) from them at run time.

❻ If unduplication is not required, call the ADD method to perform the *Insert* operation. Because of MULTIDATA:"Y" it will insert a new item from the array elements with the current _j index regardless of the items already in the table.

❼ Otherwise, if unduplication is required and the *first* duplicate for the current key must be kept, call the REF method. It automatically does what is needed because it inserts an item only if the key it accepts is not already in the table.

❽ If unduplication is required but the *last* duplicate for the current key must be kept, perform the *Update All* operation by calling the REPLACE method. It works as required here because we are attempting to add one item at a time. So, every time REPLACE is called, it can find only one item with the current key previously inserted in the table and keeps repeatedly updating it with the new values of dN[_j]. Thus, in the end, each item in the table will have the last dN[_j] value encountered for this item's key. Q.E.D.

❾ Just in case unduplication is requested, populate the arrays with missing values. This way, all duplicate-key elements being kept in the arrays will be shifted to the left, and the slack to the right will be filled with nulls.

❿ Perform the *Enumerate All* operation to retrieve the data portion values from each item into their PDV host variables and assign the latter to the corresponding elements of the corresponding arrays. After this is done, the CLEAR method is called to empty the hash table just in case the sorting routine is to be repeated – for example, in other executions of the DATA step for _N_>1, such as for new observations or new BY groups. The PUT statement is used merely to display the final state of the arrays in the SAS log.

Note that this sample program provides a basic framework for sorting arrays using the hash object. It can be further tailored to your particular needs depending on the circumstances and encapsulated accordingly. For example, a parameter can be added to indicate which arrays constitute the composite key and which constitute the data elements to be permuted in accordance with the new positions of the key-value.

10.6 Summary

The SAS hash object can be used to provide a wide range of data structures and, as shown in this chapter, it can be integrated with other SAS functionality to do that. There are any number of additional examples that are possible. Our IT users were quite pleased with the examples presented here and told us that should a follow-on project be approved, they would likely request additional examples.

Chapter 11: Hash Object Memory Management

11.1 Introduction

Anticipating performance and memory issues, the Bizarro Ball IT users have asked us to address several issues relating to performance, specifically data access times and memory management. If they should decide to proceed with the implementation of a follow-on project, they want to know what the options are as the amount of data grows. The generated data for our Proof of Concept included data for only one year. Any project will have to deal with multiple years of data.

This chapter focuses on the following three (overlapping) techniques:

1. Reducing the size of the key portion of the hash items.
2. Reducing the size of the data portion of the hash items.
3. Partitioning the data into groups that can be processed independently of one another.

This chapter addresses some of those concerns. Since the focus of this chapter is memory and performance and not the individual SAS programs, the IT and business users agreed that we need not annotate the programs to the same degree of detail as done for most of the previous programs.

Many of the examples here replicate samples in Chapters 6 and 10 by incorporating the techniques discussed here.

11.2 Memory vs. Disk Trade-Off

The data processing efficiency offered by the SAS hash object rests on two elements:

1. Its key-access operations, with the hashing algorithm running underneath, are executed in $O(1)$ time.
2. Its data and underlying structure reside completely in memory.

These advantages over disk-based structures are easy to verify via a simple test. Consider, say, such ubiquitous data processing actions as key search and data retrieval against a hash table on the one hand, and against an indexed SAS data set on the other hand. For example, suppose that we have a SAS data set created from data set Dw.Pitches and indexed on a variable RID valued with its unique observation numbers. The DATA step creating it is shown as part of Program 11.1 below.

The DATA _NULL_ step in this program does the following:

1. Loads file Pitches into a hash table H with RID as its key and Pitcher_ID and Result as its data.
2. For all RID key-values in Pitches and an equal number of key-values *not* in Pitches, searches for RID in the table. If the key-value is found, retrieves the corresponding data values of Pitcher_ID and Result from the hash table into the PDV.
3. Does exactly the same against the indexed data set Pitches using its SAS index for search and retrieval.

4. Repeats both tests above the same number of times (specified in the macro variable *test_reps* below).

5. Calculates the summary run times for both approaches and reports them in the SAS log.

Program 11.1 Chapter 11 Hash Vs SAS Index Access Speed.sas

```
data Pitches (index=(RID) keep=RID Pitcher_ID Result) ;
  set dw.Pitches ;
  RID = _N_ ;
run ;

%let test_reps = 10 ;

data _null_ ;
  dcl hash h (dataset: "Pitches", hashexp:20) ;
  h.definekey  ("RID") ;
  h.defineData ("Pitcher_ID") ;
  h.defineDone () ;
  time = time() ;
  do Rep = 1 to &test_reps ;
    do RID = 1 to N * 2 ;
      rc = h.find() ;
    end ;
  end ;
  Hash_time = time() - time ;
  time = time() ;
  do Rep = 1 to &test_reps ;
    do RID = 1 to N * 2 ;
      set Pitches key=RID nobs=N ;
    end ;
  end ;
  _error_ = 0 ; * prevent log error notes ;
  Indx_time = time() - time ;
  put "Hash_time =" Hash_time 6.2-R
    / "Indx_time =" Indx_time 6.2-R ;
  stop ;
run ;
```

Note that the value of N used in both of these loops is initialized to the number of observations in Pitches because it is the variable specified in the SET statement NOBS= option. Also note that above, automatic variable *_ERROR_* is set to zero to prevent the error notes in the log when the SET statement with the *key=RID* option fails to find the key.

When the program is run, the PUT statement prints the following in the SAS log:

```
Hash_time = 2.15
Indx_time = 21.58
```

As we see, a run with an equal number of search successes and failures shows that for the purposes of key search and data retrieval the SAS hash object works faster than the SAS index by about an order of magnitude. This is a consequence of both the difference between the search algorithms (the binary search versus hash search) and the access speed difference between memory and disk storage. The run-time figures above were obtained on the X64_7PRO platform. We told our IT users that if they are interested in trying it out under a different system, the program is included in the program file Chapter 11 Hash Vs SAS Index Access Speed.sas.

Note that the test above is strongly *biased in favor of the SAS index* because it always reads *adjacent* records which reduces paging, i.e., block I/O. Running it with the RID values picked *randomly* between 1 and $N*2$ shows that, in this case, the hash/(SAS index) speed ratio exceeds 50:1.

11.2.1 General Considerations

Holding data in memory comes at a price – the steeper, the larger the data volume handled in this manner is. At some point, it may approach or exceed the RAM capacity of the machine being used to process the data. As the IT users pointed out, the sample baseball data used in this book, though large and rich enough for our illustrative purposes, presents no challenge from this standpoint, as its largest data set of about 100 MB in size can nowadays be easily swallowed by the memory of even the most modest modern computing device. However, industrial practice, with its need to manipulate big and ever increasing volumes of data, presents a different picture. As the number of hash items, as well as the number of the hash variables and cardinality of the table keys grows, the hash memory footprint required to accomplish a task can easily reach into the territory of hundreds of gigabytes.

The sledge-hammer approach, of course, is to install more memory. Under some circumstances, it indeed makes sense. For instance, if it is vital for a large business to make crucial decisions based on frequently aggregated data (a "lead indicators" report in the managed health industry could be a good example), it appears to make more sense to invest in more RAM than to pay for coding around woeful hardware inadequacies. But as the authors have told the Bizarro Ball business users, our own experience of working on a number of industrial projects shows that this route can be blocked for a number of reasons.

Even if more machine resources can be obtained, it may turn out to be a temporary fix, working only till the time when the data volume grows or business demands change (for example, to require more hierarchical levels of data aggregation). In other words, crudely unoptimized code does not scale well. Thus, responsible SAS programmers do not merely throw a glaringly unoptimized program at the machine in hope that it will handle it and clamor for more hardware when it fails due to insufficient resources. Rather, they endeavor to avoid at least gross inefficiencies, regardless of whether or not the hardware has the capacity to cover for them at the moment, by taking at least obvious inexpensive measures to improve performance. Examples include making sure that the program reads and writes only the variables and records it needs, avoiding unnecessary sorts, etc.

Furthermore, sensible programmers aim to foresee how their code will behave if changes to the data, its volume, or business needs occur in the future – and structure their programs accordingly. Doing so requires more time and higher skill; yet, in the end, the result can prove to be well worth the price. Typical examples of this kind of effort are employing more efficient algorithms (e.g., the binary search rather than linear search) and making use of table lookup (e.g., via a format or a hash table) instead of sorting and merging, etc. This way, if (or rather when) the changes happen, chances are that the program will run successfully again in spite of them.

With the hash object, it is no different. Memory, though getting increasingly cheaper and abundant, is still a valuable resource. Hence, as the first line of defense, a programmer using the hash object should not overload it with unnecessary information for the sake of seemingly "simpler" coding and fewer lines of code. For instance, it is obvious that, to save memory, the hash table entry should contain only the variables and items it needs to do the job. Thus, using the argument tag ALL:"Y" with the DEFINEKEY and DEFINEDATA methods as a code shortcut (instead of defining only the needed variables in one way or another) must be judged accordingly. Likewise, if a hash table is used only for pure search (to find out if a key is there), using the MULTIDATA:"Y" argument tag is extraneous, as well as anything but a single short variable in the data portion.

As Bizarro Ball considers future expansion to more leagues, teams, and so on (e.g., some of the executives would like to offer the data analysis services to collegiate teams), expected data volumes may very well increase dramatically. Thus, hash memory management could be an important issue in the near term.

11.2.2 Hash Memory Overload Scenarios and Solutions

It is another question how to deal with more complex scenarios of the hash object running out of memory in spite of observing these *obvious* precautions. Typical circumstances under which it can occur include, but are not limited to, the following:

- A hash table is used as a lookup table for joining files. It can run out of memory if:
    1. The file loaded into it has too many records.
    2. The files are joined by a composite key with too many and/or too long components.
    3. The satellite variables loaded into the data portion for future retrieval are too numerous and/or too long.

- A hash table is used as a container for data aggregation. It can run out of memory if:
    1. The aggregation occurs at too many distinct key levels, consequently requiring too many unique-key table items.
    2. The statistics to be aggregated are too numerous, and so are the corresponding data portion variables needed to hold them.
    3. The keys are too numerous and/or long, and so are *both* the key and data portion variables needed to hold them (in the data portion, they are required in order to appear in the output).

Here are several tactics of resolving these hash memory overload problems which we will discuss in detail in the rest of this chapter:

- **Take advantage of intrinsic pre-grouping**. If the data input is sorted or grouped by one or more *leading components* of a *composite* key, use the BY statement to process each BY group separately, clearing the hash table after each group is processed. It works best if the BY groups are approximately equal in size.
- **Reduce the key portion length**. If the hash key is composite and the summary length of its components significantly exceeds 16 bytes, use the one-way hash function MD5 to shorten the *entire* key portion to a single $16 key. This can be done regardless of the total length of the composite key or the number of its components.
- **Offload the data portion variables (use a hash index)**. It means that only the key (simple, composite, or its MD5 signature) is stored in the key portion, and only the file record identifier (RID) is stored in the data portion. When in the course of data processing a key-value is found in the table, the requisite satellite variables, instead of being carried in the data portion, are brought into the PDV by reading the input file via the SET POINT=RID statement. Effectively, this is tantamount to using a *hash index* created in the DATA step and existing for its duration.
- **Use a known key distribution to split the input**. If the values of some composite key component (or any if its bytes or group of bytes, for that matter) *are known* to correspond to a more or less equal number of distinct composite key-values, process the file using the WHERE clause separately for each such distinct value (or a number of distinct values) of such a component one at a time. As a result, the hash memory footprint will be reduced by the number of the separate passes.

- **Use the MD5 function to split the input**. Concatenate the components of the composite key and create an MD5 signature of the concatenated result. Because the MD5 function is a *good hash function*, the value of any byte (or a combination of bytes) of its signature will correspond to approximately the same number of different composite key-values. Then process the input in separate chunks based on these values using the WHERE clause as suggested above.

Note that these methods are not exclusive to themselves and can be combined. For example, you can *both* reduce the key portion length via the MD5 function and use the data portion offloading, any method can be combined with uniform input splitting, etc.

11.3 Making Use of Existing Key Order

It often happens that the input we intend to process using a hash table is pre-sorted or pre-grouped by one or more leading variables of the composite key in question. For example, consider data set Dw.Runs where each run is recorded as a record. A subset of the file for two arbitrarily selected games is shown below. Note that the column From_obs is not a variable in the file; it just indicates the observations from which the records in the subset come.

Output 11.1 Chapter 11 Two Games from Dw.Runs

Game_SK	Top_Bot	From_obs
234B26A4DDD547C58A1A6CF33DF16E66	T	10399
234B26A4DDD547C58A1A6CF33DF16E66	B	10400
234B26A4DDD547C58A1A6CF33DF16E66	B	10401
234B26A4DDD547C58A1A6CF33DF16E66	T	10402
234B26A4DDD547C58A1A6CF33DF16E66	B	10403
234B26A4DDD547C58A1A6CF33DF16E66	B	10404
3D93DA6C9CBA8E887987F7A8E7CE0633	T	18534
3D93DA6C9CBA8E887987F7A8E7CE0633	T	18535
3D93DA6C9CBA8E887987F7A8E7CE0633	T	18536
3D93DA6C9CBA8E887987F7A8E7CE0633	B	18537
3D93DA6C9CBA8E887987F7A8E7CE0633	B	18538
3D93DA6C9CBA8E887987F7A8E7CE0633	T	18539

Suppose that we want to process it in some way using the hash object – for example, aggregate or unduplicate it – by the composite key (Game_SK,Top_Bot). In general, to do so, we would create a hash table keyed by the *entire* composite key and proceed as described earlier in the book according to the type of task at hand.

However, this file, though not sorted or grouped by the *whole* composite key, is *intrinsically sorted* by its leading component Game_SK, whose distinct values correspond to approximately an equal number of distinct key-values of the entire composite key. Since Game_SK is part of the composite key, no two *composite* key-values can be the same if their Game_SK are different. It means that no composite key-value is shared between the groups of records with different values of Game_SK. Therefore, we can make use of Game_SK in the BY statement and process each BY group independently. After each BY group is processed, we would add needed records to the output and clear the hash table to prepare it for processing the next BY group. From the standpoint of memory usage, doing so offers three advantages:

1. The hash table can now be keyed *only* by Top_Bot rather than by *both* Games_SK and Top_Bot. It shortens the key portion of the table.
2. Game_SK does not have to be replicated in the data portion. It shortens the data portion of the table.
3. The table has to hold the information pertaining to the *largest BY group* rather than the *entire file*. It reduces the maximum number of items the hash table must hold.

To see more clearly how this concept works, let us consider three very basic examples:

1. Data aggregation
2. Data unduplication
3. Joining data

11.3.1 Data Aggregation

Suppose that for each game (identified by Game_SK) we would like to find the score by each of the opposing teams. Since each record in the Runs file represents a run, it means that we would need to count the records for the top and bottom of the inning (i.e., the 2 values of Top_Bot) within each game. In other words, we would need to *aggregate* at each level of composite key (Game_SK,Top_Bot).

From the data aggregation chapter in this book (Chapter 8) we already know how a hash object *frontal attack* on this task can be carried out:

1. Use a hash table to count the records at each distinct (Game_SK,Top_Bot) key-value level.
2. Output its content when the input has been all processed.

Program 11.2 Chapter 11 Frontal Attack Aggregation.sas

```
data Scores (keep = Game_SK Top_Bot Score) ;
  if _n_ = 1 then do ;
    dcl hash h (ordered:"A") ;
    h.defineKey  ("Game_SK", "Top_Bot") ;
    h.defineData ("Game_SK", "Top_Bot", "Score") ;
    h.defineDone () ;
    dcl hiter ih ("h") ;
  end ;
  do until (LR) ;
    set Dw.Runs (keep = Game_SK Top_Bot) end = LR ;
    if h.find() ne 0 then Score = 1 ;
    else Score + 1 ;
    h.replace() ;
  end ;
  do while (ih.next() = 0) ;
    output ;
  end ;
run ;
```

For detailed explanations of the mechanics of this basic aggregation process, see any of the examples in Chapter 8. Running the program, we get the following output for the two selected games shown in Output 11.1:

Output 11.2 Chapter 11 Dw.Runs Aggregates

Game_SK	Top_Bot	Score
234B26A4DDD547C58A1A6CF33DF16E66	B	4
234B26A4DDD547C58A1A6CF33DF16E66	T	2
3D93DA6C9CBA8E887987F7A8E7CE0633	B	2
3D93DA6C9CBA8E887987F7A8E7CE0633	T	4

Note that hash table H in its final state (before its content was written out to data set Scores) has 4 items, which corresponds to the number of output records.

However, we *know* that the input is pre-sorted by Game_SK. Thus, we can reduce both the key portion and data portion lengths by aggregating the same file one Game_SK BY group at a time. The changes from the *frontal attack* Program 11.2 are shown in bold:

Program 11.3 Chapter 11 Pregrouped Aggregation.sas

```
data Scores_grouped (keep = Game_SK Top_Bot Score) ;
  if _n_ = 1 then do ;
    dcl hash h (ordered:"A") ;
    h.defineKey  ("Top_Bot") ;     ❶
    h.defineData ("Top_Bot", "Score") ;     ❷
    h.defineDone () ;
    dcl hiter ih ("h") ;
  end ;
  do until (last.Game_SK) ;     ❸
    set Dw.Runs (keep = Game_SK Top_Bot) ;
    by Game_SK ;
    if h.find() ne 0 then Score = 1 ;
    else Score + 1 ;
    h.replace() ;
  end ;
  do while (ih.next() = 0) ;     ❹
    output ;
  end ;
  h.clear() ;     ❺
run ;
```

The program generates the *same exact output* as the frontal attack Program 11.2. For our two sample games, it is identical to Output 11.2 above. Let us go over some focal points of Program 11.3:

❶   When the key portion is defined, the leading component Game_SK of the composite key

  (Game_SK,Top_Bot) is omitted. Including it is unnecessary since the aggregation occurs by the level of Top_Bot alone within each separate BY value of Game_SK.

❷   The data portion does not contain Game_SK either. Its output value comes from the PDV as populated directly by the SET statement.

❸   The DoW loop, as opposed to iterating over the *whole* file, now iterates over one Game_SK BY group at a time per one execution of the DATA step (or one iteration of its implied loop). Note that the logical structure of the program remains intact. Also note that if the data were grouped, but not sorted by Game_SK, we could add the NOTSORTED option to the BY statement.

❹   At this point, hash table H contains the distinct key-values of Top_Bot and respective sums accumulated for the BY group just processed. Hash iterator IH linked to hash table H enumerates the table, and the data portion values of its items are written as records to the output data set Scores_grouped. Every BY group that follows adds its own records to the output file in the same manner.

❺   Now that the needed information (resulting from the processing of the current BY group) is written out, the table is cleared before the processing of the next BY group commences. Thus, the memory occupied by the hash data from the current BY group is released. Therefore, the largest hash memory footprint the program will create is that for the largest BY Game_SK group available in the file.

From the standpoint of saving hash memory, the more uniformly the values of Game_SK are distributed throughout the input file and the more distinct values of Game_SK (i.e., BY groups) we have, the better this method works. Running this program back-to-back with the frontal attack program with the FULLSTIMER system option turned on shows that, in this case, making use of the intrinsic leading key order cuts memory usage roughly by half. With a rather small file like Dw.Runs it may not matter much. However, in an industrial application working against billions of input records such a reduction can make a crucial difference.

Even if the distribution of the leading key(s) values is skewed, using the existing pre-grouping still saves hash memory by shortening both the key and data portions. In the examples shown above, the savings amount to 32 bytes per hash entry. Again, it may sound trivial but can be very significant if the items in a hash summary table should number in tens or hundreds of millions, as does indeed happen in reality. For example, if Bizarro Ball offered these services to college-level teams and they all accepted the offer, there would be orders of magnitude more games and thus Game_SK values.

It is all the more true that the composite key in question may have more than just two variables and may be pre-grouped by more than one. In general, a composite key (at whose level aggregation is requested) is composed of a number of leading key variables KL (by which the file is pre-grouped) and a number of trailing key variables KT:

(KL1, KL2, ..., KLn, KT1, KT2, ..., KTm)

In such a general case, both the key and data portions of the hash table would include:

("KT1", "KT2", ..., "KTm")

and the corresponding DoW loop similar to the loop shown in Program 11.3 would look as follows:

```
do until (last.KLn) ;
  set ... ;
  by KL1 KL2 ... KLn ;
  ...
end ;
```

The more leading keys KLn (by which the file is pre-grouped) are available relative to the number of the trailing keys KTm, the more hash memory this technique saves compared to the frontal attack by the whole composite key.

11.3.2 Data Unduplication

The same technique works, almost verbatim, for data unduplication. Suppose that we want to use the same file Dw.Runs to find, for each game, the inning and its half (i.e., top vs. bottom) in which each scoring player scored *first*. Let us look, for example, at the records for the first game in the file Dw.Runs (some variables are hidden, and the column From_obs indicates the respective observation numbers in the file):

Output 11.3 Chapter 11 Records for the First Game in Dw.Runs

Game_SK	Inning	Top_Bot	Runner_ID	From_obs
0018BC1C4EA5D9A15A12D011BB93AC36	1	T	27010	1
0018BC1C4EA5D9A15A12D011BB93AC36	1	T	57764	2
0018BC1C4EA5D9A15A12D011BB93AC36	1	T	54735	3
0018BC1C4EA5D9A15A12D011BB93AC36	2	T	57764	4
0018BC1C4EA5D9A15A12D011BB93AC36	2	B	11540	5
0018BC1C4EA5D9A15A12D011BB93AC36	4	T	27010	6
0018BC1C4EA5D9A15A12D011BB93AC36	4	T	54735	7
0018BC1C4EA5D9A15A12D011BB93AC36	4	T	26278	8
0018BC1C4EA5D9A15A12D011BB93AC36	6	T	57764	9
0018BC1C4EA5D9A15A12D011BB93AC36	6	T	54735	10
0018BC1C4EA5D9A15A12D011BB93AC36	6	B	43392	11
0018BC1C4EA5D9A15A12D011BB93AC36	8	T	26278	12
0018BC1C4EA5D9A15A12D011BB93AC36	9	B	13063	13

From the data processing viewpoint, our task means that, for each game and runner, we need to output only the *first* record where the corresponding Runner_ID value is encountered. Thus, in the sample above, we would need to keep records 1, 2, 3, 5, 8, 11, and 13 and discard the rest. In other words, we would *unduplicate* the file based on the first distinct combination of key-values of composite key (Game_SK,Runner_ID).

If file Dw.Runs above *were not* pre-sorted by Game_SK, we would have to include the *entire* composite key (Game_SK,Runner_ID) in the key portion to execute a *frontal attack* as shown in Program 11.4 below. It uses the same logic as Program 10.2 in Chapter 10.

Program 11.4 Chapter 11 Frontal Attack Unduplication.sas

```
data First_scores ;
  if _n_ = 1 then do ;
    dcl hash h () ;
    h.defineKey  ("Game_SK", "Runner_ID") ;
    h.defineData ("_N_") ;
    h.defineDone () ;
    dcl hiter ih ("h") ;
  end ;
  do until (LR) ;
    set dw.Runs end = LR ;
    if h.check() = 0 then continue ;
```

```
   output ;
   h.add() ;
 end ;
run ;
```

The logic in the DO loop part requires no more than to check if the composite key-value from the current record of Dw.Runs is in table H. Therefore, the CHECK method is called to perform the *Search* operation. Hence, the data portion plays no functional role. However, as a hash table cannot exist without it, we are interested to make it as short as possible to minimize the hash memory footprint. The simplest way to do it is to define the data portion with a single numeric variable. Since in the context of the program it never interacts with the PDV, it does not matter which one we pick. The automatic variable _N_ chosen above to play the role of the dummy placeholder suits the purpose well since it is (a) numeric and (b) always present in the PDV. Therefore, no parameter type matching issues can arise.

For the input records of the game shown in Output 11.3, the program generates the following output:

Output 11.4 Chapter 11 Dw.Runs First Game Subset Unduplicated

Game_SK	Inning	Top_Bot	Runner_ID
0018BC1C4EA5D9A15A12D011BB93AC36	1	T	27010
0018BC1C4EA5D9A15A12D011BB93AC36	1	T	57764
0018BC1C4EA5D9A15A12D011BB93AC36	1	T	54735
0018BC1C4EA5D9A15A12D011BB93AC36	2	B	11540
0018BC1C4EA5D9A15A12D011BB93AC36	4	T	26278
0018BC1C4EA5D9A15A12D011BB93AC36	6	B	43392
0018BC1C4EA5D9A15A12D011BB93AC36	9	B	13063

However, if we make use of the fact that the file is intrinsically sorted by Game_SK, the program can be re-coded as shown in Program 11.5 below (the changes from Program 11.4 are shown in bold). This program generates exactly the same output as the frontal attack Program 11.4 above.

Program 11.5 Chapter 11 Pregrouped Unduplication.sas

```
data First_scores_grouped ;
  if _n_ = 1 then do ;
    dcl hash h () ;
    h.defineKey  ("Runner_ID") ;        ❶
    h.defineData ("_iorc_") ;
    h.defineDone () ;
    dcl hiter ih ("h") ;
  end ;
  do until (last.Game_SK) ;             ❷
    set dw.Runs (keep = Game_SK Inning Top_Bot Runner_ID) ;
    by Game_SK ;
    if h.check() = 0 then continue ;
    output ;
    h.add() ;
  end ;
  h.clear() ;    ❹
run ;
```

❶ Note that Game_SK is no longer included in the key portion of table H.

❷ Instead of looping through the entire file, the DoW loop now iterates through each sorted group BY Game_SK one DATA step execution at a time.

❸ This is technically made possible by the BY statement, which sets last.Game_SK=1 each time the last game record is read and thus terminates the DoW loop.

❹ After the current game is processed, the hash table is cleared of all items, ceding the memory occupied by them to the operating system. Program control is then passed to the top of the DATA step, and the next game is processed.

Just like with data aggregation, taking advantage of partial pre-sorting saves hash memory in two ways:

- It reduces the key portion length by the summary length of the pre-sorted leading keys (in this case, just Game_SK).
- The ultimate number of items in the table corresponds to the game with the largest number of distinct runners. Contrast that with the *frontal attack* program above where the ultimate number of items is the number of distinct runners in the *whole file*. Note that from the perspective of understanding the data, we know that the number of runners in a given game can never exceed the total number of players on both teams. Thus, we have a virtual guarantee on the upper bound of the number of items in the hash table object at any given point in the data.

11.3.3 Joining Data

In the cases with data aggregation and unduplication, we have dealt with a *single file* partially pre-grouped by a leading component of the composite key. An inquisitive SAS programmer may ask whether one can take advantage, in a similar fashion, of partial pre-grouping if *both* files to be *joined* by using a hash table are *identically* pre-grouped by one leading key or more. The answer to this question is "Yes". However, the implementation is a tad more involved, as in this case it requires *interleaving* the files by the pre-grouped key components and a bit more of programming gymnastics.

To illustrate how this can be done, suppose that, for each runner record in data set Dw.Runs, we would like to obtain a number of related pieces of information from those records of data set Dw.AtBats where the value of variable Runs is not zero (indicating one or more runs). For example, we might want to know:

- The *batter*, i.e., what batter was responsible for the runner scoring.
- "*How deep in the count*" (reasonably standard baseball terminology for how many pitches there were in the at bat) the at bat went and what the ball/strike count was. The Dw.AtBats fields furnishing this information are Number_of_Pitches, Balls, and Strikes.
- Whether these were two *out* runs, i.e., scored when there had already been two outs. The Dw.AtBats field for this is Outs. (The batters and runners who contribute to runs with two outs are often considered to be *clutch* players.)
- Whether the runner scored *as the result of a hit* and the *type of hit* indicated by the values of variables Is_A_Hit and Result from file Dw.AtBats.

The files Dw.Runs and Dw.AtBats are linked by the following composite key:

```
(Game_SK, Inning, Top_Bot, AB_Number)
```

For the demo purposes, we have picked three games (the same from both files) and a few rows from each game. The samples are shown below in Output 11.5 and Output 11.6. Note that the column From_Obs is *not* a variable; it is included below just to show which observations have been picked. For Dw.Runs:

Output 11.5 Chapter 11 Dw.Runs Subset for Joining

Game_SK	Inning	Top_Bot	AB_Number	Runner_ID	From_Obs
20C067AEEFBD2FD731020F3340B747DD	2	T	12	25200	9481
20C067AEEFBD2FD731020F3340B747DD	6	B	29	30733	9494
20C067AEEFBD2FD731020F3340B747DD	6	B	35	13316	9498
59C2768D26BD03738EFBD21325A7B6BA	3	T	15	29474	27138
59C2768D26BD03738EFBD21325A7B6BA	5	T	31	52914	27145
ED46E127783E76AE425FEC40885C3E2C	2	B	7	26244	72590
ED46E127783E76AE425FEC40885C3E2C	4	B	23	25142	72598
ED46E127783E76AE425FEC40885C3E2C	9	B	48	44942	72603

For Dw.AtBats with Runs>0 and non-key variables limited to Batter_ID, Result, and Is_A_Hit:

Output 11.6 Chapter 11 Dw.AtBats Subset for Joining

Game_SK	Batter_ID	Inning	Top_Bot	AB_Number	Result	Is_A_Hit	From_Obs
20C067AEEFBD2FD731020F3340B747DD	54510	2	T	12	Home Run	1	34779
20C067AEEFBD2FD731020F3340B747DD	25267	6	B	26	Single	1	34823
20C067AEEFBD2FD731020F3340B747DD	16387	6	B	29	Walk	.	34826
59C2768D26BD03738EFBD21325A7B6BA	13058	3	T	15	Home Run	1	99062
59C2768D26BD03738EFBD21325A7B6BA	29474	5	T	31	Triple	1	99086
ED46E127783E76AE425FEC40885C3E2C	50776	2	B	7	Double	1	265695
ED46E127783E76AE425FEC40885C3E2C	26244	4	B	23	Home Run	1	265721
ED46E127783E76AE425FEC40885C3E2C	33522	9	B	48	Walk	.	265767

To join the files by a hash object using a *frontal attack*, as in Program 6.6 Chapter 6 Left Join via Hash.sas, we have to:

- Key the hash table by all of the composite key variables.
- Place all of them, together with the satellite variables from Dw.AtBats, into the data portion.

Using this approach for the task at hand, we could proceed as follows.

Program 11.6 Chapter 11 Frontal Attack Join.sas

```
%let comp_keys = Game_SK Inning Top_Bot AB_Number ;      ❶
%let data_vars = Batter_ID Is_A_Hit Result ;
%let data_list = %sysfunc (tranwrd (&data_vars, %str( ), %str(,))) ;   ❷

data Join_Runs_AtBats (drop = _: Runs) ;
  if _n_ = 1 then do ;
    dcl hash h (multidata:"Y", ordered:"A") ;
    do _k = 1 to countw ("&comp_keys") ;    ❸
      h.defineKey (scan ("&comp_keys", _k)) ;
    end ;
```

```
      do _k = 1 to countw ("&data_vars") ;      ❹
        h.defineData (scan ("&data_vars", _k)) ;
      end ;
      h.defineDone() ;
      do until (LR) ;
        set dw.AtBats (keep=&comp_keys &data_vars Runs
                    where=(Runs)) end = LR ;
        h.add() ;
      end ;
    end ;
    set dw.Runs ;
    call missing (&data_list, _count) ;
    do while (h.do_over() = 0) ;
      _count = sum (_count, 1) ;
      output ;
    end ;
    if not _count then output ;
  run ;
```

❶   Assign the lists of key variables and satellite variables (from Dw.AtBats) to macro variables. Their resolved values are used later in the program instead of hard-coding the variable lists repeatedly.

❷   Create a *comma-separated* list of variable names in *data_vars* (using macro language facilities that we agreed to discuss with the IT users later) to be used in the CALL MISSING routine later.

❸   Place *all* the composite key variables in the key portion. Note that instead of passing their names to the DEFINEKEY method as a single comma-separated list of quoted names, the hash variables are added one at a time in a DO loop by passing a character expression with the SCAN function to the method call.

❹   Place the satellite variables into the data portion using the same looping technique.

In general, the program follows the principle of using a hash table to join files described in section 6.3.5 "Unique-Key Joins." Running it results in the following joined output (for the sample rows from Output 11.5 and Output 11.6):

Output 11.7 Chapter 11 Dw.Runs and Dw.AtBats Subsets Join Result

Game_SK	Batter_ID	Inning	Top_Bot	AB_Number	Result	Is_A_Hit	Runner_ID
20C067AEEFBD2FD731020F3340B747DD	54510	2	T	12	Home Run	1	25200
20C067AEEFBD2FD731020F3340B747DD	16387	6	B	29	Walk	.	30733
20C067AEEFBD2FD731020F3340B747DD	25267	6	B	35	Single	1	13316
59C2768D26BD03738EFBD21325A7B6BA	13058	3	T	15	Home Run	1	29474
59C2768D26BD03738EFBD21325A7B6BA	29474	5	T	31	Triple	1	52914
ED46E127783E76AE425FEC40885C3E2C	50776	2	B	7	Double	1	26244
ED46E127783E76AE425FEC40885C3E2C	26244	4	B	23	Home Run	1	25142
ED46E127783E76AE425FEC40885C3E2C	33522	9	B	48	Walk	.	44942

From the standpoint of hash memory usage, this head-on approach has two disadvantages:

- All components of the composite key (Game_SK,Inning,Top_Bot,AB_Number) with the summary length of 40 bytes must be present in both the key and the data portions.
- The number of items in table H equals the total number of records in the file Dw.AtBats.

However, since files Dw.Runs and Dw.AtBats are *both pre-sorted* by their common leading key variables (Game_SK,Inning), both issues above can be addressed via processing each (Game_SK,Inning) BY group separately. But since now we have to deal with *two* files, it involves the extra technicality of *interleaving* them in the proper order and treating the records from each file differently within the same BY group. The differences between the head-on approach and this one are shown in bold:

Program 11.7 Chapter 11 Pregrouped Join.sas

```
%let comp_keys = Game_SK Inning Top_Bot AB_Number ;
%let data_vars = Batter_ID Is_A_Hit Result ;
%let data_list = Batter_ID,Is_A_Hit,Result ;
%let sort_keys = Game_SK Inning ;      ❶
%let tail_keys = Top_Bot Ab_Number ;      ❷
%let last_key  = Inning ;      ❸

data Join_Runs_AtBats_grouped (drop = _: Runs) ;
  if _n_ = 1 then do ;
    dcl hash h (multidata:"Y", ordered:"A") ;
    do _k = 1 to countw ("&tail_keys") ;
      h.defineKey (scan ("&tail_keys", _k)) ;      ❹
    end ;
    do _k = 1 to countw ("&data_vars") ;
      h.defineData (scan ("&data_vars", _k)) ;
    end ;
    h.defineDone() ;
  end ;
  do until (last.&last_key) ;      ❺
    set dw.atbats (in=A keep=&comp_keys &data_vars
                          Runs where=(Runs))      ❻
        dw.runs    (in=R keep=&comp_keys Runner_ID)
    ;
    by &sort_keys ;
    if A then h.add() ;      ❼
    if not R then continue ;      ❽
    call missing (&data_list, _count) ;      ❾
    do while (h.do_over() = 0) ;
      _count = sum (_count, 1) ;
      output ;
    end ;
    if not _count then output ;
  end ;
  h.clear() ;
run ;
```

❶ Assign the list of the leading key variable names by which the files are pre-sorted to macro variable sort_keys.

❷ Assign the list of the key variables names, *except for those in sort_keys*, to the macro variable tail_keys.

❸ Assign the name of the last key variable in sort_keys to the macro variable last_key. It is needed to properly specify the *last.<variable-name>* in the UNTIL condition of the DoW loop.

❹ Define the key portion *only* with the key variables in the tail_keys list – i.e., with those *not* in the sort_keys list. Note that, just as in the frontal attack program, they are defined dynamically one at a time.

❺ Use the DoW loop to process one (Game_SK,Inning) key-value BY group per DATA step execution. Note that the UNTIL condition includes the *trailing* key variable last.&last_key from the sort_keys list.

❻ Interleave Dw.AtBats and Dw.Runs – *exactly in this sequence* – using the key variables in the sort_keys list as BY variables. Doing so forces all the records from Dw.AtBats to come *before* those from Dw.Runs within each BY group. The automatic variables A and R assigned to the IN= data set option assume binary values (1,0) identifying the records from Dw.AtBats if A=1 (while R=0) and the records from Dw.Runs if R=1 (while A=0).

❼ Within each BY group, reading the records from Dw.AtBats *before* any records from Dw.Runs allows using the IF condition to load the hash table with the key and data values from Dw.AtBats only. If some BY group has no records from Dw.AtBats, so be it – all it means is that the hash table for this BY group will have no items.

❽ If this condition evaluates true, the record is not from Dw.Runs. Hence, we can just proceed to the next record.

❾ Otherwise, we use the standard *Keynumerate* operation to look up the current (Top_Bot,AB_Number) key-value pair coming from Dw.Runs in the hash table. If it is there, we retrieve the corresponding values of the satellite hash variables (listed in data_vars) into their PDV host variables. The loop outputs as many records as there are hash items with the current (Top_Bot,AB_Number) key-value in the current BY group. Calling the MISSING routine ensures that, if there is no match, the satellite variables from the data_vars list for the current output record have missing values regardless of the values that may have been retrieved in preceding records. (This is necessary to ensure proper output in the case of a left join.) The MISSING routine also initializes the value of _count for each new (Top_Bot,AB_Number) key-value.

Note that the "*if not _count then output;*" statement is needed only if a left join, rather than an inner join, is requested. With these two particular files, there is no difference, as every composite key in the comp_keys list in one file has its match in the other.

❿ After the current BY group is processed, the *Clear* operation is performed to empty out the table before it is loaded from the next BY group. Thus, in this program the topmost number of items in the hash table is controlled, not by the *total* number of records in file Dw.AtBats but by the number of records in its largest (Game_SK,Inning) BY group.

This program generates exactly the same output (Output 11.7) as the frontal attack Program 11.6.

Note that instead of hard-coding the lists data_list, tail_keys, and last_key, they can be composed automatically using the macro function %SYSFUNC:

```
%let comp_keys = Game_SK Inning Top_Bot AB_Number ;
%let data_vars = Batter_ID Is_A_Hit Result ;
%let sort_keys = Game_SK Inning ;

%let data_list = %sysfunc (tranwrd (&data_vars, %str( ), %str(,))) ;   ❶
%let tail_keys = %sysfunc (tranwrd (&comp_keys, &sort_keys, %str())) ; ❷
%let last_key  = %sysfunc (scan (&sort_keys, -1)) ;   ❸
```

❶   Insert commas between the elements of the data_vars list by replacing the blanks between them with commas using the TRANWRD function.

❷   Take the sort keys out of the comp_keys list by replacing their names with single blanks using the TRANWRD function.

❸   Extract the *trailing* element from the sort_keys list using the SCAN function reading the list leftward (-1 means get the first element counting from the right instead of the left).

From now on, we are going to use this more robust functionality instead of hard-coded lists.

Although this program and the frontal attack Program 11.6 generate the same output, their hash memory usage is quite different. Tested under X64_7PRO, the *frontal attack* step consumes approximately 10 times more memory (about 12 megabytes) than the step-making use of the pre-sorted order. While a dozen of megabytes of RAM may be a pittance by today's standards, with much bigger data an order-of-magnitude reduction in hash memory footprint can make the difference between a program that successfully delivers its output on time and one that fails to run to completion.

11.4 MD5 Hash Key Reduction

In real-life industrial applications, data often needs to be processed by a number of long composite keys with numerous components of both numeric and character type. A typical example of such need is pre-aggregating data for online reporting at many levels of composite categorical variables. It is not uncommon in such cases to see a composite key grow to the length of hundreds of bytes. A large hash table keyed by such a multi-component key can easily present a serious challenge to the available memory resources.

It is all the more true that the entry size of a hash table in bytes does *not* merely equal the sum of the lengths of the variables in its key and data portions but usually exceeds it by a sizable margin. For example, let us consider a hash table below keyed by 20 numeric variables with only a single numeric variable in its data portion and use the ITEM_SIZE hash operator to find out what the resulting hash entry size will be. Then let us load the table with 1 million items and see if the DATA step memory footprint reported in the log is what we intuitively expect it to be.

Program 11.8 Chapter 11 Hash Entry Size Test.sas

```
data _null_ ;
  array KN [20] (20 * 1) ;
  dcl hash h() ;
  do i = 1 to dim (KN) ;
```

```
      h.defineKey (vname(KN[i])) ;
   end ;
   h.defineData ("KN1") ;
   h.definedone() ;
   Hash_Entry_Size = h.item_size ;
   put Hash_Entry_Size= ;
   do KN1 = 1 to 1E6 ;
     h.add() ;
   end ;
run ;
```

Running this step (on the X64_7PRO platform) results in the following printout in the SAS log:

Hash_Entry_Size=208

It would seem that, with 22 variables of 8 byte apiece, the hash entry size would be 176 bytes; yet in reality it is 208. Likewise, with 1 million items, we would expect the memory footprint of 167 megabytes or so, and yet the FULLSTIMER option reports that in actuality it is 208 megabytes, i.e., 25 percent larger. Though the hash entry size for given number, length, and data type of hash variables varies depending on the operating system (and whether it is 32-bit or 64-bit), the tendency of the hash memory footprint to be larger than perceived persists across the board.

Note that in order to determine the hash table entry size using the ITEM_SIZE attribute, it is unnecessary to load even a single item into the table. In fact, it makes good sense to do it against a (properly defined and created) empty table at the program planning stage if odds are that the table can be loaded with a lot of items. Estimating their future number and multiplying it by the hash entry size gives a fairly accurate idea of how much memory the table will consume when the program is actually run. In turn, the result may lead us to decide ahead of time whether any memory-saving techniques described here should be employed to reduce the hash memory footprint.

Table H used in the test above is actually rather small. The authors have encountered real-life projects where aggregate hash tables had to be keyed by composite keys with hash entry lengths upward of 400 bytes and would grow to hundreds of gigabytes of occupied memory if special measures to shorten their entries were not taken. One such measure, described above, makes use of an existing pre-sorted input order. However, in the situations where no such order exists, no variables can be relegated from the key portion of the hash table to the BY statement.

Thankfully, there exists another mechanism by means of which all key variables of the table (i.e., its composite key) can be *effectively* replaced by a single key variable of length $16 regardless of the number, data type, and lengths of the composite key components. "*Effectively*" means that from the standpoint of the data processing task *result*, using the replacement hash key is equivalent to using the composite key it replaces.

11.4.1 The General Concept

SAS software offers a number of one-way hash functions (note that, despite the name and using hashing algorithms, they are not related to the SAS hash object per se). The simplest and fastest of them is the MD5 function (apparently based on a well-known MD5 algorithm). It operates in the following manner:

signature = MD5 (<character_expression>) ;

If the defined length of character variable signature is greater than 16, MD5 populates it with 16 non-blank leading bytes followed by a number of trailing blanks. The non-blank part of the MD5 response (called the

MD5 *signature*) is a 16-byte character string whose values are all different for all different values of any possible argument. This way, any character string has its own and only one MD5 signature value. (We will briefly address a *theoretical* possibility of deviation from this rule in section 11.10.1 "MD5 Collisions and SHA256.")

The one-to-one relationship between the values of an MD5 argument and its response means that if we need to key a hash table by a composite key, we can instead:

- Concatenate its components into a character expression.
- Pass the expression to the MD5 function.
- Use its 16-byte signature in the key portion as a *single simple key* instead of the original composite key.

The advantage of this approach is a significant reduction of the key portion length and, consequently, of the entire hash entry length. Suppose, for instance, that the total length of the key portion composite key is 400 bytes, 25 times the length of its MD5 16-byte replacement key. Due to the internal hash object storage rules, it does not mean that the hash entry length will actually be reduced 25 times. For example, if the length of the keys is 400 bytes and the data portion has one 8-byte variable, replacing the composite key by its MD5 signature results in the total hash entry size of 64 bytes (as returned by the ITEM_SIZE operator on the X64_7PRO platform). While it is not the 25-fold reduction (which would be great, of course), the 7-fold hash memory footprint reduction is quite significant as well.

The concept delineated here is not unique to a particular data processing task. On the contrary, it applies to a great variety of them. As an alert IT user noticed, it is illustrated below using various data processing tasks. The intent is not to repeat what is already known but to drive home the point that the same concept works across the board and to demonstrate how it is implemented under different guises, at the same time offering the IT users a number of sample programs where specific technicalities of those guises have already been developed.

11.4.2 MD5 Key Reduction in Sample Data

Note that essentially the same technique has already been used in this book to generate sample data. To wit, in the program Chapter 5 GenerateSchedule.sas, the surrogate 16-byte character key Game_SK is created using the following expression:

```
Game_SK = md5 (catx (":", League, Away_SK, Home_SK, Date,Time)) ;
```

You can access the blog entry that describes this program from the author page at http://support.sas.com/authors. Select either "Paul Dorfman" or "Don Henderson." Then look for the cover thumbnail of this book, and select "Blog Entries." This variable is used as a universal surrogate key linking data files AtBats, Lineups, Pitches, and Runs to each other and to the data file Games. The latter cross-references each distinct value of Game_SK with the corresponding distinct value of composite key (League,Away_SK,Home_SK,Date,Time) along with other variables related to the specific game. Using this approach affords a number of advantages:

- Identifies each particular game in the data sets listed above using a single key instead of five separate keys.
- Reduces the record length of each of the data sets by (5*8-16)=22 bytes.

- Makes it easier and simpler to process (i.e., aggregate, unduplicate, join, etc.) these files anytime the program needs a game identifier, regardless of whether the hash object is used in the process or not.

- If the hash object *is* used, it needs a single $16 key Game_SK instead of the composite key with 5 components of 8 bytes each, thus reducing the hash table entry length. Also, it makes code shorter and clearer.

- This approach also provides great flexibility should the key structure need to change. For example, if the components of the key are augmented with additional variables, those fields can simply be included. For the example shown above that creates Game_SK, if an additional key (e.g., League Type: Majors, AAA, AA, A, Collegiate) is needed, the field League_Type would be included in the CATX function above and in our Games dimension table. No other program changes would be needed.

However, while having such a key as Game_SK already available as part of the data is of great utility, it does not mean that using the MD5 function to reduce the hash key portion length even further *on the fly* is not desirable or necessary. First, not all data collections have such a convenient provision as Game_SK in our sample data. Moreover, as we have seen in previous examples, data processing may involve more hash keys in addition to Game_SK. The examples in the rest of this section illustrate how this technique is used in terms of different data processing tasks.

In the following subsections we will illustrate the applicability of this approach to data aggregation, data unduplication and joining data just as we did in the section "Making Use of Existing Key Order".

11.4.3 Data Aggregation

Suppose that we would like to expand the task of finding the score for each game already discussed above and find the score for each game *and each inning* using data set Dw.Runs. In other words, we need to count the records in the bottom and top of each inning of each game. To recap, for the two selected games shown in Output 11.1, the input data with the added variable Inning looks as follows:

Output 11.8 Chapter 11 Two Games from Dw.Runs

Game_SK	Inning	Top_Bot	From_obs
234B26A4DDD547C58A1A6CF33DF16E66	2	T	10399
234B26A4DDD547C58A1A6CF33DF16E66	6	B	10400
234B26A4DDD547C58A1A6CF33DF16E66	6	B	10401
234B26A4DDD547C58A1A6CF33DF16E66	7	T	10402
234B26A4DDD547C58A1A6CF33DF16E66	8	B	10403
234B26A4DDD547C58A1A6CF33DF16E66	9	B	10404
3D93DA6C9CBA8E887987F7A8E7CE0633	1	T	18534
3D93DA6C9CBA8E887987F7A8E7CE0633	1	T	18535
3D93DA6C9CBA8E887987F7A8E7CE0633	5	T	18536
3D93DA6C9CBA8E887987F7A8E7CE0633	5	B	18537
3D93DA6C9CBA8E887987F7A8E7CE0633	7	B	18538
3D93DA6C9CBA8E887987F7A8E7CE0633	8	T	18539

Using the hash object to solve the problem *head-on* is practically the same as in the first step of Program 11.3 Chapter 11 Pregrouped Aggregation.sas. The *only* difference is that now, in addition to Game_SK, we have to include *both* Top_Bot and Inning in the key portion and data portion:

Program 11.9 Chapter 11 Frontal Attack Aggregation Inning.sas

```
data Scores_game_inning (keep = Game_SK Inning Top_Bot Score) ;
  if _n_ = 1 then do ;
    dcl hash h (ordered:"A") ;
    h.defineKey  ("Game_SK", "Inning", "Top_Bot") ;
    h.defineData ("Game_SK", "Inning", "Top_Bot", "Score") ;
    h.defineDone () ;
    dcl hiter ih ("h") ;
  end ;
  do until (LR) ;
    set Dw.Runs (keep = Game_SK Inning Top_Bot) end = LR ;
    if h.find() ne 0 then Score = 1 ;
    else Score + 1 ;
    h.replace() ;
  end ;
  do while (ih.next() = 0) ;
    output ;
  end ;
run ;
```

The output from the program (for our small visual input sample) looks as follows:

Output 11.9 Chapter 11 Frontal Attack Aggregation by (Game_SK,Top_Bot,Inning)

Game_SK	Inning	Top_Bot	Score
234B26A4DDD547C58A1A6CF33DF16E66	2	T	1
234B26A4DDD547C58A1A6CF33DF16E66	6	B	2
234B26A4DDD547C58A1A6CF33DF16E66	7	T	1
234B26A4DDD547C58A1A6CF33DF16E66	8	B	1
234B26A4DDD547C58A1A6CF33DF16E66	9	B	1
3D93DA6C9CBA8E887987F7A8E7CE0633	1	T	2
3D93DA6C9CBA8E887987F7A8E7CE0633	5	B	1
3D93DA6C9CBA8E887987F7A8E7CE0633	5	T	1
3D93DA6C9CBA8E887987F7A8E7CE0633	7	B	1
3D93DA6C9CBA8E887987F7A8E7CE0633	8	T	1

Note that all components of the composite key (Game_SK,Inning,Top_Bot) are included in the data portion of hash table H. Let us now use the MD5 function to replace the entire composite key with a single 16-byte key _MD5. The changes to the frontal attack Program 11.9 are shown below in bold:

Program 11.10 Chapter 11 MD5 Key Reduction Aggregation.sas

```
%let cat_length = 36 ;    ❶

data Scores_game_inning_MD5 (keep = Game_SK Inning Top_Bot Score) ;
  if _n_ = 1 then do ;
    dcl hash h () ;    ❷
    h.defineKey ("_MD5") ;
```

```
      h.defineData ("Game_SK", "Inning", "Top_Bot", "Score") ;
      h.defineDone () ;
      dcl hiter ih ("h") ;
   end ;
   do until (LR) ;
      set Dw.Runs end = LR ;
      length _concat $ &cat_length _MD5 $ 16 ;      ❶
      _concat = catx (":", Game_SK, Inning, Top_Bot) ;      ❸
      _MD5 = MD5 (_concat) ;
      if h.find() ne 0 then Score = 1 ;      ❹
      else Score + 1 ;
      h.replace() ;
   end ;
   do while (ih.next() = 0) ;
      output ;
   end ;
run ;
```

❶ Sizing _concat variable to accept the ensuing concatenation as $36 accounts for the actual system length of Game_SK ($16), Top_Bot ($1), 17 bytes for *numeric* variable Inning, plus 2 bytes for the CATX separator character. With only 3 key variables in question, it is not difficult to do; but with many more, this kind of hard-coding can become error-prone. So, instead of doing that, an astutely lazy programmer would notice that the key variables come from a SAS data set and use the system table Dictionary.Columns (or view Sashelp.Vcolumn) to populate the macro variable cat_length programmatically.

❷ No ORDERED argument tag is used here since the order by the _MD5 values is different from the order by the original composite key replaced by _MD5, and so ordering the output by _MD5 offers no benefit.

❸ Concatenate the components of the composite key using a colon as a separator. (Using the functions of the CAT family for the purpose is not without caveats. This is discussed in 11.10.1 MD5 Collisions and SHA256). Then feed variable _concat to the MD5 function. Its response _MD5 will be now used as the key-value.

❹ Invoke the standard data aggregation mechanism, except that now we are looking up, not the value of the original composite key, but its MD5 signature _MD5. As the relationship between them is one-to-one, it makes no difference for which one's distinct values Score is accumulated. Note that table H in this form is essentially a cross-reference between the _MD5 values in the key portion and the (Game_SK,Inning,Top_Bot) tuple in the data portion. Though the aggregation is done by the unique values of _MD5, the related values of the original composite key (Game_SK,Inning,Top_Bot) end up in the output.

For our two sample games, the program produces the following output:

Output 11.10 Chapter 11 MD5 Key Reduction Aggregation by (Game_SK,Top_Bot,Inning)

Game_SK	Inning	Top_Bot	Score
234B26A4DDD547C58A1A6CF33DF16E66	8	B	1
3D93DA6C9CBA8E887987F7A8E7CE0633	5	T	1
3D93DA6C9CBA8E887987F7A8E7CE0633	5	B	1
234B26A4DDD547C58A1A6CF33DF16E66	6	B	2
234B26A4DDD547C58A1A6CF33DF16E66	2	T	1
234B26A4DDD547C58A1A6CF33DF16E66	9	B	1
234B26A4DDD547C58A1A6CF33DF16E66	7	T	1
3D93DA6C9CBA8E887987F7A8E7CE0633	7	B	1
3D93DA6C9CBA8E887987F7A8E7CE0633	1	T	2
3D93DA6C9CBA8E887987F7A8E7CE0633	8	T	1

Note that its *record order* is different from that coming out of the *frontal attack* program. This is because the table is keyed by _MD5 whose values, though uniquely related to those of (Game_SK,Inning,Top_Bot), follow a totally different order. However, it does not present a problem because:

- Except for the inconvenience of eyeballing the unsorted result, the output is exactly the same.
- If we still want the file sorted by the original keys, it is not too much work since aggregate files are usually much smaller than unaggregated input. Also, the aggregated file can be indexed instead.
- If we wanted to display the values of the component variables used to create Game_SK, we could simply use it as a key to do a table lookup against the data set Games to get the desired values.

11.4.4 Data Unduplication

The same exact principle of using the MD5 function applies if we want to shorten the key portion when unduplicating data. Let us recall the DATA step of Program 11.4 Chapter 11 Frontal Attack Unduplication.sas in section 11.3.2. There, we used the composite hash key (Game_SK,Runner) to find and keep the records where each of its distinct key-values occurs first.

Using the MD5 function, we can now replace the composite key (Game_SK,Runner_ID) with a single 16-byte key _MD5 (the changes from Program 11.4 are shown in bold):

Program 11.11 Chapter 11 MD5 Key Reduction Unduplication.sas

```
%let cat_length = 34 ;

data First_scores_MD5 (drop = _:) ;
  if _n_ = 1 then do ;
    dcl hash h() ;
    h.defineKey  ("_MD5") ;
    h.defineData ("_N_") ;
    h.defineDone () ;
    dcl hiter ih ("h") ;
  end ;
  do until (LR) ;
    set dw.Runs end = LR ;
    length _concat $ &cat_length _MD5 $ 16 ;
```

```
    _concat = catx (":", Game_SK, Runner_ID) ;
    _MD5 = MD5 (_concat) ;
    if h.check() = 0 then continue ;
    output ;
    h.add() ;
  end ;
run ;
```

The principal differences between the head-on and MD5 approaches are these:

- With the former, we store the distinct composite key-values (Game_SK,Runner) in the hash table and search the table to see if the key-value from the current record is already in the table.
- With the latter, we use the fact that the composite key-values of (Game_SK,Runner) and _MD5 are related as one-to-one and key the table with the latter.

Since in this case the sequence of the input records defines the output sequence, the relative record orders in both are the same (in other words, the unduplication process is *stable*), and the replacement of the composite key by _MD5 does not affect it. Note that if the SORT or SQL procedure were used to accomplish the task, the same record order would not (or would not be guaranteed to) be preserved.

For data unduplication the benefit of using the MD5 key-reduction technique is aided by the fact that nothing needs to be retrieved from the data portion (the very reason it is merely filled with the dummy placeholder _N_). So, the reduction of the hash entry length as a whole is roughly proportional to the reduction of the key portion length. In the example above, the shortening of the key length from 24 bytes to 16 may look trivial. However, under different circumstances it can reach an order of magnitude or more.

One such scenario is cleansing a data set where *whole records*, i.e., the values of all variables, have been duplicated for some reason, which is a typical occurrence in ETL processes. Imagine, say, that due to either programming or clerical error, some records in the data set Dw.Games have been completely duplicated. For example, suppose that the data for a given date were loaded twice. Since that did not occur when we created the sample data, we can *emulate* such a situation as follows:

Program 11.12 Chapter 11 Full Unduplication.sas (Part 1)

```
data Games_dup ;
  set Dw.Games ;
  do _n_ = 1 to floor (ranuni(1) * 4) ;
    output ;
  end ;
run ;
```

As a result, file Games_dup would have about twice as many records as the original, some records wholly duplicated. Output 11.8 below shows the first 8 records from Games_dup. The column From_obs is not a variable from the file but indicates the original observation numbers from Dw.Games.

Output 11.11 Chapter 11 Work.Games_dup Sample

Game_SK	Date	Time	League	Home_SK	Away_SK	From_obs
006DADDD85B61F0ECA8D07835AA6D555	2017-06-13	7:00 PM	1	203	317	2
006DADDD85B61F0ECA8D07835AA6D555	2017-06-13	7:00 PM	1	203	317	2
006DADDD85B61F0ECA8D07835AA6D555	2017-06-13	7:00 PM	1	203	317	2
00A73AF3888383B73D205C02FD9E10FC	2017-08-30	7:00 PM	2	319	269	3
00DE350CB59BDFABEA1056521E23F504	2017-04-10	7:00 PM	2	115	224	4
00FC9989FF38EB3F063B2B3D3AEC1C92	2017-06-19	7:00 PM	1	342	339	5
00FC9989FF38EB3F063B2B3D3AEC1C92	2017-06-19	7:00 PM	1	342	339	5
00FC9989FF38EB3F063B2B3D3AEC1C92	2017-06-19	7:00 PM	1	342	339	5

Suppose that we need to *cleanse* the file by keeping only the *first* copy of each duplicate record while preserving the relative sequence of the input records. In other words, after the cleansing is done, our output should match the original file Dw.Games.

If we approached the task headlong using the frontal hash object attack, we would have to include all 9 variables from Games_dup (with the total byte length of 80) in the key portion, plus a dummy placeholder variable in the data portion. (Although, for this task, we do not need our key variables in the data portion since their output values come directly from the input file records, we still need at least one data portion variable due to the hash object design.) Having to include *all* input file variables in the key portion has its consequences:

- If the variables in the file are numerous, it will result in a large hash entry length.
- Even in this simple case with only 9 key portion variables (and 1 dummy data portion variable), the hash entry length on a 64-bit platform would be 128 bytes. For a real-world file with 100 million distinct records, it would result in the hash memory footprint close to 20 GB of RAM.

However, the problem can be easily addressed by using the MD5 key reduction approach because it replaces all key portion hash variables with a single 16-byte _MD5 key variable. This approach is demonstrated by the program below modified from Program 11.11 (the changes are shown in bold):

Program 11.12 Chapter 11 Full Unduplication.sas (Part 2)

```
%let cat_length = 53 ;

data Games_nodup (drop = _:) ;
  if _n_ = 1 then do ;
    dcl hash h() ;
    h.defineKey  ("_MD5") ;
    h.defineData ("_N_") ;
    h.defineDone () ;
    dcl hiter ih ("h") ;
  end ;
  do until (LR) ;
    set Games_dup end = LR ;
    array NN[*] _numeric_  ;     ❶
    array CC[*] _character_ ;
    length _concat $ &cat_length _MD5 $ 16 ;
    _concat = catx (":", of NN[*], of CC[*]) ; ❷
    _MD5 = MD5 (trim(_concat)) ;
    if h.check() = 0 then continue ;
```

```
      output ;
      h.add() ;
   end ;
run ;
```

Principally, the program is not much different from Program 11.11. The differences are noted below:

❶ The ARRAY statement in this form automatically incorporates all *non-automatic* numeric variables currently in the PDV into array *NN*. Likewise, the next statement automatically incorporates all non-automatic character variables currently in the PDV into array *CC*. Since during compile time at this point in the step the only non-automatic variables the compiler has seen are the variables from the descriptor of file Games_dup, these are the only variables the arrays incorporate. Note that it is incumbent upon the programmer to ensure that no other non-automatic variables get in the PDV before the arrays are declared; otherwise we may end up concatenating variables we do not need.

❷ Concatenate the needed variables into the MD5 argument using the shortcut array notation [*]. Note that since all the numeric variables are concatenated first followed by all the character variables (due to the order in which the arrays are declared), the concatenation order may differ from that of the variables in the file. This is okay since from the standpoint of the final result *the concatenation order does not matter*. This aspect of the technique (actually adding to its robustness) will be discussed in some more detail in the section "MD5 Argument Concatenation Ins and Outs".

For our first 8 sample records from Games_dup (Output 11.8), the program generates the following output:

Output 11.12 Chapter 11 Work.Games_dup Unduplicated

Game_SK	Date	Time	League	Home_SK	Away_SK
006DADDD85B61F0ECA8D07835AA6D555	2017-06-13	7:00 PM	1	203	317
00A73AF3888383B73D205C02FD9E10FC	2017-08-30	7:00 PM	2	319	269
00DE350CB59BDFABEA1056521E23F504	2017-04-10	7:00 PM	2	115	224
00FC9989FF38EB3F063B2B3D3AEC1C92	2017-06-19	7:00 PM	1	342	339

11.4.5 Joining Data

Let us now turn to the case of joining data already described and discussed earlier in the section "Joining Data". There, Program 11.6 Chapter 11 Frontal Attack Join.sas represents a *frontal attack* where all components of the joining composite key (Game_SK, Inning,Top_Bot, AB_Number) are included in the key portion. In the program below, files Dw.Runs and Dw.AtBats are joined by a single key variable _MD5 instead. The differences between this code and the frontal attack Program 11.6 are shown in bold:

Program 11.13 Chapter 11 MD5 Key Reduction Join.sas

```
%let comp_keys = Game_SK Inning Top_Bot AB_Number ;
%let data_vars = Batter_ID Position_code Is_A_Hit Result ;
%let data_list = %sysfunc (tranwrd (&data_vars, %str( ), %str(,))) ;
%let keys_list = %sysfunc (tranwrd (&comp_keys, %str( ), %str(,))) ; ❶

%let cat_length = 52 ;

data Join_Runs_AtBats_MD5 (drop = _: Runs) ;
   if _n_ = 1 then do ;
```

```
       dcl hash h (multidata:"Y", ordered:"A") ;
       h.defineKey ("_MD5") ;    ❷
       do _k = 1 to countw ("&data_vars") ;
         h.defineData (scan ("&data_vars", _k)) ;
       end ;
       h.defineDone() ;
       do until (LR) ;
         set dw.AtBats (keep=&comp_keys &data_vars Runs
                        where=(Runs)) end = LR ;
         link MD5 ;    ❸
         h.add() ;
       end ;
     end ;
     set dw.Runs ;
     link MD5 ;    ❹
     call missing (&data_list, _count) ;
     do while (h.do_over() = 0) ;
       _count = sum (_count, 1) ;
       output ;
     end ;
     if not _count then output ;
     return ;
     MD5: length _concat $ &cat_length _MD5 $ 16 ;    ❺
          _concat = catx ("", &keys_list) ;
          _MD5 = MD5 (_concat) ;
     return ;
run ;
```

❶ A comma-separated list of key variable names is prepared to be used in the CATX function later.

❷ Rather than defining separate key variables in a loop, a single key variable _MD5 is defined.

❸ The LINK routine MD5 is called for every record from Dw.AtBats (the right side of the join) to concatenate the components of its composite key and pass the result to the MD5 function. Its signature _MD5 is added to table H along with the values of the data_vars list variables as a new hash item.

❹ For each record read from file Dw.Runs, the LINK routine MD5 is called again, now for the *left* side of the join. It generates the value of _MD5 key to be looked up in the hash table.

❺ The MD5 LINK routine is relegated to the bottom of the step to avoid repetitive coding. This way, we have only one set of code creating the _MD5 key for both data sets involved in the join; and this ensures that they are kept in synch (e.g., if the key structure should change for some reason, we would have only one set of code to change).

The principal difference between using the MD5 key reduction technique in this program and the programs used for data aggregation and unduplication is that since in this case we have *two* files (Dw.Runs and Dw.AtBats), the composite key needs to be converted to its MD5 signature for each record from *both* input files: For Dw.AtBats with the data to be loaded into the hash table *and* for Dw.Runs before the *Keynumerate* operation is invoked.

11.5 Data Portion Offload (Hash Index)

Thus far in this chapter, we have dealt with measures of *reducing the length of the key portion* by either moving its hash variables elsewhere or replacing them with a surrogate key. In the related examples, similar measures *cannot* be used to reduce the length of the *data portion* as well because its hash variables are needed for the *Retrieve* operation to extract the data into the PDV host variables.

However, in some cases the size of the data portion can be reduced by replacing the data retrieval from the hash table with extracting the same data directly from disk. Doing so effectively relegates the hash table to the role of a searchable *hash index*. This is a tradeoff: It sacrifices a small percentage of execution speed for the possibility to remove *all* data portion variables and replace them with a *single numeric* pointer variable. Two data processing tasks that benefit from this approach most readily are (a) joining data and (b) selective unduplication.

11.5.1 Joining Data

Let us return to our archetypal case of joining file Dw.Runs to file Dw.AtBats that we have already discussed in this chapter. See the section "Joining Data" for the task outline, visual data samples, and Program 11.6 Chapter 11 Frontal Attack Join.sas.

To recap, the task is to join the two files by composite key (Game_SK,Inning,Top_Bot,AB_Number) with the intent to enrich Dw.Runs with the satellite fields Batter_ID, Is_A_Hit, and Result from Dw.AtBats. In the frontal attack program, all these data fields are included in the data portion of the hash table. As mentioned earlier, we could potentially need to bring in many more fields depending on the questions asked about the game results. In real industrial applications, data portion variables may be much more numerous – the more of them, the greater hash memory footprint they impose. The question is, can their number be reduced; and if yes, then how?

To find out, let us observe that each observation in file Dw.AtBats can be marked by its serial number – that is, the *observation number*, in SAS terminology. This number is commonly referred to as the *Record Identification Number* (or *RID Number* or just *RID*). The idea here is to include RID in the data portion and load the table from Dw.AtBats with its RID values *instead* of the values of satellite variables. Then, for each hash item whose composite key (Game_SK,Inning,Top_Bot,AB_Number) matches its counterpart from Dw.Runs, the value of RID retrieved from the table can be used to extract the corresponding data directly from Dw.AtBats by using the SET statement with POINT=RID option. With these changes in mind, the frontal attack Program 11.6 can be re-coded this way (the changes are shown in bold):

Program 11.14 Chapter 11 Hash Index Join.sas

```
%let comp_keys = Game_SK Inning Top_Bot AB_Number ;
%let data_vars = Batter_ID Is_A_Hit Result ;
%let data_list = %sysfunc (tranwrd (&data_vars, %str( ), %str(,))) ;

data Join_Runs_AtBats_RID (drop = _: Runs) ;
  if _n_ = 1 then do ;
    dcl hash h (multidata:"Y", ordered:"A") ;
    do _k = 1 to countw ("&comp_keys") ;
      h.defineKey (scan ("&comp_keys", _k)) ;
    end ;
    h.defineData ("RID") ;     ❶
    h.defineDone() ;
    do RID = 1 by 1 until (LR) ;     ❷
```

```
    set dw.AtBats (keep=&comp_keys Runs where=(Runs)) end = LR ;
      h.add() ;
    end ;
  end ;
  set dw.Runs ;
  call missing (&data_list, _count) ;
  do while (h.do_over() = 0) ;      ❸
    _count = sum (_count, 1) ;
    set dw.Runs point = RID ;       ❹
    output ;
  end ;
  if not _count then output ;
run ;
```

❶ The data portion is defined with variable RID (i.e., Dw.AtBats observation number) and *none* of the satellite variables from the data_vars list.

❷ The sequential values of RID (i.e., the sequential observation number) are created here by incrementing RID up by 1 for each next record from Dw.Runs. Each value is inserted into the hash table as a new item along with the corresponding key-values from the comp_keys list. This way, the values of RID in the table are linked with the respective records from Dw.AtBats – and thus, with the respective values of the satellite variables.

❸ The *Keynumerate* operation searches the table for the key-values coming from each record from Dw.Runs and retrieves the RID values pointing to records in Dw.AtBats.

❹ If the PDV composite key-value accepted by the DO_OVER method is found in the table, the *Keynumerate* operation retrieves the RID values from the group of items with this key-value, one at a time, into PDV host variable RID. For each enumerated item, the retrieved value of RID is then used to access Dw.AtBats directly via the SET statement with the POINT=RID option, the value of RID pointing to the exact record we need. Thus, the values of the variables from the data_vars appear in the PDV because they are read directly from the file – not as a result of their retrieval from the hash table, as in the frontal attack program. (Note that since variable RID is named in the POINT= option, it is automatically dropped despite also being used as a DO loop index.)

Running this program generates exactly the same output as the frontal attack Program 11.6. The technique used in it is notable in a number of aspects:

- **Memory-Disk Tradeoff.** Offloading the data portion variables in this manner means that the satellite values are extracted from disk. Obviously, it is slower than retrieving them from hash table memory. However, this is offset by the fact that now those same variables need not be either loaded into or read from the hash table. More importantly, the action affecting performance most – namely, searching for a key – is still done using the hash lookup in memory.

- **Performance.** Testing shows that overall, the reduction in execution performance is insignificant because POINT= access (compared to SAS index access) is in itself extremely fast. If the satellites variables involved in the process are numerous and long, the trade-off is well worth the price, as it could reduce hash memory usage from the point where the table may fail to fit in memory (which would abend the program) to where the program still runs to completion successfully.

- **Hash-Indexing.** This technique has yet another interesting aspect. Effectively, by pairing RID in the hash table with its file record, we have created, for the duration of the DATA step, a *hash index* for file Dw.AtBats defined with the composite key from the comp_keys list. Both such hash index and a standard SAS index (residing in an index file) do essentially the same job: They search for RID given a key-value, simple or composite; and if it is found, both use the RID value to locate the corresponding record and extract the data from it. The difference is that the hash memory-resident index coupled with the hashing algorithm renders both the key search and data extraction (using the POINT=) option much faster.

- **Hybrid Approach.** The hash-indexing approach can be readily adapted to a *hybrid* situation where it may be expedient to have *a few specific* satellite variables defined in the data portion *along* with RID. This way, if programming logic dictates so, the values of these *specific* hash variables can be retrieved into their PDV host counterparts from the hash table without incurring extra I/O. And yet when it dictates, at some point, that the values from more (or, indeed, all) satellite variables are needed in the PDV, they can be extracted directly from disk via the SET statement with the POINT=RID option as shown above.

11.5.2 Selective Unduplication

The hash-indexing scheme as described above can work for any data task where a hash table is used to store satellite variables with the intent of *eventually* retrieving their hash values into the PDV – for example, for data manipulation or generating output. From this standpoint:

- Using it for *simple* file unduplication is pointless. Indeed, for this task the *Retrieve* operation is not used, and the only variable in the data portion is a dummy placeholder.
- By contrast, *selective stable* unduplication illustrated by Program 10.3 Chapter 10 Selective Unduplication via PROC SORT + DATA Step.sas can benefit from data portion disk offloading because, in this case, the variables in the data portion *are* retrieved from the hash table into the PDV in order to generate the output.

Let us recall the subset of file Dw.Games used to illustrate selective unduplication in Chapter 10. (Again, From_obs is not a data set variable. It just shows the observation numbers in Dw.Games related to the records in the subset below.)

Output 11.13 Chapter 11 Subset from Dw.Games for Selective Unduplication

Game_SK	Date	Home_SK	Away_SK	From_obs
066D3D2DC65D46F1D8793F7D29311F31	2017-06-28	281	193	77
10CFAD9C990B9AA39623D1EA2B5DDD9E	**2017-10-08**	**281**	**259**	179
7F0F89E30F66297B98A7628D1D6E8841	**2017-10-11**	**203**	**259**	1419
ABA31945ED33744FA94686E025FC816A	2017-03-31	281	339	1918
B88878AC0E455E36016A423EAD4A786B	2017-06-17	281	132	2061
BA0AA3651F95B16004E9170096FB5623	2017-03-27	203	281	2080
C83E851AC9CBFE1447A69E948928B30D	2017-03-27	246	344	2245
DAD52B8138928BDE2F37EB671A8CC0DB	2017-08-28	203	348	2447
E67048EDCFE44B1CE13421427C63993B	**2017-10-04**	**246**	**317**	2584

To recap, the task lies in the following:

- Using this data, find the *away* team which each *home* team *played last*. In other words, we need to *unduplicate* the records for each key-value of Home_SK by selecting its record with the *latest* value of Date.

- Along with Home_SK, output the values of Away_SK, Date, and Game_SK from the corresponding record with the most recent value of Date.
- Note that this time, the requirements include the extra variable Game_SK in the output.
- Also note that in the data sample above, the records we want in the output are shown in bold.

To solve the task by using a *frontal attack* approach, it is enough to slightly modify Program 10.3 Chapter 10 Selective Unduplication via Hash:

Program 11.15. Chapter 11 Frontal Attack Selective Unduplication.sas

```
data _null_ ;
  dcl hash h (ordered:"A") ; *Output in Home_SK order;
  h.defineKey  ("Home_SK") ;
  h.defineData ("Home_SK", "Away_SK", "Date", "Game_SK") ;
  h.defineDone () ;
  do until (lr) ;
    set dw.games (keep = Game_SK Date Home_SK Away_SK) end = lr ;
    _Away_SK = Away_SK ;
    _Date    = Date ;
    _Game_SK = Game_SK ;
    if h.find() ne 0 then h.add() ;
    else if _Date > Date then
      h.replace (key:Home_SK, data:Home_SK, data:_Away_SK, data:_Date
                            , data:_Game_SK
                ) ;
  end ;
  h.output (dataset: "Last_games_hash") ;
  stop ;
run ;
```

This program generates the following output for the records of our sample subset from Dw.Games:

Output 11.14 Chapter 11 Results of Frontal Attack Selective Unduplication

Home_SK	Away_SK	Date	Game_SK
203	259	2017-10-11	7F0F89E30F66297B98A7628D1D6E8841
246	317	2017-10-04	E67048EDCFE44B1CE13421427C63993B
281	259	2017-10-08	10CFAD9C990B9AA39623D1EA2B5DDD9E

Note that the variables Away_SK and Game_SK are included in the data portion since their values are needed for the OUTPUT method call. Also, the temporary variables _Away_SK, and _Game_SK are included in the explicit REPLACE method call to update the data-values in table H for every input PDV value of Date greater than its presently stored hash value.

In this rather simple situation, there are only three satellite variables to be stored in the data portion and hard-coded in the explicit REPLACE method call. In other situations with many more satellite variables, two things with the frontal attack approach can become bothersome:

- With large data volumes and numerous satellite variables, hash memory can get overloaded.
- Hard-coding many arguments with the REPLACE method call is unwieldy and error-prone. In fact, even with a mere five arguments, as above, typing them in already gets pretty tedious.

However, both issues are easily addressed by using the hash index approach. In the program below, a hash table index keyed by Home_SK is used to keep the data portion variables that are not critical to the program logic out of the data portion. The program may seem to be a tad longer than the frontal attack. However, as we will see, its benefits far outweigh the cost of the few extra lines of code:

Program 11.16 Chapter 11 Hash Index Selective Unduplication.sas

```
data Last_away_hash_RID (drop = _:) ;
  dcl hash h    (ordered:"A") ;
  h.defineKey  ("Home_SK") ;
  h.defineData ("Date", "RID") ;     ❶
  h.defineDone () ;
  do _RID = 1 by 1 until (lr) ;      ❷
    set dw.Games (keep = Date Home_SK) end = lr ;     ❸
    _Date =  Date ;
    RID  = _RID ;     ❹
    if h.find() ne 0 then h.add() ;
    else if _Date > Date then
    h.replace (key:Home_SK, data:_Date, data:_RID) ;     ❺
  end ;
  dcl hiter hi ("h") ;     ❻
  do while (hi.next() = 0) ;
    set dw.Games (keep = Home_SK Away_SK Game_SK) point = RID ;     ❼
    output ;
  end ;
  stop ;
run ;
```

❶ The data portion is defined with variables Date and RID. Note the absence of other variables needed in the frontal attack Program 11.15.

❷ A temporary variable _RID instead of RID is used to increment the record identifier to safeguard its values from being overwritten by the FIND method call.

❸ No need to keep Away_SK and Game_SK in this input, as their values on disk are linked to by RID.

❹ RID is repopulated with the memorized value of _RID to ensure its value is correct, just in case it may have been overwritten in the previous iteration of the DO loop by the FIND method call.

❺ RID in this explicit REPLACE call is used instead of Home_SK, _Away_SK, and _Game_SK. If we had more satellite variables, the _RID assigned to the argument tag DATA would replace the need to code *any of them*. Essentially, all references to the satellite variables here are replaced en masse with a single RID pointer to their location on disk.

❻ The hash iterator is used to render the output instead of the OUTPUT method since satellite variables needed in the output are no longer in the data portion (the same would be true of all other satellite variables if we had more of them). The satellite variable values will be extracted directly from the input file Dw.Games via POINT=RID option.

❼ Use RID as a record pointer to retrieve Home_SK, Away_SK, and Game_SK related to the highest value of (Home_SK,Date) directly from the file; then output a record with these values for the current hash item.

This program generates the following output for the sample records shown in Output 11.10:

Output 11.15 Chapter 11 Results of Hash Index Selective Unduplication

Date	Home_SK	Game_SK	Away_SK
2017-10-11	203	7F0F89E30F66297B98A7628D1D6E8841	259
2017-10-04	246	E67048EDCFE44B1CE13421427C63993B	317
2017-10-08	281	10CFAD9C990B9AA39623D1EA2B5DDD9E	259

The output contains the same data as Output 11.14, except for the order of the variables. If it matters, it can be easily addressed by inserting a RETAIN statement listing the variables, unvalued, in the desired order immediately after the DATA statement. As we see, using a hash index instead of the frontal attack offers the following benefits:

- It reduces the data portion variables only to those required by the program logic (such as Date above) and the pointer variable RID.
- In turn, this may result in substantial savings in terms of the hash memory footprint.
- It greatly simplifies coding because it is no longer necessary to code numerous variable names in the DEFINEDATA or REPLACE method calls.
- Getting rid of hard-coding makes the program less error-prone and more robust.

11.6 Uniform Input Split

Certain industrial applications have input so large and output specifications so multi-leveled that when the hash object is used to do the processing, no memory-saving measures described in this chapter, even when combined, can fit the hash table into the available memory limits. Such situations raise the question whether there is a universal way to address the problem, even if it can be done at the expense of some reasonable trade off. "Reasonable" means an approach that would not sacrifice too much execution speed while still keeping the major benefits of the hash object.

In the quest to find such an approach, let us think why the technique of reducing the key portion length using pre-sorted order by one or more leading components of a composite key actually works. Fundamentally, it does because the existing order separates the input into distinct groups of records *sharing no composite key-values* with any other group. As the input is pre-ordered, BY group processing can be used to treat the BY groups independently and identically as if each were a separate input. And if the distinct composite key-values were distributed about equally between the BY groups, the hash memory load needed to process the file in a single chunk would be reduced roughly proportionally to their number.

11.6.1 Uniform Split Using Key Properties

Fundamentally the same scheme can be applied to an *unsorted* file if certain criteria are met. Let us call the key we are using in our data processing task – simple or composite – the *process key*. Now imagine that *some part* of our process key has such values (or ranges of values) that they *split* the input records into a *number of groups* with the following properties:

1. No value of the process key in any group is present in any other group. In other words, each group contains its own set of distinct values of the process key.
2. Each group contains roughly an equal number of distinct key-values of the process key.

Suppose we have managed to identify a part of the process key with these properties. From now on, we will call such a key part a *Uniform Key Splitter* (or a UKS, for short). If a UKS is found in the input, we can process each group of records identified by its values independently, simply combining the outputs generated from each separate group. This is true even in the case of non-additive aggregates because no key-values in different UKS-identified groups are shared.

To find a UKS candidate (if it exists), we should first remember that it *must be part of the process key* (including the entire key, if need be). That is, non-key data is irrelevant. If this requirement is met, the role of UKS can be played by a number of process key pieces, as long as the UKS values (or a number of their ranges) correspond to a more or less equal number of distinct key-values of the process key. For example:

*   One or more components (i.e., partial keys) of a composite process key. It can be used because no two composite key-values can be equal if they differ in any component.
*   A character or a combination of characters in fixed process key positions. It can be used because no two keys (simple or composite) can be equal if they differ even in a single same-position byte.

Often, a UKS can be found or is known a priori because of the knowledge of business data or its organization and structure. Consider, for example, our sample data file Dw.Runs. Even though in some innings and some inning halves no runs have been scored, we could expect from the nature of the game that overall, for each value of Inning from 1 to 9, we would have a more or less equal number of distinct values of (Game_SK,Top_Bot); or, for each value of Top_Bot ("T" and "B") we would have about the same number of unique values of (Game_SK, Inning). Therefore, both Inning and Top_Bot can be used as a UKS in hash data processing if either variable is part of the composite process key.

On the other hand, we know from the nature of our data organization that key variable Game_SK is an MD5 function signature; and since it is a uniform hash function, any byte of its response nearly *randomly* (and close to uniformly distributed) takes on the 256 values in the collating sequence. Hence, we can group these different values into 2, 4, 8, etc., range groups and use the group number as a UKS.

Some approaches based on these concepts are illustrated in the following subsections.

11.6.2 Aggregation via Partial Key Split

Let us revisit the task of obtaining the score for each game and inning from data set Dw.Runs and modify the basic frontal attack program in Program 11.9 Chapter 11 Frontal Attack Aggregation Inning.sas using the following plan:

*   Select a number of groups N_groups among which the distinct values of Inning will be split. For example, let N_groups=3.

- Segment the values of Inning from 1 to 9 arbitrarily into 3 equal groups (e.g., 1-3, 4-6, 7-9, or in any other way) and assign the groups UKS_group values from 1 to 3.
- Read Dw.Runs using the WHERE clause to select the records with UKS_group=1, do the same aggregation processing as in the frontal attack program, output the hash data, and clear the hash table.
- Do the same for UKS_group=2 and UKS_group=3.

Essentially, we are going to loop through the 3 different ranges of Inning, each time re-reading the input file with a different WHERE clause according to the selected UKS_group. This process can be organized automatically in a variety of ways instead of coding a separate block of code for each UKS_group by hand. However, before getting to this stage, let us envision what kind of code we intend to generate.

Suppose that we have decided to separate the values of Inning into 3 value groups: (3,6,9), (1,4,7), and (2,5,8). Code we would like to assemble and include in the DATA step is shown below in bold:

Program 11.17 Chapter 11 Partial Key Split By Inning Aggregation – Code to Assemble.sas

```
data Scores_inning_split_test (keep = &comp_keys Score) ;
  if _n_ = 1 then do ;
    dcl hash h (ordered:"A") ;
    do _k = 1 to countW ("&comp_keys") ;
      h.defineKey (scan ("&comp_keys", _k)) ;
      h.defineData (scan ("&comp_keys", _k)) ;
    end ;
    h.defineData ("Score") ;
    h.defineDone () ;
    dcl hiter ih ("h") ;
  end ;
  do LR = 0 by 0 until (LR) ;
    set dw.Runs (where=(Inning in (3,6,9))) end = LR ;
    link SCORE ;
  end ;
  link OUT ;
  do LR = 0 by 0 until (LR) ;
    set dw.Runs (where=(Inning in (1,4,7))) end = LR ;
    link SCORE ;
  end ;
  link OUT ;
  do LR = 0 by 0 until (LR) ;
    set dw.Runs (where=(Inning in (2,5,8))) end = LR ;
    link SCORE ;
  end ;
  link OUT ;
  return ;
  SCORE: if h.find() ne 0 then Score = 1 ;
         else Score + 1 ;
         h.replace() ;
  return ;
  OUT:   do while (ih.next() = 0) ;
           output ;
         end ;
         Num_items = h.num_items ;
         put Num_items= ;
         h.clear() ;
  return ;
run ;
```

Though this program certainly works, in this form its code is rather unwieldy because whatever facility is used to assemble the code, the respective value lists (3,6,9), (1,4,7), and (2,5,8) would have to be *hard-coded*. It is not so bad with 3 lists of 3 values each, but in the real world (as well as later in this chapter), there can be a whole lot more distinct values and ranges.

However, this is SAS; and with its cornucopia of functions, there is nearly always a remedy. In our case at hand, it comes in the guise of the MOD(N,D) function whose response is the *remainder* of the N/D quotient. Because the remainder always ranges from 0 to (D-1), the function uniformly maps the values of its first argument to the D distinct values of its response. To check how it may work towards our goal, consider the following DATA step:

Program 11.18 Chapter 11 Splitting Inning Values via MOD.sas

```
%let N_groups = 3 ;

data Groups ;
  do Inning = 1 to 9 ;
    Group     = 1 + mod (Inning, &N_groups) ;
    Group_Seq = 1 + mod (Inning + &N_groups - 1
                        ,&N_groups) ;
    output ;
  end ;
run ;
```

If we now print the output data set *Groups*, we will see the following picture:

Output 11.16 Chapter 11 Segmentation of Innings Using MOD

Inning	Group	Group_Seq
1	2	1
2	3	2
3	1	3
4	2	1
5	3	2
6	1	3
7	2	1
8	3	2
9	1	3

Evidently, the first expression maps the values of Inning (3,6,9) as Group=1, (1,4,7) as Group=2, and (2,5,8) as Group=3. The second expression also groups them uniformly but in a different order. Now if we still want to group the values of Inning in 3 groups as originally intended, we can replace the WHERE clauses with the explicitly hard-coded Inning ranges as follows:

```
where mod(Inning,3)+1 = 1
where mod(Inning,3)+1 = 2
where mod(Inning,3)+1 = 3
```

It means that the code assembling program has to generate only N_groups consecutive DO loops varying the integer on the right side of the WHERE equal sign clause from 1 to N_groups. Which particular MOD expression to use is merely a matter of choice. It can be assigned to a macro variable as a whole and will automatically appear in the WHERE clauses assembled by the code generator.

This subterfuge based on the MOD function comes in extremely handy whenever a number of distinct integer values have to be uniformly separated into a fixed number of groups. It eschews hard-coding of value ranges regardless of the hard-coding method, be it an IF-THEN-ELSE construct or the VALUE/INVALUE statement in the FORMAT procedure. (A hint: A MOD-base expression can be used to create a CNTLIN= data set for the FORMAT procedure to generate a value-grouping format for the WHERE clause.)

After this preamble, we can proceed directly to generating and executing code for splitting the input by the ranges of Inning. Below, it is done using the macro facility; however, any other code-generating technique can be used as well.

Program 11.19 Chapter 11 Partial Key Split by Inning Aggregation.sas

```
%let file_dsn  = Dw.Runs ;
%let comp_keys = Game_SK Inning Top_Bot ;
%let UKS_base  = Inning ;      ❶
%let N_groups  = 3 ;
%let UKS_group = mod (&UKS_base,&N_groups) + 1 ; ❷

%macro UKS() ;    ❸
  %do Group = 1 %to &N_groups ;
    do LR = 0 by 0 until (LR) ;
      set &file_dsn (where=(&UKS_group=&Group))
          end = LR ;
      link SCORE ;
    end ;
    link OUT ;
  %end ;
%mEnd ;

data Scores_game_inning_split (keep = &comp_keys Score) ;
  if _n_ = 1 then do ;
    dcl hash h (ordered:"A") ;
    do _k = 1 to countW ("&comp_keys") ;
      h.defineKey  (scan ("&comp_keys", _k)) ;
      h.defineData (scan ("&comp_keys", _k)) ;
    end ;
    h.defineData ("Score") ;
    h.defineDone () ;
    dcl hiter ih ("h") ;
  end ;
  %UKS()                                        ❹
  return ;
  SCORE: if h.find() ne 0 then Score = 1 ;     ❺
         else Score + 1 ;
         h.replace() ;
  return ;
  OUT:   do while (ih.next() = 0) ;
           output ;
         end ;
```

```
      * Num_items = h.num_items ;
      * put Num_items= ;
        h.clear() ;
   return ;
run ;
```

❶ Choose Inning as the UKS value base; then opt for dividing its values into N_groups=3 groups.

❷ Choose a formula to apportion the values of Inning among the 3 groups. This formula groups the values of Inning and UKS_group as (3,6,9)=1, (1,4,7)=2, and (2,5,8)=3. Of course, the values of Inning can be grouped differently. For example, formula mod(&UKS_base+&N_groups-1,&N_groups) would assign them as (1,2,3)=1, (4,5,6)=2, (7,8,9)=3. However, it does not really matter, as long as they are split equally evenly among the groups.

❸ Macro UKS is used to generate data aggregation code for each separate UKS group. In this case, it will generate 3 file-reading DoW loops (versus one in the frontal attack program). For every DoW loop it generates, it also generates the corresponding group value for the WHERE clause condition. (Also note how the automatic variable LR marking the last record in the file is auto-reinitialized for each new loop by the index construct LR=0 by 0.)

❹ Invoke macro UKS to generate the required multiple-loop block of code.

❺ The standard aggregation and output routines are relegated to the bottom of the step to avoid repetitive coding. Calls to them are also generated by macro UKS.

The SAS statements generated by macro UKS look as follows:

```
do LR=0 by 0 until (LR) ;
  set Dw.Runs (where=(mod(Inning,3)+1=1)) end=LR ;
  link SCORE ;
end ;
link OUT ;
do LR=0 by 0 until (LR) ;
  set Dw.Runs (where=(mod(Inning,3)+1=2)) end=LR ;
  link SCORE ;
end ;
link OUT ;
do LR=0 by 0 until (LR) ;
  set Dw.Runs (where=(mod(Inning,3)+1=3)) end=LR ;
  link SCORE ;
end ;
link OUT ;
```

Respectively, the SAS log reports:

```
NOTE: There were 25668 observations read from the
      data set DW.RUNS.
      WHERE (MOD(Inning,3)+1)=1;
NOTE: There were 26484 observations read from the
      data set DW.RUNS.
      WHERE (MOD(Inning,3)+1)=2;
NOTE: There were 26194 observations read from the
      data set DW.RUNS.
      WHERE (MOD(Inning,3)+1)=3;
NOTE: The data set WORK.SCORES_GAME_INNING_SPLIT
      has 28790 observations and 4 variables.
```

If we uncomment the two commented-out statements before the CLEAR method call and rerun the program, the log notes generated by the PUT statement will show for each successive pass through the file:

```
Num_items=9388
Num_items=9806
Num_items=9596
```

These numbers reflect the maximal number of items in the hash table – in other words, the number of distinct values of composite key (Game_SK,Inning,Top_Bot) – in each successive UKS group. It means that compared to the frontal attack program, the topmost number of items in the table is reduced roughly by a factor of 3.

If instead of N_groups=3, we set N_groups=9, we would reduce memory load still further, but at the expense of 9 passes through Dw.Runs for each value of Inning. With our sample data volume it is hardly needed but might be worthwhile under other circumstances. Using the uniform input split technique is always a tradeoff between the hash memory usage and extra I/O.

11.6.3 Aggregation via Key Byte Split

As already noted above, the byte values of the composite key component variable Game_SK are distributed nearly randomly due to the nature of the MD5 hash function used to create it. Thus, we can base a UKS on any byte (or bytes) of this variable. To make use of it, we need not change anything in the preceding program except for some parameterization. However, before we do it, it needs to be preceded by a few preliminary technical notes.

When the MOD function is used for value-splitting, as we have done earlier, it takes a numeric argument. In our case above, the entire first argument is &UKS_base=Inning, which is perfectly okay since the variable is numeric. But if we base the splitting on a certain byte (or bytes) of the process key, they cannot be plugged directly into the MOD function since they are character strings. However, as any character string is uniquely associated with an integer numeric equivalent, all we need to do beforehand is obtain this equivalent numeric value and pass it to the MOD function.

Suppose that we want to base our UKS on the *first* byte of Game_SK. (Due to the nature of Game_SK, any of its bytes can be used but, as we will see shortly, using the leading byte or bytes is just simpler.) The numeric equivalent of a byte, i.e., a single character, is represented by its *rank*, which is its position in the *collating sequence* – a collection of all 256 standard characters ranked from 0 to 255. Unsurprisingly, this value is returned by the SAS function RANK. So, to get the numeric equivalent of the first byte of Game_SK, we can use the expression:

```
%let UKS_base = rank (char (Game_SK, 1)) ;
```

and then plug it in as &UKS_base directly into the MOD function as its first argument. Since the RANK function returns the rank of the first (i.e., leftmost) character of any string longer than $1, the expression can be coded even more simply as:

```
%let UKS_base = rank (Game_SK) ;
```

Using such a UKS base gives us 256 rank values to work with. If our input is large enough and the composite key-values used to make up Game_SK are distributed more or less evenly, each of the 256 distinct values returned by the UKS_base expression rank will be linked to a more or less equal number of different distinct composite key-values. Grouping the ranks into N_groups much fewer than 256 (e.g., 4, 8, etc.) is likely to lead to even more even distribution of the composite key-values among the groups.

Though the RANK function, as shown above, will surely work, a better and more general way is to use standard SAS numeric informats – the tools specifically designed to convert character values into their numeric equivalents in a variety of ways.

To see what kind of advantage the informats offer, suppose that we want to base our UKS on the numeric equivalent of the two leading bytes of Game_SK rather than one (aiming, for example, at a more even UKS values distribution). It is certainly possible to get what we need using the RANK function again:

```
%let UKS_base = rank (Game_SK)+ 256 * rank (char (Game_SK, 2)) ;
```

The expression looks awkward enough, even though only 2 first bytes are involved; and with 3 bytes and more it would get increasingly long, unwieldy, and slower to execute. However, there is the PIB*w* informat in the SAS arsenal designed exactly for the purpose:

```
%let UKS_base = input (Game_SK, PIB2.) ;
```

Resulting in the same value as the RANK formula above, the informat does the job in a much more straightforward manner and much faster to boot. It achieves two goals at once:

1. Extracts the needed number of leading Game_SK bytes.
2. Produces their integer equivalent in the range from 0 to 256**2-1=65,535.

If we need to involve 3 leading bytes in the process, we can simply widen the informat to PIB3, and so on. For each extra PIB*w* byte involved, we get a 256 more wider range of the integer equivalent values, as for a given informat width *w*, the range is 256**w:

Table 11.1 Chapter 11 PIB*w* Informat Value Ranges

Informat	Maximum Number of Values
PIB1.	256
PIB2.	65,536
PIB3.	16,777,216
PIB4.	4,294,967,296
PIB5.	1,099,511,627,776
PIB6.	281,474,976,710,656

Caution: Going wider than PIB6 is *not* recommended, as it can result in a value too large to be stored accurately. As a matter of aiming at a more even distribution of key-values among UKS ranges, even going as wide as PIB5 is quite unnecessary, as the evenness practically plateaus after *w*=3.

Therefore, if we decide to base our UKS on the *w* leading bytes of Game_SK, there is no reason to use anything but the PIB*w* informat, even if *w*=1:

```
%let UKS_base = input (Game_SK, PIB1.) ;
```

On the other end of the spectrum, suppose that instead of basing the UKS on a *whole byte* with its 256 distinct numeric equivalent values, we would like to limit this range by involving only a part of the byte in the process. For example, if we base the UKS on the *first half of the byte*, we will have 16 numeric equivalent values to work with. This can be easily done with the combination of the $HEX1 and HEX1 format and informat, like so:

```
%let UKS_base = input (put (Game_SK, $HEX1.), HEX1.) ;
```

On the other hand, it is even simpler to do in one fell swoop by using the BITS*w* informat:

```
%let UKS_base = input (Game_SK, BITS4.) ;
```

In fact, by using the BITS*w* informat, we can involve any number of leading *bits* (from 1 to 8) of the first *byte* of Game_SK in the process. The resulting maximum numbers of distinct numeric equivalent values for the BITS*w* informat is $2^{**}w$, so we have:

Table 11.2 Chapter 11 BITS*w* Informat Value Ranges

Informat	Maximum Number of Values	Notes
BITS1.	2	
BITS2.	4	
BITS3.	8	
BITS4.	16	Same as HEX1
BITS5.	32	
BITS6.	64	
BITS7.	128	
BITS8.	256	Same as PIB1 or HEX2

A benefit of using *fewer than 8 leading bits* is that we may be able to segment the UKS values into a small number of groups directly without resorting to the overhead of the MOD function. For example, BITS2 generates only 4 possible equivalent numeric values. So, if we wanted only a 4-way split, the expression:

```
input (Game_SK, bits2.) + 1
```

would result in 4 possible values from 1 to 4 that naturally segment the input into 4 groups. So, this expression can be used in the 4 WHERE clauses directly as the split group identifier.

Despite such convenience of using a *shorter pa*rt of the process key for uniform splitting, the *longer the part* of the key involved, the more even the distribution of distinct key-values among the N_groups segments is likely to be. For example, if we segment data set Dw.Games into 4 groups based on the first 2, 4, 8, and 16 bits (i.e., 1/4 byte, 1/2 byte, 1 byte, and 2 bytes) using the above methods and count how many distinct values of *the whole* key Game_SK end up in each group, we will get the following picture:

Table 11.3 Chapter 11 Effect of BITS*w* Width on Game_SK Value Distribution

Group	BITS2_Count	BITS4_Count	BITS8_Count	BITS16_Count
1	693	744	729	718
2	735	716	720	734
3	710	710	706	712
4	742	710	725	716

As we see, the more bits are involved in the process, the more uniform the distribution becomes. Thus, from the practical standpoint we can just base our UKS on one or two leading bytes of the MD5 signature, and, even better, this is the simplest thing to do.

Suppose, therefore, that now we want to:

- Base our UKS on the *first* byte of Game_SK, rather than whole variable Inning.
- Split the input into 4 record groups, aiming to have approximately equal number of distinct composite key-values in each.

Correspondingly, we can merely re-parameterize UKS_base and N_groups in Program 11.19 (Chapter 11 Partial Key Split by Inning Aggregation.sas) and leave everything else intact to arrive at the following program:

Program 11.20 Chapter 11 Key Byte Split by Game_SK Aggregation.sas

```
%let file_dsn  = Dw.Runs ;
%let comp_keys = Game_SK Inning Top_Bot ;
*%let UKS_base = Inning ;   /* Program 11.19 */
*%let N_groups = 3 ;   /* Program 11.19 */
%let UKS_base  = input (Game_SK, pib1.) ;
%let N_groups  = 4 ;
%let UKS_group = mod (&UKS_base,&N_groups) + 1 ;

%macro UKS() ;
  %do Group = 1 %to &N_groups ;
    do LR = 0 by 0 until (LR) ;
      set &file_dsn (where=(&UKS_group=&Group))
          end = LR ;
      link SCORE ;
    end ;
    link OUT ;
  %end ;
%mEnd ;

data Scores_Game_SK_KeyByte_split (keep = &comp_keys Score) ;
  if _n_ = 1 then do ;
    dcl hash h (ordered:"A") ;
    do _k = 1 to countW ("&comp_keys") ;
      h.defineKey  (scan ("&comp_keys", _k)) ;
      h.defineData (scan ("&comp_keys", _k)) ;
    end ;
    h.defineData ("Score") ;
    h.defineDone () ;
    dcl hiter ih ("h") ;
  end ;
  %UKS()
  return ;
  SCORE: if h.find() ne 0 then Score = 1 ;
         else Score + 1 ;
         h.replace() ;
  return ;
  OUT:   do while (ih.next() = 0) ;
           output ;
         end ;
       * Num_items = h.num_items ;
       * put Num_items= ;
         h.clear() ;
  return ;
run ;
```

and leave the rest of the program completely intact. Running it with this new parameterization generates the following notes in the SAS log:

```
NOTE: There were 19527 observations read from the data set DW.RUNS.
      WHERE (MOD(((INPUT(Game_SK, PIB1.)+4)-1),4)+1)=1;
NOTE: There were 19307 observations read from the data set DW.RUNS.
      WHERE (MOD(((INPUT(Game_SK, PIB1.)+4)-1),4)+1)=2;
```

```
NOTE: There were 19544 observations read from the data set DW.RUNS.
      WHERE (MOD(((INPUT(Game_SK, PIB1.)+4)-1),4)+1)=3;
NOTE: There were 19968 observations read from the data set DW.RUNS.
      WHERE (MOD(((INPUT(Game_SK, PIB1.)+4)-1),4)+1)=4;
NOTE: The data set WORK.SCORES_GAME_INNING_SPLIT has
      28790 observations and 4 variables.
```

If we uncomment the two statements before the CLEAR method call and rerun the program, the SAS log notes generated by the PUT statement will show for each successive pass through Dw.Runs:

```
Num_items=7222
Num_items=7083
Num_items=7207
Num_items=7278
```

With the frontal attack approach, the topmost number of items in the hash table would equal the sum of the above numbers. As we see, using 4 value ranges of just the first Game_SK byte as a USK cuts the maximum number of items loaded into the hash table almost exactly by the factor of 4.

By varying the value of parameter N_groups and the width of the PIB*w* informat in parameter UKS_base, the number of groups into which the input is split can be selected practically arbitrarily. When the PIB1 informat is used, only the first Game_SK byte is involved, so the maximum number of split groups is 256. Using PIB*2* informat instead would allow up to a 256**2=65536 way split, and so on.

As a practical matter of hash memory to I/O tradeoff, increasing the number of splits (i.e., the value of N_groups parameter) beyond 8 or 16 may be called for only under pretty extreme circumstances. When the N_groups value is left at a moderate level (4 or 8, say), using a wider PIB*w* informat to involve more leading bytes in the process *may* lead to a more even distribution among the split groups; yet going wider than PIB2 or PIB3 will hardly make any palpable difference. Also, going above PIB6 should be *avoided*, as an integer as high as 256**7 may fail to be stored accurately due to rounding.

Note that though, with this particular approach, it is more natural to set N_groups to a power of 2 (i.e., 2, 4, 8, 16, etc.) because each byte can take on 256 distinct values, nothing precludes using a different number of groups, such as 3, 5, 6, 7, etc. The only consequence of doing so may be a *slightly* less uniform distribution of the distinct key-values among the groups.

11.6.4 Joining via Key Byte Split

If we need to join two files by a common key whose intrinsic properties allow it to be used as a good UKS candidate on both sides of the join, we can use the principles delineated above to execute the join. The only difference from data aggregation is that, in this case, we need to:

- Use a different hash object based routine.
- Split *both* inputs in the same exact manner into the same number of key-value groups.

We will illustrate it below by reusing the task of joining Dw.Runs to Dw.AtBats described earlier in the chapter and using the frontal attack Program 11.6 (Chapter 11 Frontal Attack Join.sas) as a foundation. Note that in *both* files, the variables Inning and Game_SK possess the properties needed to use either one as a UKS basis in the same fashion as it was done for split data aggregation above. The plan of attack is simple:

1. Create a WHERE clause UKS_group expression to take on the values from 1 to N_groups.
2. Read file Dw.AtBats into the hash table filtering its input for split group #1.
3. Read file Dw.Runs filtering its input for split group #1.
4. Join the groups, output the results, and clear the hash table.
5. Repeat steps 2 through 4 for every remaining split group.

Here, we will skip the rather trivial case of splitting the inputs by the straight values of Inning and concentrate on using the Game_SK variable created using the MD5 function as a UKS basis. Below, we use the *first 2 bytes* of Game_SK to split each file into 3 groups. However, the program can be re-parameterized as desired (a feature our IT users have appreciated).

Program 11.21 Chapter 11 Key Byte Split by Game_SK Join.sas

```
%let comp_keys = Game_SK Inning Top_Bot AB_Number ;
%let data_vars = Batter_ID Position_code Is_A_Hit Result ;
%let data_list = %sysfunc (tranwrd (&data_vars, %str( ), %str(,))) ;
*%let UKS_base  = Inning ;
%let UKS_base  = input (Game_SK, pib2.) ;
%let N_groups  = 3 ;
%let UKS_group = mod (&UKS_base,&N_groups) + 1 ;

%macro UKS() ;
  %do Group = 1 %to &N_groups ;
    do LR = 0 by 0 until (LR) ;
      set dw.AtBats (keep=&comp_keys &data_vars Runs
                    where=(&UKS_group=&Group and Runs))
          end=LR ;
      h.add() ;
    end ;
    do LR = 0 by 0 until (LR) ;
      set dw.Runs (where=(&UKS_group=&Group)) end=LR ;
      call missing (&data_list, _count) ;
      do while (h.do_over() = 0) ;
        _count = sum (_count, 1) ;
        output ;
      end ;
      if not _count then output ;
    end ;
    h.clear() ;
  %end ;
%mEnd ;

data Join_Game_SK_split (drop = _: Runs) ;
  dcl hash h (multidata:"Y", ordered:"A") ;
  do _k = 1 to countw ("&comp_keys") ;
    h.defineKey (scan ("&comp_keys", _k)) ;
  end ;
  do _k = 1 to countw ("&data_vars") ;
    h.defineData (scan ("&data_vars", _k)) ;
  end ;
  h.defineDone() ;
```

```
  %UKS()
  stop ;
run ;
```

Running the program results in the same output data as the frontal attack program, save for the output record order. The SAS log is telling how many records have been read into each split segment using the WHERE clauses:

```
NOTE: There were 19401 observations read from the data set DW.ATBATS.
      WHERE ((MOD(INPUT(Game_SK, PIB2.), 3)+1)=1) and Runs;
NOTE: There were 26234 observations read from the data set DW.RUNS.
      WHERE (MOD(INPUT(Game_SK, PIB2.), 3)+1)=1;
NOTE: There were 19498 observations read from the data set DW.ATBATS.
      WHERE ((MOD(INPUT(Game_SK, PIB2.), 3)+1)=2) and Runs;
NOTE: There were 26421 observations read from the data set DW.RUNS.
      WHERE (MOD(INPUT(Game_SK, PIB2.), 3)+1)=2;
NOTE: There were 18880 observations read from the data set DW.ATBATS.
      WHERE ((MOD(INPUT(Game_SK, PIB2.), 3)+1)=3) and Runs;
NOTE: There were 25691 observations read from the data set DW.RUNS.
      WHERE (MOD(INPUT(Game_SK, PIB2.), 3)+1)=3;
```

As we see, on both the Dw.AtBats and Dw.Runs sides the split is quite even with little variance. Effectively, it means that the hash table, cleared after processing each split, consumes about 1/3 of the memory required by the frontal attack program. Of course, it comes at the price of reading the data three times; but it may be a small price to pay if the frontal attack should fail due to insufficient memory.

11.7 Uniform MD5 Split On the Fly

Situations where the key used in a hash object-based data processing *happens* to have a component good enough to play the role of UKS are rather fortuitous. More often than not, the keys we have to deal with are neither sorted the way we would like them to be, nor are they known to possess any properties in terms of serving as a good UKS foundation. At the same time, the key-values of long and complex composite keys used in industrial applications are typically distributed more or less evenly throughout the data collection.

Suppose that the data volume, the nature of the task, or other factors dictate that the input must be processed in split groups with about equal numbers of distinct keys. The question then is whether there exists a method to identify the data records for such segregation *a priori* without reading the data first or analyzing the distribution of different key components beforehand. Fortunately, the answer to the above question is "yes", and again it is the MD5 function that comes to rescue.

To see how it can work, let us turn again to the sample data sets Dw.Runs and Dw.AtBats and *pretend* that we have no prior knowledge about any of the following:

- The distribution of values of any component that can be used in the process key, such as Game_SK, Inning, Top_Bot, or AB_Number.
- That the data sets are pre-sorted by any variables.
- The special way used to create variable Game_SK and the distribution of its bytes.

In other words, suppose that we know nothing special about our keys, including Game_SK – it is just another key. Of course, no special a priori knowledge is needed to process the data using a frontal hash object attack and let the memory chips fall where they may. Yet let us recall that our goal in this chapter is learning how to reduce the hash memory footprint with an eye toward real-world situations where a hash table may grow so large that it may overwhelm the available memory resources.

We already know how beneficial in terms of input splitting a variable created in the image of Game_SK can be; yet now, according to our pretense above, we cannot base our UKS on it. However, we can use the same concept to *create* a variable similar to Game_SK *on the fly*. Let us see how it can work for data aggregation and joining.

11.7.1 MD5 Split Aggregation On the Fly

Recall how we approached the data aggregation task earlier in the section "Aggregation via Key Byte Split" by using one or more bytes of Game_SK as a UKS foundation. Since we have now pretended that Game_SK is nothing but an ordinary key and cannot be used for the purpose, the solution is to use the MD5 function *on the fly* to create another 16-byte *temporary* variable (let us call it _MD5). Furthermore, in order to make the distribution of key-values between the split groups as uniform as possible, we are going to involve *all the components* of the composite key into the concatenated argument before passing it to the MD5 function.

Implementing this approach is logically and functionally similar to the scheme used in Program 11.19 Chapter 11 Partial Key Split by Inning Aggregation.sas. The principal difference is that instead of using an already existing input variable created via MD5, we are going to create it on the fly using the input key components. In the program below, it is done in a separate data view vRuns that reads Dw.Runs and creates the _MD5 key. The reference to vRuns, rather than Dw.Runs, is included in macro UKS. It makes _MD5 available to the DATA step calling macro UKS to do the actual aggregation.

The program below is derived from Program 11.20 Chapter 11 Key Byte Split by Game_SK Aggregation.sas. The differences are shown in bold.

Program 11.22 Chapter 11 MD5 Split Aggregation On the Fly.sas

```
%let comp_keys = Game_SK Inning Top_Bot ;
%let keys_list = %sysfunc (tranwrd (&comp_keys, %str( ), %str(,))) ;  ❶
%let UKS_base  = input (_MD5, pib2.) ;
%let N_groups  = 4 ;
%let UKS_group = mod (&UKS_base,&N_groups) + 1 ;

data vRuns / view = vRuns ;     ❷
  set dw.Runs ;
  length _concat $ 32 _MD5 $ 16 ;
  _concat = catx (":", &keys_list) ;
  _MD5 = md5 (_concat) ;
run ;

%macro UKS() ;
  %do Group = 1 %to &N_groups ;
    do LR = 0 by 0 until (LR) ;
      set vRuns (where=(&UKS_group=&Group)) end = LR ;     ❸
      link SCORE ;
    end ;
    link OUT ;
  %end ;
%mEnd ;
```

```
data Scores_MD5_OnTheFly_Split (keep = &comp_keys Score) ;        ❹
  if _n_ = 1 then do ;
    dcl hash h (ordered:"A") ;
    do _k = 1 to countW ("&comp_keys") ;
      h.defineKey  (scan ("&comp_keys", _k)) ;
      h.defineData (scan ("&comp_keys", _k)) ;
    end ;
    h.defineData ("Score") ;
    h.defineDone () ;
    dcl hiter ih ("h") ;
  end ;
  %UKS()
  return ;
  SCORE: if h.find() ne 0 then Score = 1 ;
         else Score + 1 ;
         h.replace() ;
  return ;
  OUT:   do while (ih.next() = 0) ;
           output ;
         end ;
       * Num_items = h.num_items ;
       * put Num_items= ;
         h.clear() ;
  return ;
run ;
```

❶   Create a comma-separated list of key variables keys_list to use in the CATX function.

❷   Create a DATA step view vRuns as input to the DATA step doing the split aggregation. This is necessary because the variable _MD5 created in it is not in Dw.Runs and so could not be used in the WHERE clause if Dw.Runs were read in directly. It is this view that creates _MD5.

❸   Macro UKS is the same as before, except that it assembles code to read view vRuns instead of data set Dw.Runs.

❹   Same as in Program 11.19 (Chapter 11 Partial Key Split by Inning Aggregation.sas).

When the program is run, macro UKS generates the following code for the 2-way split (turn the system option MPRINT on to see it in the SAS log):

```
do LR = 0 by 0 until (LR) ;
  set vRuns (where=(mod (input (_MD5, pib2.),4) + 1=1)) end = LR ;
  link SCORE ;
end ;
link OUT ;
do LR = 0 by 0 until (LR) ;
  set vRuns (where=(mod (input (_MD5, pib2.),4) + 1=2)) end = LR ;
  link SCORE ;
end ;
link OUT ;
do LR = 0 by 0 until (LR) ;
  set vRuns (where=(mod (input (_MD5, pib2.),4) + 1=3)) end = LR ;
  link SCORE ;
end ;
link OUT ;
do LR = 0 by 0 until (LR) ;
```

```
   set vRuns (where=(mod (input (_MD5, pib2.),4) + 1=4)) end = LR ;
     link SCORE ;
   end ;
   link OUT ;
```

When the assembled code is then executed in the rest of the DATA step, it generates the following log notes:

```
NOTE: There were 78346 observations read from the data set DW.RUNS.
NOTE: There were 19774 observations read from the data set WORK.VRUNS.
      WHERE (MOD(INPUT(_MD5, PIB2.), 4)+1)=1;
NOTE: There were 19351 observations read from the data set WORK.VRUNS.
      WHERE (MOD(INPUT(_MD5, PIB2.), 4)+1)=2;
NOTE: There were 19490 observations read from the data set WORK.VRUNS.
      WHERE (MOD(INPUT(_MD5, PIB2.), 4)+1)=3;
NOTE: There were 19731 observations read from the data set WORK.VRUNS.
      WHERE (MOD(INPUT(_MD5, PIB2.), 4)+1)=4;
NOTE: The data set WORK.SCORES_MD5_ONTHEFLY_SPLIT has 28790
      observations and 4 variables.
```

The PUT statement before the CLEAR method call prints the numbers of unique composite key-values in the hash table at the end of each split:

```
Num_items=7254
Num_items=7157
Num_items=7231
Num_items=7148
```

Evidently, a UKS based on the _MD5 variable splits both the input records and the numbers of unique key-values in the hash table almost evenly among the four segments.

11.7.2 MD5 Split Join On the Fly

Precisely the same principle applies to joining data, except that now we have two input files instead of one and so must have two views rather than one. Program 11.21 Chapter 11 Key Byte Split by Game_SK Join.sas that uses Game_SK for split joining provides a template for using the _MD5 key created on the fly instead. We only need to (a) add two views per Dw.AtBats and Dw.Runs, respectively, and (b) recode macro UKS to reference the views instead of the underlying data sets. The parameterization and the DATA step calling macro UKS and doing the actual joining remain. The code blocks where the alterations are made are shown below in bold.

Program 11.23 Chapter 11 MD5 Split Join On the Fly.sas

```
%let comp_keys = Game_SK Inning Top_Bot AB_Number ;
%let data_vars = Batter_ID Position_code Is_A_Hit Result ;
%let keys_list = %sysfunc (tranwrd (&comp_keys, %str( ), %str(,))) ;
%let data_list = %sysfunc (tranwrd (&data_vars, %str( ), %str(,))) ;
%let UKS_base  = input (_MD5, pib2.) ;
%let N_groups  = 3 ;
%let UKS_group = mod (&UKS_base,&N_groups) + 1 ;

data vRuns / view = vRuns ;      ❶
  set dw.Runs ;
  length _concat $ 32 _MD5 $ 16 ;
```

```
  _concat = catx (":", &keys_list) ;
  _MD5 = md5 (_concat) ;
run ;

data vAtBats / view = vatBats ;    ❷
  set dw.AtBats ;
  length _concat $ 32 _MD5 $ 16 ;
  _concat = catx (":", &keys_list) ;
  _MD5 = md5 (_concat) ;
run ;

%macro UKS() ;
  %do Group = 1 %to &N_groups ;
    do LR = 0 by 0 until (LR) ;
        set vAtBats (keep=&comp_keys &data_vars Runs _MD5  ❸
                     where=(&UKS_group=&Group and Runs))
            end=LR ;
      h.add() ;
    end ;
    do LR = 0 by 0 until (LR) ;
        set vRuns (where=(&UKS_group=&Group)) end=LR ;    ❹
        call missing (&data_list, _count) ;
        do while (h.do_over() = 0) ;
          _count = sum (_count, 1) ;
          output ;
        end ;
        if not _count then output ;
    end ;
    h.clear() ;
  %end ;
%mEnd ;

data Join_MD5_OnTheFly_Split (drop = _: Runs) ;    ❺
  dcl hash h (multidata:"Y", ordered:"A") ;
  do _k = 1 to countw ("&comp_keys") ;
    h.defineKey (scan ("&comp_keys", _k)) ;
  end ;
  do _k = 1 to countw ("&data_vars") ;
    h.defineData (scan ("&data_vars", _k)) ;
  end ;
  h.defineDone() ;
  %UKS()
  stop ;
run ;
```

❶   Create a view vRuns with Dw.Runs as input and size and populate the UKS variable _MD5.

❷   Create a view vAtBats with Dw.AtBats as input and size and populate the UKS variable _MD5. in *exactly* the same way as for view vRuns.

❸   In macro UKS, use view vAtBats instead of data set AtBats and keep variable _MD5.

❹   In macro UKS, use view vRuns instead of data set Runs.

❺   Use the same DATA step code as in Program 11.21 Chapter 11 Key Byte Split by Game_SK Join.sas.

Running this program shows that it generates the same data as the frontal attack join program, except for the different output record order because of the split input.

The ability to join files via table hash table lookup more quickly and efficiently than the SORT-MERGE or SQL was the hash object's first claim to fame. Its only major drawback is greater memory usage. The techniques demonstrated in this section (and this chapter in general) help alleviate the problem by reasonable memory and I/O tradeoffs. The input splitting methods offer the programmer a wide range of options for choosing the number of splits N_groups versus the available memory resources, data volume, and the task at hand.

Note that the MD5-based input splitting techniques described here are rather *universal*. They can be used not only when the hash object is used for data processing but also with other aggregation or joining methodologies anytime the data volume becomes problematic enough to force the programmer to think of a divide-and-conquer approach.

11.8 Uniform Split Using a SAS Index

In all input-splitting examples illustrated above, we have surrendered to the stream-of-the-consciousness apparently dictated by the nature of the method. To wit, if we need to split the input N-way to reduce the hash memory footprint (hopefully, N times), then we must generate code reading the input N times for N ranges of the UKS variable. Above, it was done by using the macro language. But if we step back a little, we may recall that, in SAS, the primary method of working with separate groups of file records is BY-group processing. In fact, we have already used it earlier to shorten the hash entry length by taking advantage of an existing order. By doing so we, in effect, were processing the input in separate groups.

In this section, though, we are dealing with situations where the input is neither pre-sorted nor pre-grouped. Let us recall, however, that if a data set is *indexed* by one or more variables, those variables can be used as BY variables, too. So, if the input had a variable created according to our UKS group-splitting expression and it were indexed, it could lend itself to input split processing in a much more *natural* way than generating SAS code for each separate split group.

The problem is, of course, that we do not have such a variable in the data file being processed, and we need to create it if we want to use a SAS index in this manner. In turn, it means that we have to create a copy of our input with the UKS variable added to it and indexed. It definitely means extra I/O and disk storage. However, under certain circumstances it may be well worth the price. It is particularly true if a data collection created with an eye toward a specific type of processing (e.g., data aggregation) has a deliberately created and indexed UKS variable.

As an example, let us recall the last data aggregation example where file Dw.Runs is aggregated by splitting the input using the leading bytes of the newly created variable _MD5 (Program 11.22 Chapter 11 MD5 Split Aggregation On the Fly.sas). Imagine that a file Dw.Runs_UKS has been created as a copy of Dw.Runs, with a UKS variable added and indexed as follows:

Program 11.24 Chapter 11 MD5 Split SAS Index Aggregation.sas (Part 1)

```
%let N_groups = 256 ;

data Dw.Runs_UKS (index=(UKS) drop = _:) ;
  set Dw.Runs ;
  length _concat $ 32 _MD5 $ 16 ;
  _concat = catx (":", Game_SK, Inning, Top_Bot) ;
```

```
  _MD5 = md5 (_concat) ;
  UKS = 1 + mod (input (_MD5, pib4.), &N_groups) ;
run ;
```

If we had this kind of indexed file, we would not need to assemble a SAS program for split processing using a macro or any other means. Instead, we could simply make use of BY group processing relying on the fact that the input file is now indexed by the UKS variable:

Program 11.24 Chapter 11 MD5 Split SAS Index Aggregation.sas (Part 2)

```
%let comp_keys = Game_SK Inning Top_Bot ;

data Scores_split_index (keep = &comp_keys Score) ;
  if _n_ = 1 then do ;
    dcl hash h() ;
    do _k = 1 to countW ("&comp_keys") ;
      h.defineKey  (scan ("&comp_keys", _k)) ;
      h.defineData (scan ("&comp_keys", _k)) ;
    end ;
    h.defineData("Score") ;
    h.defineDone() ;
    dcl hiter ih("h") ;
  end ;
  do until (last.UKS) ;
    set Dw.Runs_UKS ;
    by UKS ;
    if h.find() ne 0 then Score = 1 ;
    else Score + 1 ;
    h.replace() ;
  end ;
  do while (ih.next() = 0) ;
    output ;
  end ;
  h.clear() ;
run ;
```

The principal difference between this program and Program 11.22 (Chapter 11 MD5 Split Aggregation On the Fly.sas) is that here we no longer need to assemble code to re-read the file 256 times for each value of variable UKS. Instead, the SAS index does the job by automatically selecting a file subset with the needed value of UKS (in order ascending by UKS) for each BY group.

This is why we can afford not to worry about the number of re-reads (and increasing the number of the corresponding file buffers that also use their fair share of memory) and select the number of split groups as large as we want. Note that the value for N_groups=256 and the width for the PIB4. informat above were picked rather arbitrarily following the "big enough" principle. A rather large number of rather small split groups is of advantage in this case since: (a) it further decreases the hash memory footprint and (b) if desired, the values of UKS could be combined in a number of ranges.

One *disadvantage* of this methodology is the need to create a copy of the input file, add the UKS variable, and index the file, all of which consume system resources. However, on the bright side:

- An index on a single numeric uniformly distributed key is not too costly and performs well.
- There is no need to assemble code by using a macro or by any other means.

- The input-splitting program is simpler and easier to maintain.
- In a data warehouse environment (which is what our IT users plan to implement), a UKS variable can be created and indexed during the load time.

This technique lends itself equally well to any other type of input-splitting data processing described in this chapter – provided, of course that a file (or files) with a proper indexed UKS variable have been pre-created. In fact, if our proposed data organization is accepted by the Bizarro Ball users, we will ask them to include a specified indexed UKS variable in the data warehouse tables.

To illustrate the programmatic advantages of such an arrangement, suppose that in addition to file Dw.Runs_UKS, another file has been created from Dw.AtBats as follows (picking N_groups=16 this time around just for the sake of diversity):

Program 11.25 Chapter 11 MD5 Split SAS Index Join.sas (Part 1)

```
%let N_groups = 16 ;

data Dw.Runs_UKS (index=(UKS) keep=UKS Game_SK Inning Top_Bot AB_Number
                                Runner_ID)
     Dw.AtBats_UKS (index=(UKS) drop = _: Runner_ID) ;
  set Dw.Runs (in=R) Dw.AtBats ;
  length _concat $ 32 _MD5 $ 16 ;
  _concat = catx (":", Game_SK, Inning, Top_Bot) ;
  _MD5 = md5 (_concat) ;
  UKS = 1 + mod (input (_MD5, pib4.), &N_groups) ;
  if R then output Dw.Runs_UKS ;
  else      output Dw.AtBats_UKS ;
run ;
```

Then we could execute a N_groups-way input-split join simply by repurposing the joining code from Program 11.7 Chapter 11 Pregrouped Join.sas. Essentially, the only difference is using the UKS variable UKS as a BY variable instead of the pre-sorted variables (Game_SK,Inning):

Program 11.25 Chapter 11 MD5 Split SAS Index Join.sas (Part 2)

```
%let comp_keys = Game_SK Inning Top_Bot AB_Number ;
%let data_vars = Batter_ID Position_code Is_A_Hit Result ;
%let data_list = %sysfunc (tranwrd (&data_vars, %str( ), %str(,))) ;

data Join_Runs_AtBats_MD5_SAS_Index (drop = _: Runs) ;
  if _n_ = 1 then do ;
    dcl hash h   (multidata:"Y", ordered:"A") ;
    do _k = 1 to countw ("&comp_keys") ;
      h.defineKey (scan ("&comp_keys", _k)) ;
    end ;
    do _k = 1 to countw ("&data_vars") ;
      h.defineData (scan ("&data_vars", _k)) ;
    end ;
    h.defineDone() ;
  end ;
  do until (last.UKS) ;
    set dw.AtBats_UKS (in=A keep = UKS &comp_keys &data_vars Runs
                       where=(Runs))
        dw.Runs_UKS   (in=R keep = UKS &comp_keys Runner_ID) ;
    by UKS ;
    if A then h.add() ;
```

```
    if not R then continue ;
    call missing (&data_list, _count) ;
    do while (h.do_over() = 0) ;
      _count = sum (_count, 1) ;
      output ;
    end ;
  if not _count then output ;
  end ;
  h.clear() ;
run ;
```

Caveat: Above (shown in ***bold italics***), the UKS variable is based on (Game_SK,Inning,Top_Bot), while the processing is done by (Game_SK,Inning,Top_Bot,**AB_Number**). Note that the latter contains one more key component. It works perfectly fine – as it would if UKS were based on *all four* variables above.

However, it will ***not*** work the other way around, i.e., if UKS were based on (Game_SK,Inning,Top_Bot,AB_Number) but the processing were done by Game_SK, Inning, Top_Bot, or any of their combinations thereof. This is because the extra variable AB_Number would cause the composite key-values of (Game_SK,Inning,Top_Bot) to overlap between the split groups the UKS variable identifies.

The takeaway is that the UKS variable must be based on *the same or fewer* process key components versus those used as the hash table composite key. Of course, if the same composite key components that are used in the hash table are also used to create the UKS variable, the processing results will be correct.

11.9 Combining Hash Memory-Saving Techniques

As we have noted earlier, the hash memory-saving techniques discussed in this chapter are by no means exclusive to each other. In fact, *all of them* can be combined in one and the same program if the circumstances dictate the need to go to such lengths. Doing so can be particularly useful in a situation when the hash object must be used as the data processing tool because, due to the input data volume and the nature of the task (a) other SAS tools are not efficient enough to deliver the necessary output on time, and (b) system memory resources limit the size to which a hash table can grow. A typical situation of this sort arises in various industries when massive data needs to be pre-aggregated for online processing by tens of millions of key-value levels of numerous categorical variables.

Below, the concept of combining memory-saving techniques will be illustrated by the example of joining file Dw.Runs to Dw.AtBats using three approaches at once:

1. Exploit the pre-existing key order by key variable Game_SK and move it from the hash table to the BY statement.
2. Shorten the key portion length via the MD5 function by replacing the remaining key variables Inning, Top_Bot, and AB_Number with their MD5 function signature.
3. Offload the hash table data portion that would otherwise carry the data variables from Dw.AtBats by using the hash index keyed by the observation number from file Dw.AtBats.

Program 11.26 Chapter 11 Combined Techniques Join.sas

```
%let comp_keys = Game_SK Inning Top_Bot AB_Number ;          ❶
%let sort_keys = Game_SK ;     ❷
%let data_vars = Batter_ID Position_code Is_A_Hit Result ;     ❸
%let tail_keys = %sysfunc (tranwrd (&comp_keys, &sort_keys, %str())) ; ❹
%let last_key  = %sysfunc (scan (&sort_keys, -1)) ;   ❺
%let tail_list = %sysfunc (tranwrd (&tail_keys, %str( ), %str(,))) ;  ❻
%let data_list = %sysfunc (tranwrd (&data_vars, %str( ), %str(,))) ;  ❼

data Join_Runs_AtBats_combine (drop = _: Runs) ;
  if _n_ = 1 then do ;
    dcl hash h (multidata:"Y", ordered:"A") ;
    h.defineKey  ("_MD5") ;     ❽
    h.defineData ("RID") ;      ❾
    h.defineDone() ;
  end ;
  do until (last.&last_key) ;     ❿
    set dw.AtBats (in=A keep=&comp_keys Runs)
        dw.Runs   (in=R keep=&comp_keys Runner_ID) ;
    by &sort_keys ;
    if A = 1 then do ;     ⓫
      _RID + 1 ;     ⓬
      if not Runs then continue ;     ⓭
    end ;
    length _concat $ 37 _MD5 $ 16 ;     ⓮
    _concat = catx (":", &tail_list) ;
    _MD5 = md5 (_concat) ;
    if A = 1 then h.add(key:_MD5, data:_RID) ;     ⓯
    if R = 0 then continue ;     ⓰
    do while (h.do_over() = 0) ;     ⓱
      _count = sum (_count, 1) ;
      set dw.AtBats (keep=&data_vars) point=RID ;     ⓲
      output ;
    end ;
    if not _count then output ;     ⓳
    call missing (&data_list, _count) ;     ⓴
  end ;
  h.clear() ;     ㉑
run ;
```

❶ The common composite key by which the data sets are joined.

❷ The leading keys by which both files are pre-sorted.

❸ The variables from Dw.AtBats with which we need to enrich Dw.Runs.

❹ The list of the composite key components minus the leading keys by which the files are pre-sorted.

❺ The trailing component in the pre-sorted key list needed for the LAST.X expression in the DoW loop.

❻ Same as tail_keys list but *comma-separated* to be used as an argument to the CATX function.

❼ Same as data_vars list but comma-separated to be used in the CALL MISSING routine.

❽ The key portion of table H is keyed by a single _MD5 variable *instead* of the keys in the comp_keys list.

❾ The data portion of H is defined with a record pointer RID to Dw.AtBats *instead* of the data_vars variables.

❿ Force the DoW loop reading *interleaved* files Dw.AtBats and Dw.Runs to iterate through each BY group with the distinct key-value of sort_keys. Note that the data variables in the data_vars list are *not* kept because we are going to extract them using the POINT= option later when we need them.

⓫ A=1 indicates that the current record comes from file Dw.AtBats.

⓬ If so, increment temporary variable _RID *for every single record* coming from Dw.AtBats. This way, its values stored in the hash table will point *exactly* to the Dw.AtBats records we will eventually need to extract.

⓭ If variable Runs from Dw.AtBats has the value Runs=0, we no longer need this record, so read the next one. Note that the CONTINUE statement is executed after *every* observation number in Dw.AtBats is assigned to _RID correctly. For the same reason, this is why Dw.AtBats is filtered for Runs>0 in this manner instead of using the WHERE clause (as we have done before). If the WHERE clause were used, _RID would take on the observation numbers of the *filtered subset* of Dw.AtBats and later on point to the wrong records in the *unfiltered* file.

⓮ Size and create key variable _MD5 for (a) every composite key in the comp_keys list coming from the Dw.AtBats records with Runs>0 and (b) every such composite key for *every* record from Dw.Runs.

⓯ For every Dw.AtBats record where Runs>0, add its observation number _RID to table H along with the key _MD5.

⓰ If R=0, the BY group has no records from Dw.Runs. Hence, no joining for this BY group is necessary, and so, proceed to the next record.

⓱ Otherwise, for each record in the BY group coming from Dw.Runs, execute the standard *Keynumerate* lookup-retrieve routine. If the _MD5 value for the current Dw.Runs record is in table H (populated in the same BY group from the Dw.AtBats records), retrieve the RID values from all hash items with this value of _MD5 into the PDV one at a time in the DO WHILE loop.

⓲ For each RID value retrieved from the table, use it with the POINT=RID option to extract the values of variables in the data_vars list.

⓲ This IF statement is needed if a *left* join (rather than *inner*) is required. Because of the program logic, variable _count can have a non-missing value only if the DO WHILE loop has iterated *at least once*. The latter can happen only if the current PDV value of _MD5 from file Dw.Runs has at least one item with such a key-value in the hash table. Otherwise, the DO_OVER method call will return a non-zero code and terminate the loop before the statement with the SUM function is executed. Therefore, the condition not _count signals that there is no match. If we need a left join, we still need to output a record for the current record from Dw.Runs with all satellite Dw.AtBats variables in the data_vars list missing. If all we need is an inner join, the IF statement in question should be removed, commented out, or made dependent on a parameter.

⓴ Calling the MISSING routine here ensures that no non-missing PDV values created by the *Keynumerate* operation persist when the next record is processed (thus helping abide by the left join rules). Also, it initializes the variable _count to a missing value before the next record is read.

❹ At this point, the file joining process for the current BY group is complete, and so the hash table items inserted using the Dw.AtBat records in this BY group are no longer needed. So, they can (and should) be deleted before the next BY group is read in.

This program combining hash memory-saving approaches cuts the hash memory footprint about 10 times compared to the frontal attack program.

Other memory-conserving techniques described in this chapter but not used in the above example can be added to the mix since, as noted earlier, they are completely independent. As this example shows, it may require a degree of ingenuity, understanding of each technique, and attention to programming details.

11.10 MD5 Argument Concatenation Ins and Outs

In this chapter, as well as the rest of the book, we have made extensive use of the MD5 function that accepts a string with concatenated key components as an argument. Typically, the routine looks as follows (the key components passed to the CATX function may vary):

```
length _concat $ 37 _MD5 $ 16 ;
_concat = catx (":", Inning, Top_Bot, AB_Number) ;
_MD5 = md5 (_concat) ;
```

The *signature* returned by the function, _MD5, has been used in the book for a number of purposes, such as:

- Creating surrogate key Game_SK for our sample data collection to facilitate better linkage between its data sets, simplify coding, and conserve storage.
- Reducing hash object memory usage by shortening the key portion length.
- Creating uniform key splitter (UKS) variables to facilitate uniform input splitting.

In order for any of the above to work correctly, the routine used to obtain a signature for the process key has to satisfy two fundamental requirements:

1. It must ensure a fixed one-to-one relationship between the key and its signature. That is, the same key-value must always get the same signature; and different key-values must get different signatures.
2. It should be reasonably fast, for a routine taking too long to compute would compromise one of the rationales of using the hash object in the first place.

In order to satisfy the requirement #1, two conditions must be met:

- First, for any two distinct values of _concat, MD5 must generate two distinct 16-byte signature strings.
- Second, for any two distinct values of the process key, the concatenation routine must generate two distinct values of variable _concat.

In the subsections below, these two aspects will be addressed separately.

11.10.1 MD5 Collisions and SHA256

The MD5 function is specifically designed to generate a distinct signature for any distinct value of its argument. However, you may have heard that this MD5 property is not bulletproof. In other words there is a *small chance* that two distinct MD5 arguments may generate the same response, i.e., result in what is termed a *collision*.

This is actually true; and it worries some folks (particularly those who believe in the inviolability of Murphy's law) so much that they shy away from using the MD5 function altogether. Let us assess, though, how well those fears are grounded by putting the meaning of "small chance" into practical data processing perspective.

In the worst case scenario, the approximate number of items that need to be hashed to get a 50 percent chance of an MD5 collision is about $2{**}64 \approx 2E+19$. It means that to encounter just 1 collision, the MD5 function has to be executed against 100 quintillion distinct arguments the equal number of times, i.e., approximately 1 trillion times per second for 100 years. The probability of such an event is so infinitesimally negligible that one truly has an *enormously* greater chance of living through a baseball season where every single pitch is a strike and no batter ever gets on base. (Amusingly, some people who will confidently say that can *never, ever* happen may believe that an MD5 collision *can* happen.)

Those worriers not sufficiently convinced by this logical excursion can eschew MD5 and use the SAS function SHA256 (available since the advent of SAS 9.4) instead. SHA256 generates a *32-byte* (rather than 16-byte) signature and is deemed fully bulletproof with zero chance of collisions. Its usage exactly mirrors that of MD5; in other words, instead of MD5, one would code:

```
length _concat $ 37 _SHA $ 32 ;
_concat = catx (":", Inning, Top_Bot, AB_Number) ;
_SHA = sha256 (_concat) ;
```

So, why not do away with MD5 altogether and use SHA256 instead at all times? The answer actually dovetails with the requirement #2 listed above: The SHA256 function is *much* slower. Program file Chapter 11 MD5 vs SHA256 Test.sas includes a program for a simple programming experiment where both functions are executed 1 million times against the same non-blank arguments with different lengths. Running this experiment under the X64_7PRO platform reveals the following picture:

Output 11.17 Chapter 11 MD5 vs SHA256 Execution Times

Argument_Length	SHA_Time	MD5_Time	Speed_Ratio
2	62.37	1.81	34
4	62.14	1.77	35
8	62.27	1.81	34
16	63.18	1.75	36
32	62.02	1.79	35
64	65.97	3.33	20
128	69.08	4.76	15
256	75.40	7.80	10
512	89.95	13.93	6
1024	116.65	26.00	4

As we see, in the best case scenario, SHA256 is 4 times slower compared to MD5; and for the argument lengths we are most likely to deal with, it is 10 to 36 times slower. If this is the price one is willing to pay for the reluctance to take a 1 in 100 quintillion chance, one can always use the SHA256 function no matter what. We told our IT users that they could re-enact the experiment above, since its code, though not shown here, is included in the program file Chapter 11 MD5 vs SHA256 Test.sas.

Now that we know that the uniqueness of the one-way hash function response can be ascertained (at the extra expense of using SHA256 if desired), let us turn to various aspects of the other potential source of breaking the one-to-one relationship between the process key and _MD5 (or _SHA), i.e., to the concatenation routine.

11.10.2 Concatenation Length Sizing

The CATX function is used throughout the book to form the MD5 argument for a number of reasons:

- It is simpler to code than an equivalent expression using the || concatenation operator.
- It provides for using a concatenation separator (as its first argument).
- It automatically strips the leading and trailing blanks from the values of its arguments.
- It automatically converts numeric variable values to digit strings using the BESTw. format.

However, using the CATX function is not without caveats. To begin with, it is sensitive to the combined lengths of its arguments. If we simply code:

```
_concat = catx (":", comp1, comp2, ...,compN) ;
```

without sizing _concat variable beforehand, the compiler will set its system length to $200 since, by default, CATX (and the rest of the CAT family functions) allocates 200 bytes for its buffer. There are two problems with it:

1. First, suppose that the combined length of the key-components plus (N-1) bytes to account for the separators exceeded 200. Then, the function would return the same _concat value for all composite key-values whose concatenations do not differ in the first (200-N+1) positions, even if they differ in any position after that. Obviously, this would map different process key-values to the same _concat value and, by extension, to the same MD5 (or SHA) signature. In other words, two different key-values would be *scrambled* to the same signature value. Hence, the one-to-one relationship requirement we seek to enforce would not be met.

2. Second, even if the default 200 bytes of the CATX buffer are sufficient to cover the entire concatenation length (including the separators), it may end up hampering performance if the actual concatenation length is much shorter. This is because the time it takes the CATX function to execute is almost directly proportional to its buffer size.

A CATX buffer size different from the default length of 200 bytes can be allocated in two ways. Using the composite key (Inning,Top_Bot,AB_Number), as in the example above, it can be coded either:

```
length _concat $ 37 ;
_concat = catx (":", Inning, Top_Bot, AB_Number) ;
```

or:

```
_concat = put (catx (":", Inning, Top_Bot, AB_Number)
           , $37.) ;
```

In either case, the compiler sets the length of the CATX buffer to 37 bytes. If you wonder why the length of 37 is chosen in this case, there is method in the madness. Both Inning and AB_Number variables are numeric, so the maximum length of an integer they can accurately store under *any* operating system is 17 digits. The length of Top_Bot is $1. Adding two bytes for the colons makes a total of 37.

Therefore, it is imperative to allocate an *appropriate* buffer size for CATX instead of relying on the default value of 200. In case one might be tempted to use the "big enough" approach and allocate the buffer size at the longest possible character length of $32767 to cover all the bases and simplify coding, it would be *really* ill-advised. This is because it takes the CATX function 100 times [sic!] longer to execute with the buffer size of 32767 than with the buffer size of 200 (which again dovetails with the #2 "reasonably fast" requirement stated in the beginning of the section).

If the key component variables to be concatenated come from a specific SAS data set (as happens most often), we can use the dictionary tables to determine the exact CATX buffer size we need instead of relying on the default, calculating it by hand, or going the "big enough" route.

Our IT users have asked us to provide some code to do this auto-sizing. We have agreed to do that later. You can access the blog entry that describes this program from the author page at http://support.sas.com/authors. Select either "Paul Dorfman" or "Don Henderson." Then look for the cover thumbnail of this book, and select "Blog Entries."

11.10.4 Concatenation Delimiters and Endpoints

By taking care of the concatenation length properly, we eliminate just one, pretty obvious, reason why two distinct composite key-values may scramble into the same value of _concat. However, there exists another, less overt, possibility related to the choice of a character used to delimit the concatenated values.

The reason to use a concatenation delimiter (such as a colon ":" in our examples) in the first place is to prevent two different adjacent key components from being concatenated into the same string. For example, consider a two-component composite key (*K_num, K_char*) with two *different composite* key-values:

```
(K_num,K_char) = ( 1234, "5ABCD")
(K_num,K_char) = (12345, "ABCD" )
```

Since the composite values are *different*, they should result in *different* values after the concatenation. However, if we blindly bunch them together (for example, by using the CATS function), the result would be the same:

```
catS ( 1234, "5ABCD") = "12345ABCD"
catS (12345, "ABCD" ) = "12345ABCD"
```

That is, the difference between the two composite key-values would be *scrambled* since they concatenate to the same value of _concat. This is what we aim to prevent from occurring by using a separator character. For example, if we put a dash between the concatenated values, it will resolve this *particular* problem because now the concatenated values will differ:

```
catX ("-", 1234, "5ABCD") = "1234-5ABCD"
catX ("-",12345, "ABCD" ) = "12345-ABCD"
```

Unfortunately, it does not solve the scrambling problem *in general*. Consider, for example, another (*K_char,K_num*) pair with two *different* composite key-values:

```
(K_char,K_num) = ("ABCDE", -12345)
(K_char,K_num) = ("ABCDE-", 12345)
```

Now if the CATX function were used with the same delimiter, it would result in two *identical* concatenated values:

```
catX ("-", "ABCDE" ,-12345) = "ABCDE--12345"
catX ("-", "ABCDE-", 12345) = "ABCDE--12345"
```

Evidently, we can run into this problem with any two adjacent components (*Left* and *Right,* say) if either the *right*most character of *Left* or the *left*most character of *Right* happens to be the same as the delimiter itself. In other words, the delimiter *conflates* with the end points of the components it is intended to discriminate.

Note that though in the examples throughout this book the colon character is used as a separator, it is safe vis-à-vis our sample data since *we know* that none of our key-values contains a colon. It is very likely for other people dealing with their own data (they know better than anyone else) to encounter the same kind of situation and thus be able to judiciously select a delimiting character never present in their key-values. For instance, if an ETL cleansing process guarantees that no key ever contains low-values or high-values ("00"x or "FF"x in hex), these characters can be safely used as concatenation delimiters. In all such cases, the discourse offered below is largely of academic interest.

Yet, there are also many people (for example, consultants) whose business is to deal with *other people*'s data. And in the real data processing world, any kind of data can occur, including key components that may begin and end with *any* character available in the collating sequence. As this is the only character set from which the delimiter can be selected, no choice of it alone can prevent a potential delimiter-endpoint conflation. For the same reason, replacing an endpoint character with some other character – in hope that it will be different from the separator – will not work, either.

Fortunately, even though we have no control over the input key-values and their endpoints, there is still a solution to the problem. Before we get to it, let us make one general observation about the nature of the routine in question. Its goal is not to create an image of _concat looking exactly like the key components bunched up into one string. Rather, it is to create a unique character value of _concat for passing it to MD5 (or SHA256) for each unique value of the process key by *transforming* the latter in some manner. How the values of _concat look as a result does not matter as long as they are related to the process key-values one-to-one. Therefore, before the concatenation is done, any key component can be changed (i.e., transformed) in any manner so long as its values are uniquely (i.e., one-to-one) related to the result of the change. For example, we can do any of the following (note that when the CATX function is used, a number of these things are done automatically behind the scenes):

- Precede and/or follow a string component with any non-blank character or characters.
- Add a constant to a numeric component, multiply it by a constant, etc., if its integer precision is not compromised.
- Apply a format to any string or numeric component if the format maps its argument to its response as one-to-one. Formats of such kind include (but are not limited to) $HEXw for string components and RB8 or HEX16 for numeric components.

- Apply a function, so long as its argument and response have a one-to-one relationship. Notably, applying the MD5 function to long string components *before concatenation* can be used to reduce the CATX buffer length.

Back to the conflation problem at hand, its root cause is the possibility that some component's endpoint character may be the same as the delimiter. Hence, if we change (i.e., transform) the component in such a way that its endpoint bytes *always differ* from the delimiting character, we have a solution.

In fact, it is easy to achieve if we *surround* the key component by two extra *non-blank* characters *different* from the delimiter. Among themselves, these two characters can be different or the same: It does not matter as long as they are different from the delimiter. However, if the concatenated value is going to be viewed, it may be convenient to opt for a pair of symmetrical characters, such as parentheses or brackets, as shown below.

Returning to our simple example above, suppose that we keep the dash as the delimiter but add characters *different* from it (for example, a pair of brackets) to the beginning and the end of each component (after converting *K_num* to a string):

```
catX ("-", "[ABCDE]", "[-12345]") = "[ABCDE]-[-12345]"
catX ("-", "[ABCDE-]", "[12345]") = "[ABCDE-]-[12345]"
```

The reason this subterfuge works is that we have *forced* the *newly created* endpoint bytes to be *never* the same as the delimiting character. It renders the *original* endpoint values (i.e., A, E, -, 1, 5 above) incapable of causing a conflation with the delimiter because they no longer represent the endpoints. Also note that surrounding a component with new non-blank endpoints preserves the one-to-one relationship between the value of the component and the augmented value that goes into concatenation.

The augmentation can be done in a number of ways. If the key components are not too numerous, it can be simply hard-coded according to their data types, for instance:

```
drop _comp_ : ;
_comp_Inning   = "("||put(Inning,RB8.)||")" ;
_comp_Top_Bot  = "("||Top_Bot||")" ;
_comp_AB_Number = "("||put(AB_Number,RB8.)||")" ;
length _concat $ 19 ;
_concat = catx (":", of _comp_:) ;
```

Of course, the expressions for the _comp_: variables can be plugged in directly as arguments of the CATX function (or as operands of the concatenation operator if we decide to go that route). Alternatively, this kind of code can be generated programmatically using the dictionary tables and some method of assembling code on the fly (such as the SQL procedure, the macro facility, the CALL EXECUTE routine, etc.).

11.10.5 Auto-Formatting and Explicit Formatting

When the CATX function is used to concatenate the key components, it automatically formats them before they are concatenated according to its own internal rules for each data type (i.e., character or numeric). Let us discuss the different data types separately.

Character components are always formatted using the $CHAR*w*. format (same as $*w*.). For each particular component, the width *w* is set to the width of its actual character value. If a component is a character *variable* (rather than a character literal, such as "A1B2C3") and carries with it a different format (e.g., $HEX*w*., $OCTAL*w*., etc.), its format is *disregarded* and $CHAR*w*. is used anyway.

Such a default behavior is logical since, in terms of the number of bytes, the $CHARw. format results in the shortest possible formatted value and hence the shortest concatenation buffer size. Also, it presents no problems in terms of the one-to-one correspondence between the character component and its formatted value.

However, if there is a need to apply a different character format (for which there are conceivable use cases), it has to be done explicitly. For example, if one wanted to format Top_Bot as $HEX2., one would have to code:

```
_concat = catx (":", Inning
                   , put(Top_Bot,$hex2.)
                   , AB_Number) ;
```

and not forget to size _concat longer accordingly beforehand.

Numeric components are formatted by default using the BEST32.-L format (the modifier -L means that, after the conversion, the formatted value is left-aligned). In terms of the concatenation buffer length, the formatted value length simply depends on the number of formatted digits. With long integers, such as 10-15 digits long, it may not be optimal, as other numeric formats result in shorter strings, but it is not a big problem, either.

A meaningful problem may arise if the stored value is *fractional*. For instance, if we pass a numeric variable holding with the stored values of 1234567890.*67* and 1234567890.*68* to any function of the CAT family, we will get a formatted value "1234567890.7" in response in *both* cases. While the BEST32.2 format would preserve all the digits, this is not what the CATX function uses internally using BEST32. instead. Of course, one may object that the practice of using fractional numeric values as key components is questionable. Yet it does occur in the real world, and so the possibility must be accounted for as well.

The solution is to use a format whose formatted value always represents the numeric variable value being formatted *exactly*. The shortest (and fastest) SAS format among those that *guarantee* this to happen is RB8. (The format HEX16. will do the job, too, but its formatted value is twice as long, and it is not as fast.) So, if we used RB8. to convert *any* numeric variable to an 8-byte string and then pass the latter to the CATX function, it would totally ensure that no too-different numeric variable values could be ever scrambled to the same formatted value. It would be very convenient if a specific numeric-to-character conversion format could be specified with the CATX function. As no such provision exists, the conversion has to be done explicitly.

In terms of our concatenation routine, the RB8. format is much more preferable to any other numeric format (especially the default BEST*w.*) for a variety of reasons, such as:

- RB8. maps the value of a numeric variable to an 8-byte character string *exactly* as it is stored, never losing precision regardless of whether the value is integer or fractional.
- Thus, it maintains a strictly one-to-one relationship between the numeric value and its formatted 8-byte string.
- The value formatted by RB8. never contains endpoint blanks, so there is no need to strip them.
- Of all numeric formats, RB8. is the fastest because this is the way SAS numbers are stored in memory. So, when the format is called, no manipulations are involved: The 8-byte RB8. numeric value string image is copied directly from the physical memory address of the numeric variable in question. As a result, RB8. formats numeric values a whopping 6 times faster than BEST*w.*

If CATX is used, its default formatting behavior can be changed by formatting numeric components explicitly. For example:

```
_concat = catx (":", put(Inning,RB8.)
                   , Top_Bot
                   , put(AB_Number,RB8.)) ;
```

Or, if the concatenation operator ‖ is used, we would code:

```
_concat = put(Inning,RB8.)
            ||":"
            ||Top_Bot
            ||put(AB_Number,RB8.) ;
```

If the concatenation operator method is used, we have to format the numeric components explicitly using *some* format, for otherwise:

- The BEST12. format will be used by default, which can lead to scrambling meaningful digits.
- The log will be peppered with pesky numeric-to-character conversion notes.

But if we need to use a format anyway, it is better to select one doing its job more safely and faster than BEST*w*., and RB8. is the best tool for the job. In fact, its superiority is another good reason to concatenate via the ‖ operator (in addition to the automatic length sizing this approach offers).

However, there is another format worth considering in the specific situation when the number being formatted is an integer less than $256**4=4,294,967,296$. In this case, the PIB4. format can be used to write the number as a 4-byte formatted value, thus saving 4 bytes of the buffer length for every component it formats, while its formatting speed is practically the same as that of RB8.

11.10.6 Concatenation Order and Consistency

Simply put, from the standpoint of our goal of mapping composite key-values to their _concat counterparts in a one-to-one manner, the order in which the components are concatenated *does not matter*. Obviously, different concatenation orders produce different _concat values, but this is of no concern so long as the key-values and the values of the resulting _concat are related uniquely.

Therefore, if it is deemed more expedient or convenient to concatenate all character components first and numeric components – second (or the other way around), so be it. We have seen an example of doing just that in the program included in the program file Chapter 11 Full Unduplication.sas (Program 11.12) where arrays are used to segregate the numeric and character components into two separate collections and feed them to the CATX function in that order.

A final note concerning concatenation *consistency* is due, lest the above be misunderstood in the sense that one concatenation order can be used to populate _concat in one record and a different order – in another. It needs to be *emphasized* that throughout the entire data processing task, not only the concatenation order but also the *entire routine* of creating an MD5 (or SHA256) signature *must be exactly consistent* for every data element (a file record, a hash item, etc.) the signature is supposed to identify.

For example, if two files are joined by keying a hash table with _MD5 variable (e.g., as in Program 11.13, Chapter 11 MD5 Key Reduction Join.sas), the routine of valuing _MD5 must be exactly the same for every record on both sides being joined. The concatenation order in the routine itself can be a matter of choice, but once it is selected, it must be persisted throughout the task. The same pertains to every other element of the routine.

11.11 Summary

This chapter provides a deep dive into a number of alternative methods to deal with sizing and performance issues. It is targeted to our IT users to provide a foundation for alternative approaches they could investigate to address sizing and performance issues. The IT users have expressed their appreciation for this detailed discussion, adding that it would take them some time to fully digest it.

Part Four—Wrapping up: Two Case Studies

It is always nice to finish a project on a positive note, which is the point of Part Four. This part concludes with several samples in the hope and expectation that the headquarters office of Bizarro Ball decides to use SAS, and specifically, the SAS hash object as a key component of their data management and reporting requirements.

Part Four contains two chapters that represent two different use cases—addressing ad-hoc questions and research needs.

1. Chapter 12 builds on previous examples that calculated metrics to highlight what can be done with the hash object given that it has the rich set of DATA step programming facilities. This case study provides an example of researching and comparing alternative metrics.
2. Chapter 13 provides an example of how existing programs could be repackaged and re-used to address the typical ad-hoc questions that come up in virtually every project.

A secondary goal of Part Four is to remind the users of our Proof of Concept that it is just that - a proof of concept. In other words, SAS has many tools that can be leveraged in order to build a series of tools that can be packaged as needed to address a given question.

Chapter 12: Researching Alternative Pitching Metrics

12.1 Overview

Our business users and most baseball analysts agree that quantifying the performance of pitchers is much harder than for batters, and our users want to investigate a number of alternatives.

Earned Run Average (ERA) has long been a default metric to evaluate pitchers despite its limitations. For starting pitchers, if they are replaced and the relief pitcher gives up a hit that scores any of the runners who are already on base, the starting pitcher is charged for those runs.

Likewise, if a relief pitcher allows "inherited runners" (i.e., those runners already on base when they entered the game) their ERA does not reflect those runs.

We reminded the users that our mock data could be used to demonstrate alternative metrics but cautioned them to not make any inferences from the results since our mock data did not allow for such scenarios because pitchers can be replaced only at the beginning of an inning. Likewise, we did not distinguish between runs that are earned vs. unearned.

A popular alternative metric is Walks plus Hits per Inning Pitched (abbreviated as WHIP). Calculating WHIP was one of our examples in section 9.5.2 "Multiple Split Calculations."

The challenge for this case study is to create a data set that has a number of other metrics that can be analyzed in a number of ways (for example, perhaps correlation analysis). The initial set of requested alternative metrics they asked for include:

- For the WHIP calculation, if a pitcher hits a batter with a pitch that does not count against their WHIP. Since hitting a batter with a pitch produces results similar to a walk, they would like to see what impact that has on the WHIP calculation.
- One other criticism of the WHIP metric is that, for example, a home run counts the same as a single. They want to investigate using the number of bases allowed (e.g., 4 for a Home Run and 1 for a single) instead of the number of hits. This would be similar to the difference between how a batter's Batting Average (BA) is calculated vs. his Slugging Average (SLG).

We agreed to use one of the sample *splits* programs from Chapter 9 and modify it to create desired metrics.

12.2 The Sample Program

This use case example required just a few modifications to Program 9.8 Chapter 9 HoH Multiple Splits Batter and Pitcher.sas. The key lines of code we had to add to create the program below are noted in bold and annotated.

We also pointed out to the business and IT users that once these programs were modularized as part of the long-term project, it would be easier to use pieces and parts to address ad-hoc and research questions like this one.

Program 12.1 - Chapter 12 Pitcher Metrics What-If.sas

```
data chapter9splits;
 set template.chapter9splits;      ❶
 by hashTable;
 output;
 if last.hashTable;
 Column = "IP";
 output;
 Column = "ERA";
 output;
 Column = "WHIP";
 output;
 Column = "_Runs";
 output;
 Column = "_Outs";
 output;
 Column = "_Walks";
 output;
 Column = "_HBP";    ❷
 output;
 Column = "WHIP_HBP";
 output;
 Column = "BASES_IP";
 output;
run;

data _null_;
 dcl hash HoH(ordered:"A");
 HoH.defineKey("hashTable");
 HoH.defineData("hashTable","H","calcAndOutput");    ❸
 HoH.defineDone();
 dcl hiter HoH_Iter("HoH");
 dcl hash h();
 dcl hiter iter;
 /* define the lookup hash object tables */
 do while(lr=0);
    set template.chapter9lookuptables
        chapter9splits(in=CalcAndOutput)
    end=lr;
    by hashTable;
    if first.hashTable then
    do;   /* create the hash object instance */
       if datasetTag ne ' ' then h = _new_ hash(dataset:datasetTag
                                      ,multidata:"Y");
```

```
         else h = _new_ hash(multidata:"Y");
   end; /* create the hash object instance */
   if Is_A_key then h.DefineKey(Column);
   h.DefineData(Column);
   if last.hashTable then
   do;  /* close the definition and add it to our HoH hash table */
      h.defineDone();
      HoH.add();
   end; /* close the definition and add it to our HoH hash table */
end;
/* create non-scalar fields for the lookup tables */
HoH.find(key:"GAMES");
dcl hash games;
games = h;
HoH.find(key:"PLAYERS");
dcl hash players;
players = h;
HoH.find(key:"TEAMS");
dcl hash teams;
teams = h;
dcl hash pitchers(dataset:"dw.pitches(rename=(pitcher_id = Player_ID))");  ❹
pitchers.defineKey("game_sk","top_bot","ab_number");
pitchers.defineData("player_id");
pitchers.defineDone();

if 0 then set dw.players
            dw.teams
            dw.games
            dw.pitches;
format PAs AtBats Hits comma6. BA OBP SLG OPS 5.3
      IP comma6. ERA WHIP WHIP_HBP BASES_IP 6.3;   ❺

lr = 0;
do until(lr);
   set dw.AtBats end = lr;
   call missing(Team_SK,Last_Name,First_Name,Team_Name,Date,Month,DayOfWeek);
   games.find();
   pitchers.find();
   players_rc = players.find();   ❻
   do while(players_rc = 0);
      if (Start_Date le Date le End_Date) then leave;
      players_rc = players.find_next();
   end;
   if players_rc ne 0 then call missing(Team_SK,First_Name,Last_Name);
   teams.find();

   do while (HoH_Iter.next() = 0);
      if not calcAndOutput then continue;
      call missing(PAs,AtBats,Hits,_Bases,_Reached_Base
                  ,_Outs,_Runs,_Bases,_HBP);   ❼
      rc = h.find();
      PAs           + 1;
      AtBats        + Is_An_AB;
      Hits          + Is_A_Hit;
      _Bases        + Bases;
      _Reached_Base + Is_An_OnBase;
```

```
      _Outs          + Is_An_Out;
      _Runs          + Runs;
      _Walks         + (Result = "Walk");
      _HBP           + (Result = "Hit By Pitch");   ❽
      BA = divide(Hits,AtBats);
      OBP = divide(_Reached_Base,PAs);
      SLG = divide(_Bases,AtBats);
      OPS = sum(OBP,SLG);
      if _Outs then
      do;  /* calculate pitcher metrics suppressing missing value note */
         IP = _Outs/3;
         ERA = divide(_Runs*9,IP);
         WHIP = divide(sum(_Walks,Hits),IP);
         WHIP_HBP = divide(sum(_Walks,Hits,_HBP),IP);   ❾
         BASES_IP = divide(_Bases,IP);
      end; /* calculate pitcher metrics missing value note */
      h.replace();
   end;
 end;
 do while (HoH_Iter.next() = 0);
    if not calcAndOutput  then continue;
    h.output(dataset:hashTable||"(drop=_:)");
 end;
 stop;
run;
```

❶ We are calculating pitcher metrics only, so we don't need to read in the data set twice to create pitcher and batter hash tables.

❷ Three additional variables are needed in the data portion of our splits hash tables to calculate these alternative metrics.

❸ The data item corresponding to the hash iterator is no longer needed.

❹ A hash object that can be used to find and retrieve the Pitcher_ID for each at bat. In the examples in Chapter 9 an SQL step was used to remove the duplicates (i.e., multiple pitches per at bat). Given the challenge to use only hash object methodology, we simply need to omit the MULTIDATA argument tag in order to force only a single hash table item for each pitcher. Note also that we renamed Pitcher_ID to Player_ID in order to facilitate the lookup to get the pitcher name.

❺ Add our output metrics to the format statement.

❻ Use the FIND method to look up the pitcher's name and team. Since we need to look up only the pitcher name, we moved this lookup to before the loop that enumerates through the splits hash objects.

❼ The columns _Bases and _HBP were added to the CALL MISSING call routine.

❽ Count the number of at bats that resulted in an HBP event as well as the number of pitches that were thrown.

❾ Calculate our two additional metrics: WHIP_HBP which includes at bats for which the result was a Hit by Pitch: and BASES_IP which replaces the total number of hits with the total number of bases.

The program created the same 5 split tables plus the additional metrics WHIP including Hit by Pitch (WHIP_HBP) and Bases Allowed per Inning (BASES_IP) as was shown in Output 9.22.

Output 12.1 First 5 Rows of byPlayer Splits Output

Last Name	First Name	Player ID	PAs	AtBats	Hits	BA	OBP	SLG	OPS	IP	ERA	WHIP	WHIP HBP	BASES IP
Davis	Harold	51972	1,079	981	361	0.368	0.425	0.885	1.310	195	14.077	2.821	3.036	4.451
Scott	Joseph	28568	1,488	1,346	496	0.368	0.429	0.871	1.300	270	13.100	2.441	2.615	4.344
Alexander	Jason	48313	1,029	931	344	0.369	0.430	0.903	1.333	185	13.378	2.568	2.730	4.546
Reed	Terry	21799	999	922	352	0.382	0.429	0.910	1.339	180	14.350	2.433	2.572	4.661
Wilson	Christian	40826	973	895	313	0.350	0.402	0.811	1.213	180	11.850	2.450	2.594	4.033

12.2.1 Adding More Metrics

The users were surprised at how simple the changes were to make and wanted confirmation that more metrics could be added quickly. We pointed out that it depended on what the metrics were, adding that metrics that are calculated from values in the AtBats data could be that simple. We updated our program and saved it with a new name (Chapter 12 Pitcher Metrics What-If Updated.sas) to illustrate this.

The following lines of code were added to Program 12.1 that is highlighted in ❷.

```
Column = "BASES_PA";
output;
Column = "PAs_IP";
output;
Column = "_Pitches";
output;
Column = "PITCHES_IP";
output;
```

The variables BASES_PA, PAs_IP, and PITCHES_IP were added to the FORMAT statement in ❺.

The variable _Pitches was added to the CALL MISSING function in code described in ❼ and was calculated immediately after ❽.

These lines were added to the code described in ❾.

```
    PITCHES_IP = divide(_Pitches,IP);
    PAs_IP = divide(PAs,IP);
end; /* calculate pitcher metrics suppressing missing value note */
BASES_PA = divide(_Bases,PAs);
```

The modified program produced the same output as Program 12.1 Chapter 12 Pitcher Metrics What-If.sas plus the additional metrics Bases per Plate Appearance (BASES_PA), Plate Appearances per Inning Pitched (PAs_IP), and Pitches per Inning Pitched (Pitches_PA).

Output 12.2 First 5 Rows of byPlayer Splits Output

Last Name	First Name	Player ID	PAs	AtBats	Hits	BA	OBP	SLG	OPS	IP	ERA	WHIP	WHIP HBP
Davis	Harold	51972	1,079	981	361	0.368	0.425	0.885	1.310	195	14.077	2.821	3.036
Scott	Joseph	28568	1,488	1,346	496	0.368	0.429	0.871	1.300	270	13.100	2.441	2.615
Alexander	Jason	48313	1,029	931	344	0.369	0.430	0.903	1.333	185	13.378	2.568	2.730
Reed	Terry	21799	999	922	352	0.382	0.429	0.910	1.339	180	14.350	2.433	2.572
Wilson	Christian	40826	973	895	313	0.350	0.402	0.811	1.213	180	11.850	2.450	2.594

Last Name	First Name	Player ID	BASES IP	BASES PA	PAs IP	PITCHES IP
Davis	Harold	51972	4.451	0.804	5.533	16.559
Scott	Joseph	28568	4.344	0.788	5.511	16.963
Alexander	Jason	48313	4.546	0.817	5.562	17.173
Reed	Terry	21799	4.661	0.840	5.550	16.944
Wilson	Christian	40826	4.033	0.746	5.406	16.567

12.2.2 One Output Data Set with All the *Splits* Results

Our users then asked if it was possible to create a single output data set with the results from all of the splits. We agreed to do this but cautioned them to be careful when doing this since all the key variables and all the analysis variables would need to be included in a single output data set. The sample program below creates one output data set regardless of how many splits are being calculated and includes all the variables (both key variables and analysis variables) that are included in any of the splits. Variables that do not apply for a given split will have missing values.

The changes to the above program (Chapter 12 Pitcher Metrics What-If Updated.sas) are highlighted in bold.

Program 12.2 - Chapter 12 Pitcher Metrics What-If One Dataset.sas

```
data chapter9splits;    ❶
   .
   .
   .
run;

proc sql noprint;    ❷
 select distinct column into:keepList separated by ' '
 from chapter9splits
 /* where substr(column,1,1) ne '_' */;    ❸
 select distinct column into:missingList separated by ','
 from chapter9splits;
quit;

data combined;    ❹
 keep hashTable &keepList;
 dcl hash HoH(ordered:"A");
 HoH.defineKey ("hashTable");
```

```
HoH.defineData ("hashTable","H","ITER","calcAndOutput"); ❺
HoH.defineDone();
dcl hiter HoH_Iter("HoH");
dcl hash h();
dcl hiter iter;
/* define the lookup hash object tables */
do while(lr=0);
   set template.chapter9lookuptables
       chapter9splits(in=CalcAndOutput)
   end=lr;
   by hashTable;
   if first.hashTable then
   do;  /* create the hash object instance */
      if datasetTag ne ' ' then h = _new_ hash(dataset:datasetTag
                                          ,multidata:"Y");
      else h = _new_ hash(multidata:"Y");
   end; /* create the hash object instance */
   if Is_A_key then h.DefineKey(Column);
   h.DefineData(Column);
   if last.hashTable then
   do;  /* close the definition and add it to our HoH hash table */
      h.defineDone();
      iter = _new_ hiter("h");    ❻
      HoH.add();
   end; /* close the definition and add  it to our HoH hash table */
end;
/* create non-scalar fields for the lookup tables */
HoH.find(key:"GAMES");
dcl hash games;
games = h;
HoH.find(key:"PLAYERS");
dcl hash players;
players = h;
HoH.find(key:"TEAMS");
dcl hash teams;
teams = h;
dcl hash pitchers(dataset:"dw.pitches(rename=(pitcher_id = Player_ID))");
pitchers.defineKey("game_sk","top_bot","ab_number");
pitchers.defineData("player_id");
pitchers.defineDone();

if 0 then set dw.players
            dw.teams
            dw.games
            dw.pitches;
format PAs AtBats Hits comma6. BA OBP SLG OPS 5.3
       IP comma6. ERA WHIP WHIP_HBP BASES_IP BASES_PA PAs_IP PITCHES_IP 6.3;

lr = 0;
do until(lr);
   set dw.AtBats end = lr;
   call missing(Team_SK,Last_Name,First_Name,Team_Name,Date,Month,DayOfWeek);
   games.find();
   pitchers.find();
   players_rc = players.find();
   do while(players_rc = 0);
      if (Start_Date le Date le End_Date) then leave;
```

```
            players_rc = players.find_next();
        end;
        if players_rc ne 0 then call missing(Team_SK,First_Name,Last_Name);
        teams.find();

        do while (HoH_Iter.next() = 0);
            if not calcAndOutput then continue;
            call missing(PAs,AtBats,Hits,_Bases,_Reached_Base,_Outs,_Runs
                        ,_Bases,_HBP,_Pitches);
            rc = h.find();
            PAs             + 1;
            AtBats          + Is_An_AB;
            Hits            + Is_A_Hit;
            _Bases          + Bases;
            _Reached_Base   + Is_An_OnBase;
            _Outs           + Is_An_Out;
            _Runs           + Runs;
            _Walks          + (Result = "Walk");
            _HBP            + (Result = "Hit By Pitch");
            _Pitches        + Number_of_Pitches;
            BA = divide(Hits,AtBats);
            OBP = divide(_Reached_Base,PAs);
            SLG = divide(_Bases,AtBats);
            OPS = sum(OBP,SLG);
            if _Outs then
            do;  /* calculate pitcher metrics suppressing missing value note */
                IP = _Outs/3;
                ERA = divide(_Runs*9,IP);
                WHIP = divide(sum(_Walks,Hits),IP);
                WHIP_HBP = divide(sum(_Walks,Hits,_HBP),IP);
                BASES_IP = divide(_Bases,IP);
                PITCHES_IP = divide(_Pitches,IP);
                PAs_IP = divide(PAs,IP);
            end; /* calculate pitcher metrics suppressing missing value note */
            BASES_PA = divide(_Bases,PAs);
            h.replace();
        end;
    end;
    do while (HoH_Iter.next() = 0);
        if not calcAndOutput  then continue;
        call missing(&missingList);      ❼
        do while (iter.next() = 0);      ❽
            output;
        end;
    end;
    stop;
run;
```

❶ No changes to the creation of the parameter file.

❷ Create two macro variables to use: as the list of variables to include in the output data set (a blank-separated list); and the list of variables to be used in the CALL MISSING routine (a comma-separated list). Both lists include all the variables referenced in our *splits* definitions parameter files. We could have used hard-coded %LET macro statements for these two lists in order to be fully compliant with the user's request to not use other SAS facilities. They agreed that this was an appropriate exception to the *use only hash object facilities* restriction.

❸ If the accumulator variables are not desired in the output this line can be uncommented.

❹ Define the output data set containing the combined results in the DATA statement and specify the list of variables to be kept using the macro variable created in the previous SQL step.

❺ Include a non-scalar field of type iterator in the Hash of Hash that can be used to loop through the splits hash tables. This will be used to create the combined output data set.

❻ For each *splits* hash table create an instance of the iterator object.

❼ Use the MISSING call routine to set all the fields to null before iterating through each hash table. This ensures that no values will be carried down from the previous hash table.

❽ Instead of using the OUTPUT method which creates a separate data set for each hash table, use the iterator to enumerate all the data items in each hash table and output each item to the combined SAS data set.

The output below shows selected rows in our combined output object. The output is ordered alphabetically by the hashTable name (i.e., the *split* name). The CALL MISSING routine is what ensures that, for example, the value 6 for the variable DayOfWeek is not carried down to the observations for the BYMONTH *split* and the value of 8 is not carried down to the observations for the BYPLAYER *split*.

Output 12.3 Multiple Splits As a Single Output Data Set

Obs	hashTable	Player ID	Team SK	First Name	Last Name	Team Name	Month	Day Of Week	PAs	AtBats	Hits
2	BYDAYOFWEEK	.	.				.	1	48,298	43,790	16,348
3	BYDAYOFWEEK	.	.				.	3	47,880	43,526	16,121
4	BYDAYOFWEEK	.	.				.	7	47,717	43,338	15,895
5	BYDAYOFWEEK	.	.				.	4	47,854	43,493	16,079
6	BYDAYOFWEEK	.	.				.	6	47,711	43,245	15,815
7	BYMONTH	.	.				5	.	43,067	39,056	14,399
8	BYMONTH	.	.				9	.	41,754	37,829	14,074
9	BYMONTH	.	.				3	.	15,743	14,266	5,135
10	BYMONTH	.	.				7	.	42,827	38,883	14,143
11	BYMONTH	.	.				4	.	41,651	37,894	14,108
12	BYMONTH	.	.				10	.	20,934	18,965	7,118
13	BYMONTH	.	.				6	.	39,885	36,247	13,400
14	BYMONTH	.	.				8	.	41,443	37,664	13,925
15	BYPLAYER	51972	.	Harold	Davis		.	.	1,079	981	361

Obs	BA	OBP	SLG	OPS	IP	ERA	WHIP	WHIP HBP	BASES IP	BASES PA	PAs IP	PITCHES IP	Bases
2	0.373	0.432	0.885	1.317	8,640	14.010	2.248	2.416	4.486	0.802	5.590	17.016	38757
3	0.370	0.428	0.877	1.305	8,640	13.607	2.200	2.370	4.418	0.797	5.542	16.913	38171
4	0.367	0.425	0.873	1.298	8,640	13.458	2.184	2.348	4.380	0.793	5.523	16.879	37846
5	0.370	0.427	0.873	1.300	8,640	13.420	2.198	2.366	4.392	0.793	5.539	16.864	37949
6	0.366	0.425	0.871	1.296	8,640	13.508	2.193	2.358	4.359	0.789	5.522	16.811	37658
7	0.369	0.427	0.881	1.308	7,776	13.655	2.198	2.370	4.423	0.799	5.538	16.859	34393
8	0.372	0.431	0.883	1.314	7,488	13.867	2.239	2.407	4.461	0.800	5.576	17.020	33402
9	0.360	0.420	0.850	1.269	2,880	12.969	2.140	2.315	4.208	0.770	5.466	16.617	12119
10	0.364	0.422	0.867	1.290	7,776	13.361	2.162	2.331	4.337	0.787	5.508	16.761	33722
11	0.372	0.429	0.879	1.308	7,488	13.632	2.225	2.387	4.447	0.800	5.562	16.911	33302
12	0.375	0.434	0.885	1.319	3,744	13.974	2.277	2.438	4.481	0.801	5.591	17.097	16776
13	0.370	0.427	0.882	1.309	7,200	13.641	2.200	2.367	4.440	0.801	5.540	16.860	31967
14	0.370	0.427	0.872	1.299	7,488	13.520	2.196	2.364	4.387	0.793	5.535	16.960	32849
15	0.368	0.425	0.885	1.310	195	14.077	2.821	3.036	4.451	0.804	5.533	16.559	868

Obs	Reached Base	Outs	Runs	HBP	Pitches	Walks
2	20856	25920	13450	1451	147018	3076
3	20475	25920	13063	1468	146127	2886
4	20274	25920	12920	1414	145833	2976
5	20440	25920	12883	1459	145703	2908
6	20281	25920	12968	1423	145246	3131
7	18410	23328	11798	1341	131098	2692
8	17999	22464	11537	1256	127447	2695
9	6612	8640	4150	505	47858	1028
10	18087	23328	11544	1313	130331	2666
11	17865	22464	11342	1214	126627	2552
12	9087	11232	5813	604	64010	1406
13	17038	21600	10913	1197	121394	2443
14	17704	22464	11249	1258	126999	2521
15	459	585	305	42	3229	189

12.3 Summary

The point of this case study was to illustrate how metrics could be added to our sample programs.

Upon presenting these results, we informed the business and IT users that we were moving on to the next case study and could not entertain any more modifications to this one. They responded that while they had additional requests that it was clear that using the hash object in a DATA step offered a lot of flexibility. They could see the advantages of this approach given their interest in researching additional metrics.

We impressed upon the business and IT users that *stealing* snippets of code from other programs is not an approach that we would suggest as a long-term solution. The SAS macro language, among the many available tools provided by SAS software, provides a framework to build a set of re-usable tools that can be parameterized and mixed and matched as needed in order to address ad-hoc questions. That approach can also be combined with custom coding as needed.

Chapter 13: What If the Count Is 0-2 After the First Two Pitches

13.1 Overview

Pace of play (i.e., how long the games last) is becoming a major concern for Bizarro Ball. There is a perception that very long at bats make games last a lot longer. Some have suggested that the rule be changed that says a foul does not count as the third strike. Purists really object to such a change and have asked how frequent long at bats are - specifically for those where the first two pitches result in strikes. The debate among the business users focused on the breakdown of when such at bats resulted in a run or not as well as whether it was specific to teams/batters/runners that were able to do this (foul off lots of pitches). So they asked for a program that finds all the at bats where the first two pitches were strikes and had at least 10 pitches. They requested two reports:

- A listing of the relevant data fields for all the rows in the AtBats data set that resulted in a run.
- A listing of the relevant data fields for all the rows in the AtBats data set that did not result in a run being scored.

On the assumption that the subsetting was done using the SAS hash object and the listing could be produced by one of the SAS reporting procedures, we agreed to provide such an example.

We further decided to also produce a listing of the counts for how many times such at bats occurred. They did not ask for this output, but we expected that as soon as they saw the results they would ask for it. We had already demonstrated how to create frequency tables (i.e., a distribution) using hash objects and felt that anticipating such a request would buy us some brownie points.

13.2 The Sample Program

This program is made up of quite a few snippets of code that we "stole" from other programs – either a simple cut-and-paste or a simple modification of other code.

Program 13.1 - Chapter 13 Long At Bats with Initial 0-2 Count.sas

```
%let Number_of_Pitches = 10;     ❶
```

```
data _null_;
 if 0 then set dw.pitches(keep = Pitcher_ID)     ❷
                dw.players
                dw.games
                dw.teams
                dw.runs
 ;

 /* define the lookup hash object lookup tables */     ❸
 dcl hash HoH(ordered:"A");
 HoH.defineKey ("hashTable");
 HoH.defineData ("hashTable","H");
 HoH.defineDone();
 dcl hash h();
 do while(lr=0);
    set template.chapter9lookuptables end=lr;
    by hashTable;
    if first.hashTable then h = _new_ hash(dataset:datasetTag,multidata:"Y");
    if Is_A_Key then h.DefineKey(Column);
    h.DefineData(Column);
    if last.hashTable then
    do;  /* close the definition and add it to our HoH hash table */
       h.defineDone();
       rc=HoH.add();
    end; /* close the definition and add it to our HoH hash table */
 end;

 HoH.find(key:"GAMES");     ❹
 dcl hash games;
 games = h;
 HoH.find(key:"PLAYERS");
 dcl hash players;
 players = h;
 HoH.find(key:"TEAMS");
 dcl hash teams;
 teams = h;

 dcl hash Count_0_2(dataset:"DW.Pitches(where=(Strikes = 2 and Balls = 0))");     ❺
 Count_0_2.defineKey("game_sk","top_bot","ab_number");
 Count_0_2.defineData("Pitcher_ID");
 Count_0_2.defineDone();

 dcl hash results(multidata:"Y",ordered:"A");     ❻
 results.defineKey("Date");
 results.defineData("Date","Team_AtBat","Team_InField","AB_Number","Result"
                    ,"Number_of_Pitches","Runs","Batter","Pitcher","Runner");
 results.defineDone();

 dcl hash runners(dataset:"DW.Runs",multidata:"Y");     ❼
 runners.defineKey("Game_SK","Top_Bot","AB_Number");
 runners.defineData("Runner_ID");
 runners.defineDone();
```

```
dcl hash distribution(ORDERED:"A");      ❽
distribution.defineKey("Result");
distribution.defineData("Result","Count");
distribution.defineDone();

lr = 0;
do until(lr);
    set dw.atbats end = lr;
    where Number_of_Pitches >= &Number_of_Pitches;      ❾
    if Count_0_2.find() > 0 then continue;      ❿
    if distribution.find() gt 0 then Count = 0;      ⓫
    Count = Count + 1;
    distribution.replace();
    games.find();      ⓬
    players_rc = players.find(Key:Batter_ID);      ⓭
    link players;
    Batter = catx(', ',Last_Name,First_Name);      ⓮
    Team_AtBat = Team_Name;
    players_rc = players.find(Key:Pitcher_ID);      ⓯
    link players;
    Pitcher = catx(', ',Last_Name,First_Name);
    Team_InField = Team_Name;
    if runs then      ⓰
    do;  /* if runs scored - get runner data */
       rc = runners.find();      ⓱
       do while(rc=0);
           players_rc = players.find(Key:Runner_ID);      ⓲
           link players;
           Runner = catx(', ',Last_Name,First_Name);
           results.add();      ⓳
           rc = runners.find_next();
       end;
    end; /* if runs scored - get runner data */
    else
    do;  /* no runs scored */
       Runner = ' ';      ⓴
       results.add();
    end; /* no runs scored */
end;
results.output(dataset:"Runs_Scored(where=(Runs))");      ㉑
results.output(dataset:"No_Runs_Scored(where=(not Runs))");
distribution.output(dataset:"Distribution");      ㉒
stop;
return;
players:      ㉓
    do while(players_rc = 0);
        if (Start_Date le Date le End_Date) then leave;
        players_rc = players.find_next();
    end;
    if players_rc ne 0 then call missing(Team_SK,First_Name,Last_Name);
    teams.find();
return;
stop;
run;
```

❶ A macro variable is used to define the threshold for the number of pitches for our subset criteria. The rationale is simply to make it easy to update this value should there be a request to use a different value (which, of course, there will be).

❷ Use the IF 0 THEN SET construct to define the needed PDV host variables. We added a KEEP data set option to one of the data sets to illustrate that we need not define all of the variables. Such refinements would be good examples of work that can be done later.

❸ We copied and modified the code starting with this comment through the end of the DO WHILE – END block from the examples in Chapter 9 to define the same needed lookup tables using a Hash of Hash approach. The changes made were to: only keep the data set name in the SET statement that defines the lookup tables; and removing the logic used to determine what kind of hash tables were being creating (lookup tables vs. result tables).

❹ Just as for the *splits* example in Chapter 9, we need a non-scalar variable of type hash object for each lookup table. We use the FIND method to populate non-scalar variables whose values point to the instances of the lookup table hash objects for Games, Players, and Teams.

❺ Create a lookup table that has two functions. First, it identifies all the at bats that meet the criteria of the first two pitches being strikes (i.e., the count is 0 balls and 2 strikes after the second pitch of the at bat). Second, it provides the id value for the pitcher, which is one of the variables for the output report.

❻ Create a hash table for the results of all the at bats that meet our criteria. We do not need a unique key so we decided to just use the date of the game as the key so the data will be ordered by the date of the game (thanks to the ORDERED:"A"attribute value). And if multiple such events occur on the same date, they will be added to the hash table in order. The data portion contains all the variables for the output report.

❼ We define a hash table that contains all the rows for the runs scored. The variables that are the keys, Game_SK, Top_Bot, and AB_Number uniquely define each at bat in the entire season. The data portion is the Runner_ID, one of the variables requested in the output. The MULTIDATA argument tag is used because multiple runners can score in a single at bat (e.g., if the result of the at bat is a Home Run and there are runners on base).

❽ Create the hash table that will tabulate the distribution of how many rows in the AtBats data set meet our filter criteria. The result field is the key for the table. Since the MULTIDATA argument is not used, the hash table will contain one row for each value of the Result variable.

❾ Read our data in a DO WHILE loop and use a WHERE clause to include only those at bats that have a number of pitches that meet our threshold criteria.

❿ This statement performs two functions. First, if the FIND method returns a non-zero value, this at bat did not start with an 0-2 count. So we use the CONTINUE statement to exit the DO-END block and read the next observation. Second, if the FIND method finds a value it retrieves the id for the pitcher.

⓫ Update the distribution hash table with the cumulative count of the result for the AtBats row. If the FIND method returns a non-zero value, the value of Result is not found in the hash table so the PDV host variable Count is initialized to 0. If it is found, the PDV host variable is updated with the value of Count from the distribution hash table. The Count field is then incremented and the hash table is updated in order to update the cumulative distribution.

⓬ Use the FIND method to retrieve the data from the GAMES hash table and update the PDV host variables. Specifically, we want the Date column.

⑬   We now have host variables that have the Batter_ID and Pitcher_ID values. We need to look up their names and their teams. Since the PLAYERS hash table is the Type 2 Slowly Changing Dimension table created in Chapter 7, it is not just a simple FIND method call. In earlier examples we just assigned a value to the PDV host variable for the key value to look up. What we do here is use an explicit FIND method call to specify the key and point to the first item for that key. In this case it is the Batter_ID. We then use a LINK statement to execute the search logic (since we need to do it multiple times for different id values).

⑭   We now have the data for the batter and we create a field that is their name and another field for their team.

⑮   Do the same thing to get the pitcher name and team.

⑯   If any runs were scored in this at bat we need to get the runner data. Note that since the variable Runs was read from our input data set, we don't need to look it up.

⑰   Use the FIND and FIND_NEXT methods in a loop to read all the runner data.

⑱   Retrieve the runner name information from the PLAYERS hash table just as we did for batters and pitchers.

⑲   Use the ADD method to add an item to our results table as we have all the data needed for those at bats that resulted in runs being scored.

⑳   If no runs were scored for this at bat, we still want an item in our hash table (i.e., a row in our output SAS data set) with a blank value for the runner name. The ADD method is used here as well.

㉑   Create our output data sets using the OUTPUT method. Using just Runs as the WHERE clause, any item with a non-zero, non-missing value results in the WHERE clause being true. Likewise, Not Runs is true only when no runs were scored.

㉒   Create the output data set that contains the count of the number of at bats that met our filter criteria for each Result value.

㉓   This is the same Type 2 Slowly Changing Dimension table lookup that was shown in Section 7.3.3.1 "Performing Table Lookups Using an SCD Type 2 Table."

Output 13.1 Distribution of the Results of the 0-2 At Bats with 10 Pitches

Result	Count
Double	2
Home Run	2
Out	10
Single	3
Strikeout	18
Triple	2
Walk	18
	55

Output 13.2 - Runs Scored - At Bats with 10 or More Pitches, Starting with a Count of 0-2

Date	Batter	Team AtBat	Runner	Runs	Pitcher	Team InField	AB Number	Number of Pitches	Result
2017-03-31	Phillips, Nicholas	Owls	Young, Jerry	1	Miller, Jordan	Owls	35	11	Walk
2017-04-02	Jenkins, Kevin	Cardinals	Sanchez, Jeremy	1	Hayes, Ryan	Cardinals	14	10	Walk
2017-04-21	Bailey, Joseph	Saints	Simmons, Ethan	1	Alexander, Benjamin	Saints	48	10	Triple
2017-04-21	Sanchez, Harold	Grizzlies	Watson, Nathan	1	Miller, Kenneth	Grizzlies	32	13	Walk
2017-04-22	Taylor, Gary	Bobcats	Nelson, Willie	2	Hernandez, Justin	Bobcats	23	10	Home Run
2017-04-22		Bobcats	Taylor, Gary		Hernandez, Justin	Bobcats	23	10	Home Run
2017-04-25	Foster, Alexander	Wolves	Cook, Justin	1	Rivera, John	Wolves	43	11	Walk
2017-08-23		Wolves	Alexander, Henry		Hill, Alexander	Wolves	41	11	Walk
2017-07-18	Simmons, Steven	Rebels	Williams, Jeffrey	2	Edwards, Anthony	Rebels	44	10	Double
2017-07-18		Rebels	Baker, Alan		Edwards, Anthony	Rebels	44	10	Double
2017-07-25	Cooper, Eric	Bobcats	Cooper, Eric	1	Cooper, Eric	Bobcats	9	10	Home Run
2017-08-12	Anderson, Brian	Mavericks	Moore, Eugene	1	Thompson, Douglas	Mavericks	28	10	Walk
2017-09-05	Parker, Kenneth	Titans	Rivera, Douglas	2	Edwards, Justin	Titans	14	10	Walk
2017-09-05		Titans	Peterson, Lawrence		Edwards, Justin	Titans	14	10	Walk

Output 13.3 – No Runs Scored - At Bats with 10 or More Pitches, Starting with a Count of 0-2

Date	Batter	Team AtBat	Pitcher	Team InField	AB Number	Number of Pitches	Result
2017-03-20	Adams, Jack	Blazers	Adams, Joseph	Blazers	17	10	Out
2017-03-22	Hughes, Jerry	Hurricanes	Hughes, Eric	Hurricanes	33	10	Out
2017-03-25	Scott, Ryan	Bluejays	Wilson, Carl	Bluejays	23	10	Strikeout
2017-03-31	Morgan, Nicholas	Bears	Lee, Richard	Bears	41	11	Strikeout
2017-04-09	Robinson, Douglas	Bears	Collins, David	Bears	7	12	Out
2017-04-11	Walker, Willie	Titans	Hall, Frank	Titans	39	11	Walk
2017-04-25	Ross, John	Chiefs	Ross, John	Chiefs	18	10	Out
2017-05-03	Lee, Andrew	Bearcats	Ward, Keith	Bearcats	26	11	Out
2017-05-09	Gonzales, William	Red Sox	Gonzales, William	Red Sox	18	10	Out
2017-05-09	Griffin, Jordan	Tigers	Ramirez, Dylan	Tigers	29	10	Strikeout

Date	Batter	Team AtBat	Pitcher	Team InField	AB Number	Number of Pitches	Result
2017-05-19	Gonzalez, Benjamin	Storm	Phillips, Timothy	Storm	15	10	Strikeout
2017-05-19	Alexander, Jerry	Mountaineers	Alexander, Jason	Mountaineers	28	10	Strikeout
2017-05-23	Cook, Anthony	Pirates	Barnes, Willie	Pirates	16	10	Strikeout
2017-05-24	Rogers, Dylan	Tigers	Ramirez, Dylan	Tigers	4	10	Single
2017-05-26	Stewart, Brandon	Tigers	Martin, Zachary	Tigers	12	11	Walk
2017-06-05	Harris, Austin	Redskins	Murphy, Gerald	Redskins	20	11	Walk
2017-06-14	Turner, Keith	Patriots	Hughes, Thomas	Patriots	8	12	Strikeout
2017-06-16	Gray, Joshua	Red Raiders	Thomas, George	Red Raiders	25	14	Strikeout
2017-06-18	Powell, Terry	Vikings	Hayes, Bobby	Vikings	42	11	Walk
2017-06-19	Sanchez, Jesse	Broncos	Lopez, Bruce	Broncos	7	10	Strikeout
2017-07-01	Carter, Jesse	Patriots	Gray, Bobby	Patriots	25	10	Strikeout
2017-07-12	Sanchez, Jeremy	Cardinals	Foster, Lawrence	Cardinals	15	10	Strikeout
2017-07-17	Patterson, Benjamin	Owls	Miller, Roy	Owls	10	10	Out
2017-07-19	Clark, Carl	Bobcats	Washington, Mark	Bobcats	35	10	Walk
2017-07-26	Wood, Wayne	Monarchs	Smith, Mark	Monarchs	29	10	Double
2017-07-28	Edwards, Michael	Bluejays	Watson, Dennis	Bluejays	11	10	Out
2017-08-08	Peterson, Dennis	Titans	Flores, Brian	Titans	3	10	Strikeout
2017-08-09	Watson, Nathan	Grizzlies	Adams, Frank	Grizzlies	38	10	Walk
2017-08-12	Torres, Dylan	Saints	Richardson, Louis	Saints	30	10	Strikeout
2017-08-12	Miller, Ryan	Comets	Perez, Scott	Comets	4	10	Single
2017-08-15	Hughes, Jerry	Hurricanes	Hughes, Eric	Hurricanes	26	10	Walk
2017-08-18	Carter, Jonathan	Chiefs	Ramirez, Roger	Chiefs	52	10	Walk
2017-08-28	Russell, Randy	Bears	Hernandez, Donald	Bears	1	10	Strikeout
2017-09-02	Washington, Benjamin	Storm	Hill, Billy	Storm	3	10	Strikeout
2017-09-03	Flores, Richard	Comets	Rodriguez, Benjamin	Comets	40	10	Out
2017-09-12	Griffin, Jordan	Tigers	Martin, Zachary	Tigers	17	10	Out
2017-09-16	Adams, Jacoby	Grizzlies	Miller, Kenneth	Grizzlies	20	11	Strikeout
2017-09-22	Jackson, Howard	Gladiators	Martinez, Jerry	Gladiators	6	10	Strikeout
2017-09-22	Cook, Howard	Red Sox	White, Bruce	Red Sox	10	11	Single
2017-09-30	Bryant, Paul	Gladiators	Martinez, Jerry	Gladiators	8	10	Triple
2017-10-03	Wilson, Louis	Owls	Miller, Roy	Owls	21	10	Walk
2017-10-03	Reed, Henry	Pirates	Bennett, George	Pirates	21	10	Strikeout
2017-10-06	Adams, Jacoby	Grizzlies	Adams, Frank	Grizzlies	31	13	Walk
2017-10-06	Mitchell, Carl	Titans	Hernandez, Howard	Titans	6	12	Walk

Both the business and IT users liked these reports and asked about the code that was used to produce them. We told them that Output 13.1 was created using PROC PRINT, and PROC REPORT was used to produce Output 13.2 and 13.3. They seemed quite happy when we told them that the sample programs included the code to produce the reports. We recommended the business and IT users review *Carpenter's Complete Guide to the SAS REPORT Procedure* by Art Carpenter to learn more helpful tricks for using PROC REPORT.

13.3 Summary

For the purposes of this sample, we wrote and documented all the code (as opposed to highlighting differences from other programs), as that approach would provide more insights on how the SAS hash object can be used for such requests.

We reinforced the comment we made about the first case study: that *stealing* snippets of code from other programs is not an approach that we would suggest as a long-term solution. SAS software, including the macro language, provides a framework to build a set of re-usable tools that can be parameterized and mixed and matched as needed in order to address both their ongoing reporting requirements as well as ad-hoc questions.

Given that this Proof of Concept is now complete we suggested that the business and IT users consider prioritizing their requirements and proposed that an iterative approach to building their solution was an ideal way to proceed. We suggested that a good place to start would be to develop a list of all the reports they want- both on an ongoing basis as well as a number of ad-hoc questions.

The League office thanked us for our efforts and said that we would definitely be hearing from them regarding an implementation project.

Index

Ready to take your SAS® and JMP®skills up a notch?

Be among the first to know about new books,
special events, and exclusive discounts.
support.sas.com/newbooks

Share your expertise. Write a book with SAS.
support.sas.com/publish

sas.com/books
for additional books and resources.

§sas.

THE POWER TO KNOW.

www.ingramcontent.com/pod-product-compliance
Lightning Source LLC
Chambersburg PA
CBHW081042220326
41598CB00038B/6960